Springer Series in Design and Innovation

Volume 33

Springer Series in Design and Innovation (SSDI) publishes books on innovation and the latest developments in the fields of Product Design, Interior Design and Communication Design, with particular emphasis on technological and formal innovation, and on the application of digital technologies and new materials. The series explores all aspects of design, e.g. Human-Centered Design/User Experience, Service Design, and Design Thinking, which provide transversal and innovative approaches oriented on the involvement of people throughout the design development process. In addition, it covers emerging areas of research that may represent essential opportunities for economic and social development.

In fields ranging from the humanities to engineering and architecture, design is increasingly being recognized as a key means of bringing ideas to the market by transforming them into user-friendly and appealing products or services. Moreover, it provides a variety of methodologies, tools and techniques that can be used at different stages of the innovation process to enhance the value of new products and services.

The series' scope includes monographs, professional books, advanced textbooks, selected contributions from specialized conferences and workshops, and outstanding Ph.D. theses.

The volumes of the series are single-blind peer-reviewed.

Keywords: Product and System Innovation; Product design; Interior design; Communication Design; Human-Centered Design/User Experience; Service Design; Design Thinking; Digital Innovation; Innovation of Materials.

How to submit proposals

Proposals must include: title, keywords, presentation (max 10,000 characters), table of contents, chapter abstracts, editors'/authors' CV.

In case of proceedings, chairmen/editors are requested to submit the link to conference website (incl. relevant information such as committee members, topics, key dates, keynote speakers, information about the reviewing process, etc.), and approx. number of papers.

Proposals must be sent to: series editor Prof. Francesca Tosi (francesca.tosi@unifi.it) and/or publishing editor Mr. Pierpaolo Riva (pierpaolo.riva@springer.com).

Nuno Martins · Daniel Brandão ·
Adérito Fernandes-Marcos
Editors

Perspectives on Design and Digital Communication IV

Research, Innovations and Best Practices

Springer

Editors
Nuno Martins
Research Institute for Design
Media and Culture, School of Design
Polytechnic Institute of Cávado and Ave
Barcelos
Barcelos, Portugal

Daniel Brandão
Communication and Society Research
Centre
Institute of Social Sciences
University of Minho Braga
Braga, Portugal

Adérito Fernandes-Marcos
Faculty of Arts and Humanities
University of Saint Joseph
Macau, China

ISSN 2661-8184 ISSN 2661-8192 (electronic)
Springer Series in Design and Innovation
ISBN 978-3-031-41769-6 ISBN 978-3-031-41770-2 (eBook)
https://doi.org/10.1007/978-3-031-41770-2

This Springer imprint is published by the registered company Springer Nature Switzerland AG
The registered company address is: Gewerbestrasse 11, 6330 Cham, Switzerland

Preface

Over the past 15 years, we have witnessed the rise of a visual aesthetic brought forth by digital media, serving as a means of artistic and cultural expression that embraces the multifaceted nature of design. This process has become integral to the ongoing global digital technological revolution. As time has progressed, this aesthetic has evolved to encompass the emergence of new materiality and the rematerialisation of digital artistic, cultural and design practices. In this evolution, tangible physical materials have assumed an equal role alongside the digital realm, expanding and repositioning it within an aesthetic framework that has come to be known as post-digital. From a post-digital perspective, the digital domain is perceived as omnipresent, resembling a fluid medium that permeates and engulfs everything and everyone. It possesses a sense of transparency, being ubiquitous and, to a large extent, invisible. Additionally, it empowers us by existing as an underlying and inherent element within the physicality of objects and materials [1, 2, 3].

Like digital media art, the design process is inherently centred around the artefact itself, accompanied by cycles of action and reflection that contribute to developing a new digital aesthetic rooted in the hybridity and physicality of these artefacts. While the creation-research process in digital media art differs from the commonly adopted methodology in design, which focuses on problem-solving, and guides the entire process, both processes share the common goal of systematisation (as opposed to ad hoc procedures) towards creating a final artefact. This artefact aims to provide a meaningful and aesthetically significative fruition experience and fulfil its utilitarian dimensions, thus becoming an integral part of the overall process.

In the digital era, design expands its utilitarian dimensions to encompass (post)digital aesthetics, merging and enhancing the digital and physical aspects of materials and objects, whether tangible or intangible. This aesthetics possesses an inevitable volatility, creating a collective imagination that expresses cultural references within society while also incorporating the individual perspectives of the designers/artists. Such contemporary aesthetics inherently prioritises the experience on the end-user's side in scenarios that are intertwined with the fluidity of the digital realm. It challenges and blurs the boundaries between physical and digital materialities, prompting new avenues for experimentation and research in terms of stylistic innovation, representation of reality, the dynamic between society and the individual, and the role of design/art and the identity of the designer/artist. Furthermore, this exploration examines user behaviours towards emerging technologies, acknowledging their impact on the creative process [2, 4].

As we navigate through the book's various chapters, we encounter significant progress in the realm of creation-research within design and digital communication. These advancements provide us with crucial insights into the best practices of research and innovation. Oftentimes, these practices leverage new and appropriate technologies, making a profound impact on optimising design projects, enhancing education and fostering enriching learning experiences. Within these contexts, the book also showcases compelling evidence of new approaches in (post)digital aesthetics, leaving a lasting impression on the field.

This book includes original contributions by authoritative authors based on the best papers presented at the 6th edition of International Conference on Digital Design and Communication (Digicom 2022), held on 3–5 November 2022, in Barcelos, Portugal, and some invited chapters written by leading international researchers. The 20 chapters cover the following topics:

- User Experience and Interface Design;
- Pedagogy, Society and Design Practice;
- Branding Design;
- Game Design and Virtual Environments;
- Design Strategies and Challenges.

The discoveries showcased in this book possess the potential to inspire and provide valuable support to individuals in the field of design and digital communications. The content caters to both researchers and practitioners, offering insights that can be beneficial to their respective pursuits.

Barcelos, Portugal Nuno Martins
Braga, Portugal Daniel Brandão
Macau, China Adérito Fernandes-Marcos

References

1. Fernandes-Marcos A (2017) Computer artefact: the crucial element in artistic practice in digital art and culture. Revista Lusófona de Estudos Culturais 4(1):129–. https://doi.org/10.21814/rlec.182
2. Cramer F, Jandrić P (2021) Postdigital: a term that sucks but is useful. Postdigital Sci Educ 3:966–989. https://doi.org/10.1007/s42438-021-00225-9
3. Manovich L (2001) Post-media aesthetics. http://manovich.net/content/04-projects/032-post-media-aesthetics/29_article_2001.pdf. Accessed on June 20, 2023

4. Fernandes-Marcos A, Pereira S (2022) Post-digital textile aesthetics: a critical reflection on the process of digital re-materialisation of artefacts. In: Joseph M (ed) Proceedings of third international conference on digital creation in arts, media and technology (ARTeFACTo2022Macao), University of Saint, 2022, pp 1–9. https://doi.org/10.1109/ARTeFACTo57448.2022.10061256

Contents

User Experience and Interface Design

Achieving Privacy Through UX Design in Instant Messaging Apps

Davide M. Parrilli and Rodrigo Hernández-Ramírez

Abstract Protecting and enhancing informational privacy is crucial in user-to-user interactions, particularly through instant messaging apps. Careless and malicious handling of personal information can harm users' well-being and dignity. However, there is a need for a better understanding of the relationship between informational privacy and UX design in both design practice and academic research. This paper aims to bridge this gap and explain how privacy can be achieved through meaningful UX design. The research presented in this paper focuses on Confide, an instant messaging app promoted as the 'Confidential Messenger'. In a lab test involving the app's actual usage, we first subjected it to a privacy assessment. Then, we analyzed Confide's UX through a heuristic assessment followed by an authors' UX report. The research revealed that users' freedom to manage their privacy settings leads to a meaningful UX. The methodology proposed in this paper is appropriate for investigating how designed products combine informational privacy and good UX design due to its replicability and scalability.

Keywords Confide App · Informational privacy · Interaction design · UX design

1 Introduction and Scope

Interaction is a crucial aspect of human personality, and human exchanges are fundamental moments of human existence [1]. Nowadays, technology often mediates these exchanges, facilitating and significantly influencing when and how they occur [2]. Instant messaging apps are ubiquitous means of interaction, used for personal and professional purposes, such as communicating with doctors, lawyers, professors, and

D. M. Parrilli (✉) · R. Hernández-Ramírez
UNIDCOM/IADE, Faculdade de Design, Tecnologia e Comunicação, Universidade Europeia, Lisbon, Portugal
e-mail: davide.parrilli@universidadeeuropeia.pt

R. Hernández-Ramírez
e-mail: rodrigo.ramirez@universidadeeuropeia.pt

N. Martins et al. (eds.), *Perspectives on Design and Digital Communication IV*, Springer Series in Design and Innovation 33, https://doi.org/10.1007/978-3-031-41770-2_1

loved ones. These apps have become one of the most direct, universal, and widespread ways in which human beings engage and share their experiences with others.

Interactions between humans imply exchanging potentially valuable information; hence, it is crucial to be aware of privacy and confidentiality concerns when sharing that information. Sharing personal data in free interactions relies on trust [3]—albeit it is essential to note that not all personal data deserves the same level of confidentiality. For instance, the expected privacy threshold of an intimate picture exchanged between partners is higher than that of an image of the sender surrounded by a multitude in a public event.

The exchange of personal information between interacting parties may involve the personal data of third parties [4]. Informal privacy is the right of individuals and groups to determine who, how, and when their personal data can be accessed and processed by the recipient [5]. Those whose personal information has been shared without consent cannot exercise their privacy. The most severe and damaging abuses of people's private information imply a lack of awareness and consent from the victims. Instant messaging applications often facilitate such violations of privacy.

Instant messaging apps are technological artifacts that often claim neutrality, but as with any technology, this neutrality is just a mirage [6]. Human goals and intentions cannot exist independently of technological mediation [7]. A technology that allows the instant sharing of sensitive personal information to countless recipients and, thus, irreparable damage to someone's reputation creates and reinforces malicious behavior.

Instant messaging apps must be designed to empower senders and other data subjects to exercise their privacy rights and claims effectively. The sender should have complete control over the privacy settings. For instance, the sender should be able to restrict the recipients from forwarding received messages and media to third parties and taking screenshots. This would safeguard the personal information of other data subjects too.

This paper builds upon previous research on informational privacy and instant messaging apps [8, 9]. Through a meticulous analysis of the privacy features of popular instant messaging apps such as WhatsApp, Telegram, and Signal, we have identified design patterns that can effectively reduce or minimize privacy risks for the sender of personal information. However, there is still a gap in the literature regarding privacy, design, and instant messaging apps. To address this gap, there is a pressing need to investigate the relationship between privacy and user experience (UX) design and comprehend the design patterns and approaches that can simultaneously guarantee an adequate level of informational privacy and a meaningful UX.

Research in privacy notices and cookie consent banners has shown that notifications are often lengthy and overly complex, making it difficult for the average user to understand their content and meaning [10–12]. For example, some websites force users to accept all cookies without allowing them to browse through privacy policies and settings, all in the name of simplicity (i.e., clicking a single button instead of trudging through an inextricably complex privacy policy agreement). This creates a simple interaction through dark patterns, which are necessarily unethical [13–16].

Designers could alternatively create a simple interaction by nudging users to *reject* all cookies. However, this practice would threaten the very reason cookies are widely used in surveillance society [17] and capitalism [18–20], which is to gather people's personal data for profit.

In this broad context, personal data are significant not only ontologically [1, 21] but also financially and politically: storing personal data is profitable and allows those with access to it to obtain all sorts of intimate aspects of people's lives. This paper explores the relationship between informational privacy and UX design in instant messaging apps. The main objective is to explain how a UX design that is both satisfactory and meaningful can achieve informational privacy in instant messaging apps. It also aims to prove that a frustrating UX is a barrier to users recognizing the value of privacy on a large scale.

1.1 Methodology and Content

This paper intends to fill a gap in the research about privacy and UX design by answering the following research questions: *is it possible to assess the level of privacy protection of a digital interface through a heuristic evaluation of its UX design?* In case of an affirmative answer, *which other methods can complement the heuristic evaluation to offer a comprehensive privacy assessment of a digital interface?*

The authors identified an instant messaging app, Confide [22], that allegedly allows users to "communicate digitally with the same level of privacy and security as the spoken world." The first step in our methodology was critically using Confide. The authors used the app for seven days, during which they interacted through Confide, exchanging messages and attachments and experimenting with its privacy features. Confide was chosen for this research due to its patented ScreenShield technology that prevents screenshots of messages. Other instant messaging apps designed for privacy and confidentiality include Three-ma, The Secure Messenger, Kryptet, LiteWire Messaging, and Sircle—Private Messenger. However, our previous research concluded that screenshot prevention is an appropriate means to protect the informational privacy of all parties engaged in digital interactions [8, 9]. Therefore, testing an instant messaging app that includes this feature is consistent with investigating the relationship between privacy and UX design.

The authors installed the Confide app on their devices[1] for the lab test and interacted by sending texts, pictures, and videos in peer-to-peer chats.[2] The next step in the methodology was a qualitative privacy assessment. We answered a list of research questions aimed at understanding how the app protects and enhances users' informational privacy—the authors tested this method in previous research [9] and

[1] Confide version 10.0.3 installed on iPhones with iOS 16.3.1.

[2] Confide permits also group chats. The privacy features of group chats are equivalent to those of peer-to-peer chats.

reproduced it with some slight adjustments in the investigation discussed in this contribution[3]:

1. Does Confide allow sending temporary messages (i.e., messages that are erased after they are read or once a certain amount of time set by the sender or group administrator expires)?
2. If so, does this mode prevent recipients from saving media (e.g., audio or images) or files included in the message on their device?
3. Does this mode prevent recipients from taking screenshots or making screen recordings of temporary messages?
4. Does this mode prevent the recipient from forwarding the message, media, and files to third parties?
5. Can the sender delete messages and files for everyone once they have been sent?
6. Can the sender prevent the recipient from forwarding messages or attachments to other users?
7. Can the sender prevent the recipient from copying and saving messages and attachments?
8. Can the sender prevent the recipient from making screenshots or screen recordings of messages?
9. Is it possible to send hidden text that requires the recipient to carry out some type of action to read the message?
10. Can the sender be notified if the recipient forwards, saves, takes a screenshot, or makes a video recording of a message and its attachments?
11. Is the caller notified if the other participant takes a screenshot or records the screen?

In the next section, we will report on the results of the privacy assessment, providing a qualitative and critical evaluation. The third stage in the methodology is a qualitative UX analysis. For the UX assessment, we first conducted a qualitative heuristic analysis based on Nielsen's 10 Usability Heuristics for User Interface Design [23–26]. The outcome of the heuristic analysis is described in the third section of this paper, followed by a critical report of the authors' personal UX experience. Applying the heuristics provided a valuable and objective means of evaluating the quality of the app's interaction design. However, the authors' experience report serves as a helpful complement, documenting their personal experiences with the app.

A multidisciplinary literature review on privacy, UI design, and UX design supports the remaining steps in the methodology. In the paper's final section, we present some concluding remarks and describe the developments of our research.

[3] Although Confide prevents users from taking screenshots, we decided to keep the questions about screenshots and screen recording and answer them, to get a detailed overview of the design and impact of this feature.

2 Privacy Assessment of Confide App

Confide instant messaging app is built for confidentiality; the product website states that "Confidential means more than just encryption"—messages exchanged on other popular instant messaging apps such as WhatsApp, Telegram, and Signal are encrypted. However, they offer ample opportunities for recipients to misuse and abuse information without the sender's knowledge and consent [8, 9]. For our assessments, the unpaid version of Confide was tested. Confide Plus for premium features offers some extra tools: though they have not been tested, in the privacy assessment discussed in this section we report these extra features when relevant.

2.1 Outcome of the Test

Our research begins with a privacy assessment of the Confide app. We relied on the 11 research questions introduced in Sect. 1.1 to evaluate how Confide protects users' privacy in interactions involving texts, files, and media such as pictures and videos.

1. RQ1—Does confide allow sending temporary messages?
   Confide allows users to send messages with the Confidential Mode ON or OFF. When Confidential Mode is ON (which is the default setting), messages are destroyed as soon as they are read and covered with a message text overlay. When Confidential Mode is OFF, messages are visible for 24 h after being sent before being permanently deleted. In summary, all messages sent on Confide are temporary.
2. RQ2—Does this mode prevent recipients from saving media or files?
   Regardless of the Confidential Mode selected, recipients of images, videos, or files cannot save them on their devices. Users can only view them (for 24 h with Confidential Mode OFF or only once with Confidential Mode ON).
3. RQ3—Does this mode prevent recipients from taking screenshots or screen recordings of temporary messages?
   Confide does not allow users, including message senders, to take screenshots. If a screenshot is attempted by any party in the conversation, the user is expelled from the chat, and all messages are deleted. The other party is also notified of the attempted screenshot. This feature is triggered whenever a party tries to take a screenshot of the chat or an attached file.

   Interestingly, if one attempts to take a screenshot, a file is recorded on the device, but the screen's content is replaced by a message stating that Confide prevents screenshots with the patented ScreenShield technology. This technology also prevents recording the screen.
4. RQ4—Does Confide prevent recipients from forwarding messages, media, and files to third parties?
   Messages, media, and files cannot be forwarded to third parties.

5. RQ5—Can the sender delete messages and files for everyone?
Users of the free version can delete messages but only for themselves. The Message Retraction feature, available for Confide Plus subscribers, allows messages to be unsent for all users.
6. RQ6—Can the sender prevent recipients from forwarding messages, media, and files to other users?
Message forwarding is not allowed by default in Confide.
7. RQ7—Can the sender prevent recipients from copying and saving messages, media, and files?
The sender and the recipient cannot copy and save messages, media, and files exchanged in a chat.
8. RQ8—Can the sender prevent recipients from taking screenshots or screen recordings?
As mentioned above, Confide has an absolute ban on screenshots and screen recordings.
9. RQ9—Is it possible to send hidden text to recipients?
All messages are hidden when the default Confidential Mode is ON, and a black strip covers the text. To read the text, the sender and recipient must tap on it with their finger. Media and files are not immediately visible, and a message with a clip and the text Confidential Attachment appears instead. The attachment requires action to be seen, such as a tap-and-hold gesture.
10. RQ10—Can the sender be notified if the recipient forwards, saves, takes a screenshot, or makes a video recording of the message?
Regardless of which party (the sender or the recipient) tries to take a screenshot or make a screen recording of the message, the other party receives a notification. There is no notification mechanism for forwarding and saving messages and attachments.
11. RQ11—Is the caller notified if the other participant takes a screenshot or records the screen?
Confide does not allow voice and video calls on the platform.

2.2 Discussion

Confide provides a robust level of informational privacy. It offers a very high default level of privacy and confidentiality for both the sender and the receiver of information, dramatically reducing the risk of information abuse and misuse. Screenshots, message forwarding, and message saving are not possible, and the only way to record chat content is by taking pictures and videos using another device.

Incognito Mode (only for Confide Plus subscribers) offers additional privacy for users. It prevents users from being found by others unless they want to be found; e-mail and phone number searches do not work with Incognito users. Users can manage existing friends who can see and message them. Confide only tries to find friends from the contact list of Incognito users if they explicitly request to do so.

However, these features do not impact the privacy and confidentiality of exchanged messages.

Although Confide's privacy features are robust, two dimensions of privacy are missing from the app: freedom and trust. Users are forced to adhere to strict privacy settings, with the only notable exception of rejecting Confidential Mode. However, even with this mode off, all messages are destroyed after 24 h. Confide does not allow users to keep memories shared with other app users. Freedom is a critical component of privacy because there is no freedom without privacy [5, 17], but the opposite is equally valid.

The freedom to choose the most appropriate privacy settings allows users to make ethical and careful choices. Users can make meaningful decisions when they are free to choose. In the case of Confide, users' privacy decisions are fully delegated to the app, which decides that the most stringent privacy settings are suitable for all users. One can legitimately claim that users are not forced to use Confide unless they work for an organization that uses it for their business. However, this is not an excuse for not allowing users to develop their privacy virtues when using the app.

Freedom is closely related to trust [3]. Confide users cannot decide by themselves which recipients deserve their trust. The design of Confide's privacy features dictates that no recipient can be trusted. Users cannot keep received messages or reread them because they are emotionally valuable or just beneficial for the receiver. Nobody can record and store a picture or video sent by a loved one. In this sense, default privacy settings that do not allow trust, intended as the freedom to trust other users and memory, are dystopic. Despite being allegedly built for privacy and confidentiality, Confide inscribes itself onto the (digital) panopticon phenomenon [27]. It takes for granted that all other users are watchers in the Panopticon, where we are forced to live without recognizing the possibility of being free and engaging in meaningful and equal interactions with other human beings.

3 UX Analysis of Confide App

To link informational privacy and UX design, it is crucial to demonstrate that robust privacy can be achieved through good UX design without sacrificing user experience. The first step in this process is to conduct a UX analysis of the product under examination. By assessing the privacy and UX of products already on the market, designers can gain insights into creating products that respect users' privacy while providing an enjoyable experience.

In the case of Confide, a twofold assessment was conducted. The first assessment was based on Nielsen's usability heuristics, while the second involved summarizing customers' UX experiences. The latter complemented the former and took the form of a report detailing users' feelings, challenges, frustrations, and satisfactions when using the app. This analysis was inspired by a customer journey map, which helped detail the interactions and emotions in a clear and accessible manner [28].

Customers' UX reports, as self-observation exercises, provide valuable insights into what users need and want from a human-centered design perspective. This is important because a deep understanding of people is necessary to design usable and understandable products [29]. The objective of this analysis is not to provide a final assessment of Confide's quality but to reveal a sound methodology that can be easily replicated on a larger scale to evaluate digital products.

3.1 Results of the Heuristic Analysis for UX

To ensure objective results in the UX analysis of Confide, we use a methodology extensively discussed and tested in the last decade: a heuristic evaluation of the Confide interface [24, 25] based on the 10 Nielsen's usability heuristics [23]. An interface is evaluated heuristically "by simply looking at the interface and passing judgment according to one's own opinion [...] trying to come up with an opinion about what is good and bad about the interface." [25]

The authors performed the heuristic evaluation to assess the UX level of Confide qualitatively and neutrally without defining a scale for the severity of the identified problem and setting specific recommendations. Although more evaluators would be required to provide a more reliable analysis, our goal is not to make a final evaluation of Confide but to test a replicable method that allows us to obtain preliminary results. The following paragraphs identify critical and positive issues for each heuristic.

1. Heuristic 1: Visibility of System Status
   No specific issues are found regarding the first heuristic. The interface keeps users informed about what is going on through feedback messages. In particular, when users attempt to take forbidden actions, such as screenshots, they immediately receive clear feedback.
2. Heuristic 2: Match Between System and the Real World
   The interface design uses no specific unfamiliar or incomprehensible words, phrases, or concepts, but please refer to heuristic 6 to know more about a particular consistency and standard issue we identified.
3. Heuristic 3: User Control and Freedom
   In general, Confide allows users to adjust the privacy settings of the interface only to a minimum extent. A specific critical issue that undermines user control and freedom arises when a user tries to take a screenshot or make a screen recording of a Confide chat. The user cannot redo the action but only click the "I understand" button (see Fig. 1). The consequences of the attempt are inevitable: all messages exchanged in the affected chat are deleted, and it will be necessary to start a new conversation from scratch with the other user.

   Another concern affecting user control and freedom is that only Confide Plus users can delete messages once they have been sent. Free users cannot delete messages sent by mistake—which is an issue also under the perspective

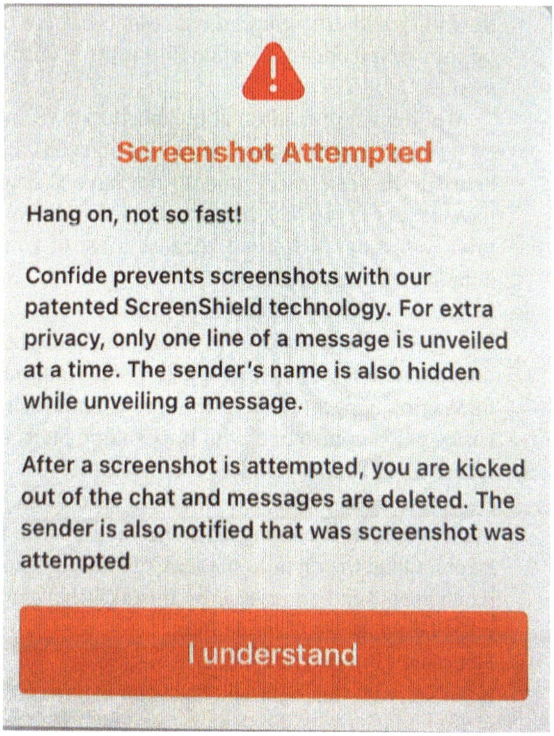

Fig. 1 A user who tries to take a screenshot of the chat receives a severe warning message

of heuristic 9. By comparison, this option is available for all users of WhatsApp, Telegram, and Signal [8, 9].

4. Heuristic 4: Consistency and Standards
Contrary to other popular instant messaging apps like WhatsApp, Telegram, and Signal [8, 9], which use plain and standard language, the function to delete messages in Confide is called message retraction. This language inconsistency can generate misunderstandings among users, who are more used to dealing with the word 'delete'.
5. Heuristic 5: Error Prevention
A contradiction with this heuristic is found when a user tries to take a screenshot of the chat. No confirmation option is presented to the user before committing the action and experiencing its consequences. No constraints are in place either: no obstacles prevent users from taking screenshots, such as deactivating interactive elements.
6. Heuristic 6: Recognition Rather than Recall
In the Confide interface, menus and menu items are visible and easily retrievable.
7. Heuristic 7: Flexibility and Efficiency of Use
One feature that ensures efficiency of use is the possibility to activate the Confidentiality Mode directly in the user-to-user chat without the need to browse through the privacy or chat settings. On the contrary, an aspect that reduces

flexibility and efficiency of use, especially for users with special needs, is the option offered only to Confide Plus users to customize the interface with custom themes.

We identify the automatic cancellation of messages and attachments after 24 h (with Confidentiality Mode disengaged) as a feature that undermines this heuristic because users who do not have access to the app for a day will not have access to the messages sent to them. In general, flexibility to regulate the privacy settings is limited because most of Confide's privacy features are by default and unalterable.

8. Heuristic 8: Aesthetic and Minimalist Design
   The Confide interface is minimal and straightforward, and only commonly understandable icons are used. Users familiar with other popular instant messaging apps and their interfaces are not likely to experience any adaptation issues when confronted with the Confide interface.
9. Heuristic 9: Help Users Recover from Errors
   As commented under heuristic 3, in case of an attempt to take a screenshot or make a screen recording of a chat, there is no possibility of recovering from errors. Once the error is made, Confide does not offer any way back. Further, when users send a message by mistake, they can delete it for both parties only if they sign up for Confide Plus.
10. Heuristic 10: Help and Documentation
    No issue with this heuristic has been identified.

It emerges that all heuristics are essential for assessing the privacy level of a digital interface through its UX. However, some heuristics contribute to a larger extent to a good privacy management experience. We refer, in particular, to heuristic 3—because use control and freedom are critical for meaningful and ethical privacy for users; heuristic 7—because users should have the choice to adapt the privacy settings flexibly; and heuristic 9—because users who send information by mistake should be able to recover for their error. Heuristic 8 indirectly plays a vital role since a clear, straightforward design ensures users can easily choose their preferred privacy settings.

3.2 Report of the Authors' UX

Heuristic analysis is valuable for identifying issues affecting Confide's user experience. However, it has limitations and may not catch all usability issues. Additionally, it may not give specific suggestions on how to address those issues [25]. Relying solely on heuristic analysis is insufficient to fully understand how a system or interface can be improved to enhance the user experience.

One effective method to achieve this goal is a customer journey map, which maps the emotions, feelings, satisfactions, and frustrations experienced by users in the real world. Service designers have successfully used this method [28], which can be usefully extended beyond service design.

Interestingly, the heuristic assessment provides valuable insights about the expected users' feelings likely to be reflected in the customers' UX report. It is important to note that the heuristic analysis and the customers' UX assessment were conducted by the same research team in the case examined. This, combined with the limited number of researchers involved, may lead to biased results. We suggest that the two evaluations should be conducted by different teams with an adequate number of testers to ensure sound results. However, the scope of our research is not to elaborate final analyses about Confide but to establish a link between privacy and UX design by suggesting a methodological approach to assess the relationship between the two.

Based on the authors' experience, three aspects of Confide's design have caused frustration:

1. Users cannot decide on privacy settings for their interactions.
2. The consequences of taking a screenshot are extremely harsh.
3. Messages and attachments are deleted after 24 h when Confidential mode is off, penalizing users who are offline or cannot access the app or their device for a day.

We understand that preventing users from taking screenshots, saving files on their devices, and forwarding messages and attachments effectively protects users' privacy. However, automatically deleting messages and files after 24 h may not be immediately understandable as a privacy-protecting function. Users may sometimes want to apply less strict settings if they trust the other party. Enforcing stringent privacy and confidentiality settings can make using Confide less enjoyable for everyday communication. However, based on our experience, Confide can be helpful when conducting business interactions that require secrecy or when sensitive information must be exchanged under dire circumstances, such as whistleblowing or talking to the press within a repressive state.

When we attempted to take a screenshot of a chat, the content was erased, and the offender was expelled from the chat, which deleted all written interactions and required starting a new conversation from scratch. There is no recovery option when a user accidentally takes a screenshot, which can be frustrating. While WhatsApp, Telegram, and Signal allow users to adjust privacy settings, they lack some of the essential default privacy features that Confide has, making it difficult to fine-tune privacy protection for users [8, 9].

4 Conclusions

The analysis of the privacy and UX design features of the instant messaging app Confide indicates that more research is necessary to fully understand the relationship between informational privacy and UX design. However, this paper establishes a link between privacy and UX design. Our research shows that it is possible to heuristically measure the privacy level offered by a digital interface through its UX. This method is adequate but insufficient to get the complete picture: the heuristic analysis should be supported by a privacy assessment, where the evaluators test the interface and answer a list of predefined questions, and a real users' UX report. As a result, the proposed multistep method led to a satisfactory privacy assessment of a digital interface from which operative insight can be extracted.

We believe that users' privacy is best protected and enhanced through UX design that allows them to make ethical choices about protecting their personal information. Privacy can be viewed from different perspectives—as a value, claim, or right—but not as an obligation. The same applies to trust: people have the right to trust other users but cannot be forced to mistrust all other users.

These considerations should be reflected in how an app that allows interactions between people is designed. That is, in how the conditions that shape customers' UX are designed. These conditions should rely on users' freedom and control to set their privacy preferences without undue constraints from the app provider. Through heuristic analysis, we have shown that a lack of users' freedom and control, together with other aspects of an interface, can be observed and categorized. However, heuristics are not fully adequate to grasp the complexity of UX issues from the users' perspective. Heuristics should be paired with customers' UX reports to understand how users in the real world experience a designed product.

The methodology tested in the research presented in this paper is new but has the potential to be easily replicated on a larger scale to assess how a designed product enhances both users' privacy and UX. The authors intend to improve this method in the further steps of their research. Indeed, the analysis presented in this paper is a crucial steppingstone to conceptualizing and prototyping an instant messaging app that protects and amplifies privacy through a design that ensures users' freedom and control. In other words, an instant messaging app designed for privacy and a satisfactory, meaningful UX. The app's prototype will undergo a customer's UX evaluation to understand if the intended developers' goals have been achieved.

From a more holistic perspective, the success of all instant messaging apps designed for users' privacy relies on an adequate awareness of the value and importance of privacy among users. Educating younger and older users is essential to have more privacy—that is, users who understand how and when to control and protect their personal information. As designers, we recognize that our means to educate are limited. Nevertheless, we can create tools that facilitate the process for users to claim and exercise control over their information. This perspective on how things are and how they could and should be designed differently places designers in a privileged position: that of starting much-needed conversations and envisioning solutions.

Acknowledgements This study was supported by UNIDCOM under a Grant by the Fundação para a Ciência e Tecnologia (FCT) no. UIDB/DES/00711/2020 attributed to UNIDCOM/IADE—Universidade Europeia, Lisbon, Portugal.

References

1. Floridi L (2010) Information: a very short introduction. Oxford University Press, Oxford; New York
2. Verbeek P-P (2006) Materializing morality: design ethics and technological mediation. Sci Technol Human Values 31:361–380. https://doi.org/10.1177/0162243905285847
3. Waldman AE (2018) Privacy as trust: information privacy for an information age. Cambridge University Press, Cambridge, United Kingdom; New York, NY
4. Vallor S (2016) Technology and the virtues: a philosophical guide to a future worth wanting. Oxford University Press, New York, NY
5. Westin AF (1967) Privacy and freedom. Scribner, New York, NY
6. Bowles C (2018) Future ethics. NowNext Press, London
7. Verbeek P-P (2015) Cover story beyond interaction: a short introduction to mediation theory. Interactions 22:26–31. https://doi.org/10.1145/2751314
8. Parrilli DM, Hernández-Ramírez R (2023) Protecting users' information and dignity through privacy-enhancing design. In: Martins N, Brandão D, Paiva F (eds) Perspectives on design and digital communication III. Springer International Publishing, Cham, pp 69–88. https://doi.org/10.1007/978-3-031-06809-6_5.
9. Parrilli DM, Hernández-Ramírez R (2023) Does signal APP's design protect users' informational privacy? A critical assessment. In: Martins N, Brandão D (eds) Advances in design and digital communication III. Springer Nature Switzerland, Cham, pp 236–245. https://doi.org/10.1007/978-3-031-20364-0_21
10. Litman-Navarro K (2021) We read 150 privacy policies. They were an incomprehensible disaster. https://www.nytimes.com/interactive/2019/06/12/opinion/facebook-google-privacy-policies.html?mtrref=undefined&assetType=REGIWALL. Accessed 16 Apr 2021
11. Pater R (2016) The politics of design: a (not so) global manual for visual communication. BIS Publishers, Amsterdam, The Netherlands
12. Santos C, Bielova N, Matte C (2020) Are cookie banners indeed compliant with the law? https://doi.org/10.26116/TECHREG.2020.009
13. Narayanan A, Mathur A, Chetty M, Kshirsagar M (2020) Dark patterns: past, present, and future: the evolution of tricky user interfaces. Queue 18:67–92. https://doi.org/10.1145/3400899.3400901
14. Gray CM, Kou Y, Battles B, Hoggatt J, Toombs AL (2018) The dark (patterns) side of UX design. In: Proceedings of the 2018 CHI conference on human factors in computing systems. ACM, Montreal QC Canada, pp 1–14. https://doi.org/10.1145/3173574.3174108
15. Luguri J, Strahilevitz L (2019) Shining a light on dark patterns. The University of Chicago, Chicago
16. Nelissen L, Funk M (2022) Rationalizing dark patterns: examining the process of designing privacy UX through speculative enactments. https://doi.org/10.57698/V16I1.05
17. DeBrabander F (2020) Life after privacy: reclaiming democracy in a surveillance society. Cambridge University Press, Cambridge, United Kingdom; New York, NY, USA
18. Doctorow C (2020) How to destroy surveillance capitalism. Medium editions
19. Foster JB, McChesney RW (2014) Surveillance capitalism: monopoly-finance capital, the military-industrial complex, and the digital age. Month Rev 66
20. Zuboff S (2019) The age of surveillance capitalism: the fight for a human future at the new frontier of power. Public Affairs, New York
21. Floridi L (2014) The 4th revolution: how the infosphere is reshaping human reality. Oxford University Press, New York; Oxford
22. Confide|Your confidential messenger (2023). https://getconfide.com. Accessed 15 Mar 2023
23. Nielsen J. 10 Usability heuristics for user interface design. https://www.nngroup.com/articles/ten-usability-heuristics/
24. Nielsen J. How to conduct a heuristic evaluation. https://www.nngroup.com/articles/how-to-conduct-a-heuristic-evaluation/

25. Nielsen J, Molich R (1990) Heuristic evaluation of user interfaces. In: Proceedings of the SIGCHI conference on Human factors in computing systems empowering people—CHI '90. ACM Press, Seattle, Washington, United States, pp 249–256. https://doi.org/10.1145/97243.97281
26. Nielsen J (1994) Enhancing the explanatory power of usability heuristics. In: Proceedings of the SIGCHI conference on human factors in computing systems. Association for Computing Machinery, New York, NY, USA, pp 152–158. https://doi.org/10.1145/191666.191729
27. Foucault M (2003) Surveiller et punir: naissance de la prison. Gallimard, Paris
28. Stickdorn M, Schneider J (2011) This is service design thinking: basics, tools, cases. WILEY, Hoboken, New Jersey
29. Norman DA (2013) The design of everyday things. Basic Books, New York, New York

Entutela 2.0: Enhancing Legal User Experiences Through Understanding the Narrative of Facts and Claims

Santiago de Francisco Vela, Camila Padilla Casas, and Andrés Polanía

Abstract *Entutela* is a prototype legal interface built to allow anyone to draft and file a *tutela* (a writ of protection of constitutional rights) without legal assistance or any specific knowledge. The *tutela* was implemented in 2021 as a digital tool to improve citizens' experience and contribute to the judicial branch's digital transformation. This article describes the second iteration of *Entutela*, where we focus on constructing content by citizens and developing the interface at the front-end and back-end levels. Both developments involve a transdisciplinary approach to access to justice, combining design, law, and engineering to define, analyze, propose, and implement solutions that respond to the real needs of the context, taking into account regulatory constraints, stakeholder visions, and impacts within the system. The article also emphasizes the need to adapt legal artifacts so as to make them available to everyone, regardless of their complexity, exploring the potential of legal interfaces within a digitized legal system. Overall, the article highlights the importance of developing user-friendly legal interfaces to promote access to justice and close the gap between legal knowledge and the public.

Keywords Access to justice · Legal interface · Storytelling · Transdisciplinary

S. de Francisco Vela (✉) · C. Padilla Casas
Departamento de Diseño, Universidad de los Andes, Cra. 1 Este No. 18A—70, Bogotá, Colombia
e-mail: s.defrancisco@uniandes.edu.co

C. Padilla Casas
e-mail: c.padilla12@uniandes.edu.co

A. Polanía
Facultad de Artes y Humanidades, Universidad de los Andes, Cra. 1 Este No. 18A—70, Bogotá, Colombia
e-mail: af.polania212@uniandes.edu.co

N. Martins et al. (eds.), *Perspectives on Design and Digital Communication IV*, Springer Series in Design and Innovation 33, https://doi.org/10.1007/978-3-031-41770-2_2

1 Introduction: Giving Context to *Tutelas*

The *tutela* is the primary judicial defense mechanism introduced by the Colombian Political Constitution of 1991 to protect the fundamental rights of foreigners or locals living on Colombian territory. It is an agile, informal, and simple legal instrument that anyone can use without needing to comply with the formalities of other judicial processes. For example, *tutelas* can be presented in written form on a document or napkin, and they can also be presented orally. They can be submitted to any of the more than 4,500 judges in the country, and the judge must resolve it within ten days of its submission. These characteristics make the *tutela* so attractive and viable that since its implementation, more than nine million *tutelas* have been filed [1]. However, during the pandemic, the Colombian judicial system faced the challenge of guaranteeing access to justice for all citizens [2, 3]. During this period, most courts were closed, and an online tool had to be improvised so people could file *tutelas*. The first version of *Tutela en línea* was launched in June 2020. This tool works with a form that allows citizens to fill out contact information and attach their *tutela* file in PDF. This information is sent to the email of a distribution judge, who allocates the *tutela* to one of 4500 judges. The tool centralizes the receipt of documents but does not address other challenges related to the digital transformation of justice. Moreover, although forced by the pandemic, the opportunity arose to build a system that would drive this digital transformation. The solution presented does not impact any of the critical indicators of the justice system. These indicators include decongesting the judicial branch, improving the *tutela* selection process, or protecting fundamental rights. Although *Tutela en línea* could have been a milestone in the progress of digital transformation, it fell short at a technical, functional, and experiential level because it only focused on the reception of a document rather than on the complete flow of the user experience. These lessons learned led us to propose *Entutela* as a digital tool. *Entutela* is a legal interface prototype built based on the experiences of citizens, lawyers, judges, and justices [4]. It is designed to allow anyone to draft and file a *tutela* without legal assistance, whether or not they have legal knowledge. A functional version was launched with a court between April and June 2021. The results of *Entutela* yielded some exciting work opportunities concerning user experience and legal interfaces.

This article has two purposes. The first is to explore narrative structures that facilitate the construction of the content of a *tutela* to establish the facts and claims of the citizen. The second is to define the appropriate technological package to develop this tool within the parameters of a legal interface. For content creation, we furthered the writing of facts and claims using storytelling tools. To develop the interface, we advanced in the database architecture, functionalities, parameterization, and visualization deployment in different devices. The development of legal interfaces becomes very relevant within the growing field of Legal Design since the digitization of legal artifacts needs translation and visualization so that different people can operate them [4]. This article contributes to expanding the technological and design resources that allow the development of solutions to improve access to justice and advance the digital transformation of the judicial system.

1.1 Access to Justice from a Transdisciplinary Perspective

The law determines access to justice. Typically, laws are attributed to the practice of law, as defined, and operated by lawyers. However, laws cover all citizens, with or without legal knowledge. Leaving such a decisive issue as access to justice in the hands of lawyers alone may generate a risk of gaps being formed in the legal system, even more so when considering the modernization of justice systems.

The union of disciplinary knowledge has made it possible to broaden the means of guaranteeing access to justice to more people. Today, it is no longer just about norms but about the touchpoints a citizen has to access any legal mechanism. The digital revolution has broadened the spectrum from physical touchpoints, such as courts, to digital ones. We see cases where legal innovations have made it possible to overcome crises and create new ways of accessing justice [5].

This trend has occurred, among many other factors, thanks to the cooperation between design and law. The last ten years have witnessed the forging of Legal Design as a new branch of design that allows user-centered design methodologies to solve problems in the legal context. Prior to this, other disciplines have been interested in enriching the practice of law. For instance, engineers and lawyers have common interests. Like engineers, lawyers are trained to solve problems following a strict process [6]. The difference lies in the type of input and output results they deal with.

In Legal Design, users are involved in observations, explorations, and dynamic prototyping of ideas. Meanwhile, in Legal Engineering, fields of action are defined, within the regulatory framework, to deploy a measurable solution that impacts the system. In Systemic Design, on the other hand, future scenarios are projected based on the new design proposals, evaluating the most relevant of these [7]. This vision of transdisciplinary work makes it possible to define, analyze, cocreate, propose, and implement solutions that respond to the context's real needs, considering regulatory constraints, stakeholders' visions, and impacts within the system [7].

1.2 Legal Interfaces in a Digitize Legal System

Law, as a discipline, has focused on the formation, creation, modification, and composition of laws and rules framed in the legal environment. Contracts, legal documents, and legal forms are the tangible representations of these laws and rules. These elements, like any object, have been created by individuals with a specific purpose and materiality. In this sense, contracts or legal documents can be called artifacts. These artifacts serve different purposes and inhabit the social context of each country. They are intended to be used by people to regulate possible human behaviors and resolve conflicts [8]. Compared to regular artifacts, the legal artifacts that exist today are so complex that, in most cases, they cannot be used by people who have not legal knowledge. This situation is challenging because laws and rules have to operate in contexts and realities different from those of the law. However, the complexity of the legal system is not an excuse for the judiciary to adapt legal artifacts so that anyone can use them, regardless of their complexity.

In Colombia, the Political Constitution is responsible for issuing the laws, rules, codes, and regulations that make up the legal system. Accordingly, it also defines the *tutela* mechanism. Both the Constitution and the *tutela* are complex legal artifacts that have been constructed with specific techniques, intentions, and goals in mind [9]. As a legal artifact, Colombia's Political Constitution organizes social behaviors, norms, and sanctions for citizens and foreigners. However, its interface, which consists only of technical texts, makes it inaccessible to a large number of people. The legal contents and procedures are complex for citizens, requiring advanced knowledge to read and decipher.

On the other hand, the *tutela,* as a legal artifact, is the primary judicial defense mechanism to protect fundamental rights in Colombia. It is an agile, informal, and simple instrument that anyone can use without legal assistance or representation. *Tutelas* do not require formalities for their presentation; they can be written, oral, formal, or informal. They can be submitted to any of the more than 4,500 judges in the country and must be resolved within ten days. The somewhat ambiguous interface for the *tutela* has been standardized by the judicial operators, taking away the beauty of the mechanism.

The digital transformation program for the judicial branch in Colombia, announced by the National Planning Department (DNP), provides the opportunity to imagine the design of new legal interfaces. Tools like this program can be useful in addressing the issue of unequal access to information in legal documents and enhancing the overall accessibility metrics of justice. This fertile landscape leads to the development of technological projects that help modernize the judicial branch and bring it closer to citizens. Interfaces are one of these technological developments that make it possible to decode the contents of legal artifacts. However, to develop these interfaces, we must understand the relationship between human beings, their needs, and technology.

1.3 The Process of Telling (Legal) Stories

Storytelling involves a process of rationalizing who is involved, what happens during the story, and when and where events occur. The best stories are those that identify a common thread that allows us to understand the relationships between those elements. This is why stories are composed of formal, technical, and structural aspects that can generate a message in a specific language [10]. Formally, storytelling, as a communication model, requires a sender and a receiver. For Lothe [11], this implies understanding the relationships given by Narrator-Narration-Reader. This model describes the importance of the narrator and the content of the message for the reader to decompose. At technical level, the system of signs, symbols, and signals used in a narrative must be comprehensible to both the narrator and the reader. Moreover, at the structural level, the coherence of the content of the message is determined by the way the narrator organizes the elements of the story (the who, the what, the when, and the where).

In the case of *tutelas*, these documents are legal stories that evince violations of fundamental rights. In essence, a *tutela* can be seen as a legal narrative, which means it must contain the essential elements of a story. Many legal documents, such as the *tutela*, identify requirements that regulate procedures and minimum legal content. However, they do not refer to the narrative or how the story is told. They only ask for the facts to be stated. This practice triggers argumentative writing that, beyond highlighting the elements of a story, justifies some actions under legal concepts. This practice results in documents with a very high percentage of technical topics. Most of these documents or artifacts are templates and pro forma formats that do not allow other narrative structures, preventing the narrator from communicating his or her intention more clearly.

In most cases, legal artifacts are constructed in a textual support —a letter size document and 12 pt font—, with a structure determined by a hierarchy of information specific to these documents. Considering that most of these artifacts are used for a person, in this case, a judge, to decide, the reading process should favor making a situation understandable rather than technical over-argumentation. Additionally, such a narrative must consider that the reading process involves different legal actors who intervene and identify the elements in the constructed narrative. In the end, judges are trained to understand legal technicalities, and their decision should be motivated by understanding the facts of a case.

From this perspective, it could be advantageous to identify within the structure of a *tutela* the type of characters, events, and other narrative elements within the stories to mark how information is shaped and communicated. It is, therefore, possible to apply rules and dynamics centered on the legal context when shaping the information. In this order of ideas, the narrator is expected to determine the story's standpoint and the perspective shown to the reader. Thus, the conception of the characters, the events, the chronological order, and other narrative elements within the stories mark their sense of interpretation, verisimilitude, and feasibility within a legal context. From a narrative perspective, it is possible to propose mechanisms for the construction of legal artifacts that allow the creation of specific legal instruments, such as *tutelas*, without legal assistance.

Another critical factor that precedes the *Entutela* project is the overwhelming number of *tutelas* filed each year. In 2022, there were around 600,000 *tutelas*, and this number is expected to grow or remain the same [1]. There is a possibility that the number of *tutelas* will continue to increase if the means to file *tutelas* digitally are facilitated. However, it is essential to note that the goal of the *Entutela* project is not to reduce the number of *tutelas* filed. Instead, it proposes a way to structure the writs so that the input of information in the databases allows for a better analysis of the types of *tutelas* and, therefore, the types of decisions to be made.

2 Methodology: Combining Creative Research with Technical Development

The *Entutela* 2.0 development methodology was based on creative research techniques [12, 13]. The research was developed taking as a starting point the results of the *Entutela* prototype, where the process of writing a *tutela* was simplified using a digital form, establishing a primary database, and generating a document that met all the regulatory requirements to receive a *tutela* [4]. As in *Entutela* version 1.0, the project's focus emphasized the transformation process through research rather than the outcome itself [13]. However, in this case, significant progress was also made in the design and development of the interface.

The research process developed around two themes. The first was related to the experience of the affected users or the need to draft a *tutela*; the second was related to the technical design and development of the interface, understanding the *tutela* as a legal artifact. To assess the user experience, prototype navigation validations were performed with individuals possessing varying levels of legal knowledge (high, medium, or low). These validations were guided by different tools proposed as solutions to help individuals understand and draft a *tutela*. In addition, two storytelling workshops were held at the Legal Clinic of Universidad de los Andes engaging citizens needing legal advice. Meanwhile, for the interface design, efforts were focused on constructing the database, refining tool navigation, and enhancing visual components.

2.1 User Experience Exploration

The legal experience of an individual facing a *tutela* procedure was divided into two explorations. The first analyzed the different stages of filing a *tutela* and the second inquiring about different techniques of drafting the facts and claims of the *tutela*. For the first exploration, five tools were developed to support the secuential process from recognizing the need to file a *tutela* to tracking its progress. The first tool was an informational video to help individuals determine whether a *tutela* might be the proper judicial mechanism for their case. The second tool was a checklist for individuals they fulfilled the necessary criteria for a *tutela*. The third tool was *Entutela's* enhanced digital form, designed to enable individuals without legal expertise to create and visualize a *tutela* in order to improve the experience of access to justice. The fourth tool involved a set of cards explaining, in *clear language*, the fundamental rights in Colombia, what they are about, and provide some examples. The fifth tool was a calendar to allow the claimant to follow up on the stages of the *tutela* after filing it (see Fig. 1).

The first exploration consisted of eight validations of the prototype conducted with people with high-level legal knowledge (4), average legal knowledge (1), and low or no legal knowledge (3). These people resided in different departments of Colombia (Bogotá, La Guajira, Casanare, Cundinamarca, and Norte de Santander). All interviews were conducted virtually and lasted 2 h. To validate the prototype and explore the five tools, participants had access to a legal case, which was used to observe how those with high, average, and low or no legal knowledge made decisions on how to file a *tutela*.

The second exploration consisted of two co-creation workshops focused on building the basis for drafting the facts and petitions of a *tutela*. The workshops were held at the Legal Clinic at Universidad de los Andes and lasted 90 min. A total of 8 people participated, 3 in the first workshop and 5 in the second. All participants were citizens who were seeking legal advice regarding their cases. The workshops were held outside office hours, and each participant was offered a voucher equivalent to 20 Euros.

The workshops began with a presentation by the team members and the participants. An introduction to the workshop activities followed this. The first activity was called "identifying your story" and was designed to identify the requests to be made in the case story. In this activity, based on their case, the participants had to identify (i) the characters involved in the case (protagonists and antagonists), (ii) the request they wanted to make to the antagonist, (iii) the moment when the situation occurred that led them to set up the case, and (iv) the place where the events occurred. This activity lasted between 15 and 20 min. For the second activity, "How to tell the story," each participant could narrate the story with the four elements identified in the previous activity. Once they had socialized the story, they were given a guide to structuring the story in the form of a template. In the template, they had to include "who," "what," "when," and "where" for each sentence they wrote. Once they finished restructuring the story, they returned to socialize it by reading the template. This activity lasted between 40 and 60 min. The workshop closed with a space for reflection where participants were asked to comment on how they felt, what was the most accessible

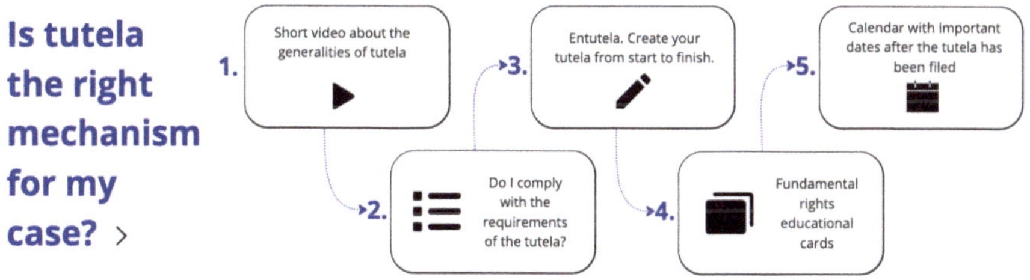

Fig. 1 Overview of the navigation process to present the tools to the participants

and most challenging part of the story to tell, what helped them most in structuring the story, and what they would change about the activity.

2.2 Technical Requirements of the User Interface

The user interface was divided into three layers to ensure a comprehensive and effective design, and encompassed a digital platform to add more tools besides *Entutela*. These included some of the ones that were prototyped in the user experience exploration phase. The first layer was the data layer, consisting of a database manager (Mysql) and a tool to parameterize the data (JSON). The second layer was the logic layer, where we explored how to input the information for each tool and how the data could be stored (data layer) and used (web layer). The third layer was the web layer, where the navigation paths of the digital tool were proposed. To ensure the functioning of all the technological infrastructure layers, several characteristics were defined. These included hosting, web browsing protocols that allow authentication and data encryption, content manager, database software, SEO, navigation analysis system, and source code repository. By incorporating these features, the user interface was able to function efficiently and provide a seamless user experience in the construction of the legal stories for *Entutela* users (see Fig. 2).

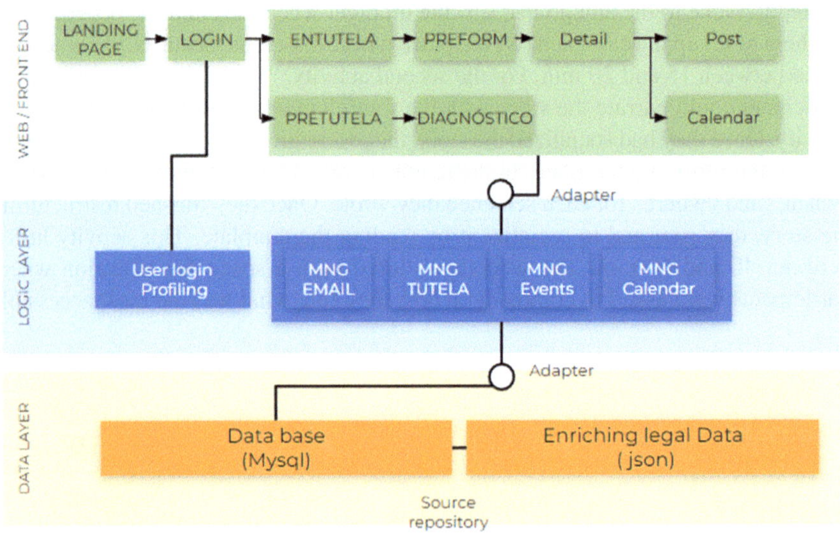

Fig. 2 Diagram of the 3-layer technology infrastructure and how the layers interconnect

3 Results: Legal Narratives Versus Life Stories

The results of the scans can be divided into three parts. The first relates to the navigation and connection of the five tools. In this exploration, we wanted to identify the gaps and differences between profiles with high legal knowledge versus those with average, low, and no legal knowledge. The second relates to the user experience and facts and claims content creation. In this exploration, storytelling structures [10, 11], visual thinking tools [14], and ways of rationalizing information [15] were used. The objective was to help simplify, organize, and prioritize the contents of the facts that led the violation of a person's rights, and to simplify, organize, and prioritize the response expected by that person as part of their reparation. The third is related to the design and development of the user interface. It relates to more technical aspects concerning database architecture, operation, parameterization, and deployment. The idea here was to create a digital platform to store several legal digital tools for citizen use; however, this process was only undertaken for *Entutela.*

3.1 Understanding Gaps Through User Prototyping

Overall, there was a notable difference between the three groups of people, not only in terms of legal knowledge classification but also in terms of differences in literacy, especially for the group with little or no legal knowledge. Those with high-level legal knowledge (HLK) found the toolkit to be an exciting initiative to promote education in legal issues and access to justice. This perception was given because they felt that the content was adequately focused on citizens (especially those in urban areas). On the other hand, they proposed adjustments that they considered could improve citizens' experience. These adjustments included the possibility to access the tools through social networks, have downloadable guides for the different tools, and ways to automate the content to link to the official platforms of the Judicial Branch. Those with average legal knowledge (ALK) found great value in using the tools, so they intended to recommend them to their acquaintances. This perception was due to the fact that they felt that the information presented, and the use of the tools was clear, and because they felt legally supported while using the tools. The only comment that stood out was to keep the website as simple as possible, i.e., not to include more tools than strictly necessary, as it could generate confusion. Finally, for those who had little or no legal knowledge (LNLK), the navigation experience was not as positive as for the other profiles. First, they needed help to understand the terminology and chronology of events. These difficulties were still evident despite using other media, such as video or infographics, to present the information. Another aspect that was mentioned was the process' lack of humanization, especially in terms of the texts and descriptions of each tool. Some participants required clarification as to what kind of tools they could use or whether they could only use them online. This comment was made in relation to areas where there is no access to the internet or a computer.

Table 1 The participants' responses after navigating the tools to create and file a *tutela*

No.	Cat	Video	Checklist	*Entutela*	Calendar	Others
1	LNLK	It needs to display more information	It needs to be more empathetic and human. Include real situations	Needed assistance. Mixed up the current situation with facts	Found it enjoyable	Access barrier to digital tools
2	LNLK	It needs to provide more security for legal actions	It is easy to understand. Less technical more colloquial	Difficult to fill it in CHRONO order. The rest was easy to use	Found it useful	Favorite tool: *Entutela*
3	ALK	It needs to be more precise with the *tutela* process timing	Confusion due to certain terms	Email verification can be a burden	More visual	Recommend the tools
4	HLK	N/A	Adjust in terms, dates, and legal considerations	It needs to be more precise with facts and evidence		Not sure if it will work in rural areas
5	HLK	N/A	It should suggest alt legal mechanism	Provide examples of how to initiate the claims	Agrees with the dates	It could foster legal education
6	HLK	N/A	Validation system on fundam. rights	Suggest to reverse the order of some questions	It should be sent with the PDF	Filling instructions should be included in the PDF
7	LNLK	Acknowledges the need for a *tutela* in her case	Needed assistance with some terms	Struggled to understand how to make requests	N/A (Iteration process)	It is still difficult to understand
8	LNLK	N/A	Technically OK, but difficult for seniors	Mixed up the current situation with facts	N/A (Iteration process)	N/A

Each participant was asked about each tool on a case-by-case basis. Table 1 lists the responses and perceptions of each participant.

3.2 Structuring Content as Stories

In *Entutela* version 1.0 [4], the Path of Expressions [16] was used as a reference to understand the pattern of a *tutela*. The current problem represents the "today" field. The events and facts that led them to their current situation represent the "past" and the "memories." The claims and what they expect to be resolved represent the "future" and the "dreams" [4]. However, the wording of the facts and demands was still very terse. In the cases presented in *Entutela* 1.0, long texts were still evident, with no grammatical structure or precise chronology. To better structure their stories, the participants performed several exercises to work on a new way of telling their stories.

During the initial exercise, participants were tasked with identifying the elements necessary to construct their stories before sharing them. Of the eight participants, seven required assistance in identifying these elements. This was mainly due to the complexity of their stories, which often involved multiple locations or significant time gaps. In other cases, there needed to be more clarity about the characters. For example, in matters related to health, it was important to determine which entities (health-promoting organizations, health service providers, the Ministry of Health, or other relevant entities) should be included. After completing the first exercise and drawing up the list of elements with the assistance of the facilitators, the participants shared their stories. However, in most cases, the list did not prove to be helpful in structuring their stories. In a few instances, the facilitators attempted to refer to the list to try to help structure the stories.

In the second exercise, participants were asked to rewrite the story chronologically following a template (see Fig. 3). The template consisted of five sections that provided the structure for the narrative:

1. Identify themselves and identify the problem, associating it with a fundamental rights issue from a list of options.
2. Identify the facts by reporting when the event occurred, what happened, and where it happened. For this item, three lines were left to be completed with more events.

Fig. 3 A participant filling out the chronological structure template during the storytelling workshop

3. Identify the reasons why the situation is affecting them.
4. Enlist the person's actions to solve this situation and the answers they obtained.
5. Identify who or what entity is violating their fundamental right.

The participants had to name the person or entity responsible for responding to their claim. At this point, it was suggested that they begin drafting the part of the request with verbs such as "authorize," "order," "inform," or "respond," among others. At the end of the form were three examples of how to write these requests. All eight people filled out the forms and left with more precise ideas about how to tell their stories, despite significant challenges in understanding the forms.

3.3 Shaping the (Legal) Interface

The three layers of the digital platform are interconnected and work together to ensure a seamless and reliable user experience. From a user perspective, the interface must fulfill a goal, it must be able to make it easy, and it must be able to give rise to an emotion [17]. In the case of *Entutela*, the platform was prioritized for developing these layers because it had the most information and made it possible to build the most robust content. This allowed for a comprehensive and thorough approach to the data structure, interactions, and presentation layers.

User explorations helped us to refine the tool's architecture and navigation, which, in turn, ended up defining the main and essential functions that an interface needs to generate content, store data, and authenticate information. Thirty-seven requirements were created, prioritized, and classified according to complexity through *Entutela's* decision tree. These requirements were used to define the visible modules, which, unlike the first form, allow users to go back and forth in the fields to be filled in. including this functionality was crucial as the construction of history rarely occurs linearly.

The initial module is dedicated to user registration, which involves the creation of a personal profile, acceptance of the tool's terms and conditions, and agreement to an oath confirming that no previous *tutela* actions have been filed. This oath serves as a legal requirement to validate a *tutela*.

The second module consists of five navigable options. The first option involves defining one's role in the *tutela* process. This allows individuals to indicate whether they are filing the *tutela* on their own behalf, on behalf of someone else, or with legal representation. The second option involves defining the specific petition and how the issue is expected to be resolved. The third option involves selecting the specific, which should be aligned with the petition stated earlier. The fourth option is dedicated to narrating the facts of the case. In this section, an updated format from the workshops is used, which includes subcategories for facts, impacts, and solutions. Participants are encouraged to provide one idea per sentence under these subcategories. The fifth option allows users to review the document and generate a PDF version of the *tutela* for further use (see Fig. 4).

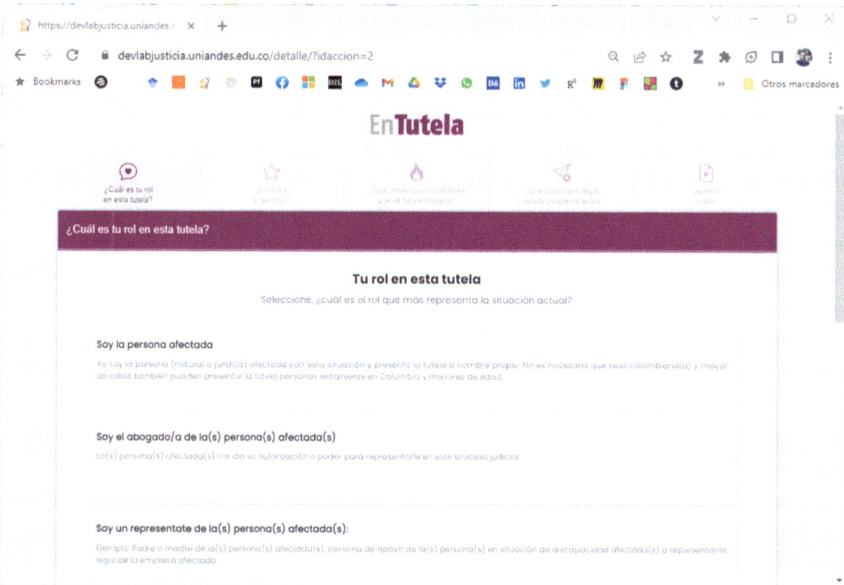

Fig. 4 *Entutela*'s second module interface for writing and reviewing the content of a *tutela*

Considering that this tool serves as an interface for the legal document known as the *tutela*, the PDF form was redesigned taking into account the insights gained from the judges' readings [4]. Once the PDF is created, the user receives an email notification containing instructions on how to upload the document to "*Tutela en línea*." The email also provides information on important dates for process monitoring. The option to edit the *tutela* is also available, as it allows for possible additions to be made before submitting the document in the future.

4 Discussion: Challenges and Opportunities

Legal artifacts require legal interfaces, just as regular artifacts require interfaces [18]. However, there is a significant difference when it comes to the normative framework surrounding legal artifacts. There is a perceived resistance among lawyers, judges, and other judicial actors towards embracing new ways of materializing these artifacts. This makes it challenging to imagine legal interfaces that go beyond traditional judicial parameters. *Entutela* 1.0 and *Entutela* 2.0 are designed to break these paradigms by presenting legal information in a way that makes sense to the user, specifically to citizens with low or no legal knowledge.

The project encountered more challenges than anticipated in its pursuit. The gap in understanding legal terms proved to be a more complex issue than simply presenting them in plain language. Despite the significant linguistic development achieved by

Entutela, many participants still struggled to comprehend the actions and decisions they needed to make in order to craft their stories. Individuals are accustomed to receiving legal assistance from various legal actors, such as students from the legal aid clinic. Typically, citizens would approach the legal aid clinic, present their case, and rely on the clinic staff to solve their situation. This experience has both advantages and disadvantages for citizens. On the one hand, they can delegate their cases to someone who "knows" what to do. On the other hand, they feel restricted because they lack the tools to express their situations themselves.

Following the workshops with the participants, numerous questions emerged, outnumbering the solutions. One possible explanation is linked to the medium employed during the workshops. Paper templates have inherent limitations compared to the capabilities that digital tools offer. For instance, paper cannot correct spelling errors or automatically arrange facts chronologically. However, when considering the overall development of the process, paper prototyping proved to be the most appropriate choice for conducting the workshops.

Nevertheless, the majority of participants found the experience to be refreshing. The opportunity to write and narrate their story using different methods allowed them to refine their narrative. However, it remains unclear whether this refinement helped them in achieving their goal of safeguarding their fundamental rights. Based on the project findings, a hypothesis emerges that suggests that if we expect citizens to embrace this tool, facilitators may need to address the onboarding experience. This notion aligns with the recognition that digital tools can potentially pose another barrier to ensuring access to justice [4]. Therefore, facilitators can contribute to expanding access to and utilization of these tools by using them as educational resources or conducting legal brigades to reach more people in rural areas.

However, in future iterations, attention must be given to the back-end components of the platform to ensure proper storage of information and expose the data stored in each tool. Strengthening security and access protocols is also crucial to prevent system vulnerabilities. Despite the effort to translate the content into a more colloquial version, the current display of information on the platform remains complex. This complexity may stem not only from the more precise language, but also from the meaning of the terms themselves. Future collaboration with linguists and semioticians should be considered to improve the language aspect. It is important to note that the core of this tool lies in the legal content made available to citizens. The functionalities of filing petitions and identifying fundamental rights elevate this tool to a higher level, as they minimize the need for legal knowledge and reduce uncertainty when individuals encounter legal situation.

Nevertheless, further exploration is needed to enhance these functionalities. For example, there is room for improvement in incorporating metadata derived from judgments and other digital legal artifacts. An exciting idea is to develop artifacts such as personas and user stories before starting the development process. Finally, the mobile experience needs to be reconsidered with a focus on generating value for the user. Currently, the exact definition of what this experience entails is yet to be fully defined.

Entutela 2.0 represents a groundbreaking leap in legal interfaces for *tutelas*, positioning itself as one of the most advanced tools available. Its introduction marks a transformative moment in shaping the future of justice in Colombia. The tool revolutionizes how citizens engage with the justice system, offering relatable means of access. It redefines the role of data in the system, emphasizing its use to enhance efficiency. Rather than focusing solely on the quantity of *tutelas* filed, the emphasis shifts towards how the justice system can effectively safeguard fundamental rights.

References

1. *Corte Constitucional de Colombia* (2022) Tutelas filed at the Constitutional Court. https://www.corteconstitucional.gov.co/lacorte/estadisticas.php. Accessed 10 Mar 2023
2. Moreno Torres MC (2021) Los nuevos entornos tecnológicos y su impacto en la garantía del derecho humano al acceso a la justicia: Un enfoque en época de covid-19. Revista chilena de derecho y tecnología 10(1):309–327
3. Ariza-Sosa GR, Agudelo-Galeano JJ, Saldarriaga-Quintero LA, Ortega-Mosquera MC, Cecilia D (2021) Crisis humanitaria de emergencia en Colombia por violencia contra las mujeres durante la pandemia de COVID-19. Revista de la Facultad de Derecho y Ciencias Políticas 51(134):125–150
4. Padilla Casas C, de Francisco Vela S (2022) Entutela. A digital tool to facilitate the drafting and creation of tutelas in Colombia. In: Advances in design and digital communication III: Proceedings of the 6th international conference on design and digital communication, Digicom 2022, November 3–5, 2022, Barcelos, Portugal, pp 366–376. Springer Nature, Cham
5. Sourdin T, Li B, McNamara DM (2020) Court innovations and access to justice in times of crisis. Health Policy Technol 9(4):447–453
6. Howarth D (2013) Law as engineering: thinking about what lawyers do. Edward Elgar Publishing
7. De Francisco Vela S, Guzman-Abello L (2023) Systemic design of justice: transdisciplinary approach to access to justice. (Legal Design Roundtable 2023) [Manuscript submitted for revision]
8. Harwood J (2011) The interface: IBM and the transformation of corporate design, 1945–1976. U of Minnesota Press
9. Kroes P (2002) Design methodology and the nature of technical artefacts. Des Stud 23(3):287–302
10. Wei SL, Wei H (2006) Uncovering hidden maps: Illustrative narratology for digital artists/designers. Comput Compos 23(4):480–502
11. Lothe J (2000) Narrative in fiction and film. An introduction. OUP, 0xford
12. Candy L (2006) Practice based research: a guide. CCS report 1(2):1–19
13. Daza SL (2009) Investigación—Creación, un acercamiento a la investigación en las artes. horizontes pedagógicos. https://horizontespedagogicos.ibero.edu.co/article/view/339/303. Accessed 11 May 2022
14. Kernbach S (2019) Visual thinking in life design: a conceptual framework of visual tools and templates. In: 2019 23rd International conference information visualisation (IV), pp 329–334. IEEE
15. Roam D (2009) Back of the napkin: Solving problems and selling ideas with pictures. Marshall Cavendish International Asia Pte
16. Sanders EBN, Stappers PJ (2014) Probes, toolkits and prototypes: three approaches to making in codesigning. CoDesign 10(1):5–14
17. Walter A (2020) Designing for emotion by Aarron Walter. A Book Apart. https://abookapart.com/products/designing-for-emotion. Accessed 4 Apr 2022

18. Polanía A, de Francisco Vela S (2022) Designing interfaces for legal artifacts. using storytelling to enhance digital legal experiences. In: Advances in design and digital communication III: Proceedings of the 6th international conference on design and digital communication, Digicom 2022, November 3–5, 2022, Barcelos, Portugal, pp 137–147. Springer Nature, Cham

The Impact of Visual Design on Public Participation. Case Study in Cyberjournalism

Sara Alves and Andreia Pinto de Sousa

Abstract Indispensable for the proper functioning of democratic societies, journalism assumes a pivotal role in democracy and the community. Although information is the primary element in this area, the visual communication design is equally relevant. Placed on the same level as language and communication, cyberjournalism and visual communication contribute to creating the social world. Moreover, the effects of visual communication appear before one realises or reflects on them, influencing greater or lesser public participation. This manuscript seeks to understand how visual communication and the design adopted by online news platforms can affect public participation in cyberjournalism. To conduct this study, a literature review and methodological analysis were carried out, which focus focused on three online news platforms and the public. The analysis corpus is divided into five moments that address themes of cyberjournalism, participation design, design principles, design elements, credibility, and confidence. The results we obtained are relevant for distinct areas: Human-Computer Interaction, allowing the adoption of more conscious techniques in terms of interface and communication design; for the public, since what is at stake is the knowledge they derive from news content; and for news producers, encouraging quality consumption and informed participation in increasing digital literacy through the adoption of meaningful techniques towards more conscious forms of interaction.

Keywords Visual communication · Human-computer Interaction · Participation

S. Alves · A. P. de Sousa (✉)
Lusófona University, Porto, Portugal
e-mail: andreia.pinto.sousa@ulusofona.pt

A. P. de Sousa
HEI-Lab: Digital Human-Environment Interaction Labs, Lisbon, Portugal

N. Martins et al. (eds.), *Perspectives on Design and Digital Communication IV*, Springer Series in Design and Innovation 33, https://doi.org/10.1007/978-3-031-41770-2_3

1 Cyberjournalism and Participation: The Relationship with Visual Communication Design

Indispensable for the proper functioning of democratic societies, journalism plays a structural role in democracy and society. Based on five central pillars: "the notion of public service, autonomy, objectivity, ethics, and immediacy" [1–3], allow the acquisition of knowledge necessary for a life in society [4] and protect citizens from possible abuses of power [3].

On the other hand, the widespread use of the Internet and new technologies has caused a structural change in all sectors—at the social level, transforming how individuals communicate and relate, and at the business level, modifying how they communicate with the public. This new space, made up of networks, is also called cyberspace, "a communication space opened by the worldwide interconnection of computers and computer memories that includes the set of electronic communication systems" [, p. 92].5

The area of journalism was one of the sectors where this transformation was felt, and due to the need to adapt to a new reality, a new form of journalism arose—cyberjournalism. As [1] indicates, "innovative narratives and unprecedented practices emerged, new genres were born (...) they led the news media to reconfigure themselves to respond to the demands of the new medium, the trends of the moment and the growth and sophistication of online audiences".

This audience, built around information networks, also known as the "network society" [6–8]. At the same time, studies [9] also demonstrate that the news media sought to encourage and support public participation in the news context.

Although information is the primordial element of journalism and online journalism, the accompanying visual communication plays an equally relevant role. The design creates the social and cultural world on the same footing as language and communication [10]. As [11] explains, vision precedes perception, which means that the effects of visual communication appear before you realise or reflect on them. In the case of visual design, there is an impact on how readers receive and interpret information, on trust, usability, and even on the value attributed to themes [12].

Nevertheless, not only that, the impact of visual design can be felt at several levels, such as: increased public involvement with information as well as greater attention [10], it can contribute positively or negatively to emotional and cognitive processes [13, 14], affects the understanding and knowledge that comes from reading the news [15, 16], lead to greater or lesser confidence [17, 18], or even influence the way readers navigate platforms [19].

Given the range of cases influenced by visual communication, it is understandable that the visual design adopted by online news platforms can positively affect or negatively affect participation. However, it remains to deepen the way this impact is processed. As some authors [20–22] indicate, participation and design in journalism have received little attention compared to other studies.

2 Elements for Visual and Information Architecture Interface Dimensions

As mentioned, design impacts how the reader receives and interprets the message. Several authors have studied the impact of design principles and elements on interface perception. The Interaction Design Foundation defines design principles, which determine how the features that build visual literacy are used, as broadly applicable design guidelines or considerations. As [18] points out, these guidelines allow products, services, or systems to be helpful, desirable, and easy to use, which can determine the success of that same product, service, or system. Analysing those elements and principles on online news platforms is crucial to understanding how they can influence public participation. Due to o the nature of these platforms in this study and based on the interface analysis framework proposed by [18], the focus is on two interface dimensions proposed by the researcher—the Visual Dimension and the Information Architecture Dimension.

Design elements build visual literacy and play a key role in visual communication. Therefore, understanding these elements is crucial to understanding how they affect readers and how they can influence public participation in a news context. In this context, the study focused on five elements that make up visual literacy: typography, colours, images, dimensions, and composition.

2.1 Typography

In the case of online news platforms, which work essentially with information, typography is one of "the most important elements in the graphic design of the online newspaper (…) it must be simple, effective and profound" [23, p. 271]. Having these characteristics as objective to attract the public and allow the information to be skimmed, the main point is adopting good typography techniques not to compromise the understanding and knowledge from this reading. Therefore, typography is "a design discipline that involves the use of typefaces and the organisation of those typefaces to create readable, usable, and user-friendly interfaces or experiences" [24, p. 2].

The ease of use of an online news platform is related to the ease with which the reader can read and understand the information, and typography can help in this task. Palomo [21] refers to a set of stable characteristics in these platforms: the use of italics and underlines is avoided, font sizes are similar; use of one or two fonts; combine serif and sans serif fonts; texts are left-aligned; and the design is simple.

2.2 Colours

In a journalistic context, colours significantly impact the reader. When used correctly, they can increase the success of the message [25]. Of the various repercussions on the reader, the following can be listed: the impact on trust, satisfaction, and loyalty [19, 26], provoking more significant learning and motivation [27], affect how the public retains information [19], or cause a positive or negative psychological impact [28, 29].

Studies [19, 26] indicate that cold colours are seen more positively than warm colours, with blue being considered the most favourite colour and grey being the least favourite. On online news platforms, [21] and [30] found the following patterns: a combination of cold and warm colours, text in black or grey colour; distinct colours to draw the reader's attention; and recurrent use of blue, red, grey, and black.

2.3 Images

The use of images on the homepages of online platforms is also one of the elements that significantly impact the reader [21]. Through eye-tracking studies [31], there is value in including these images. This element improves the reader's experience in several ways: it influences satisfaction, trust, loyalty, and attention [19], encourages greater public involvement [32], and the space occupied by the image indicates the importance of the news or theme.

The position in which these images appear on the interface should also be considered based on two fa̧ctors [19]: when there are faces, visualisation is more dispersed, and when these are placed above the midpoint of the page, longer fixations are recorded, leading to a diversion of attention to relevant information such as titles or texts.

2.4 Dimension

The dimensions of the elements present in the interface play an increasingly important role, especially for news brands that project the same content to different devices, from smartphones to computers or tablets. For example, [33, p. 509] state, "phone-sized screens that support multi-touch interactions require onscreen objects to be large enough that they can be activated easily with fingers, without the user accidentally triggering other interactions while doing so".

Studies [15, 16, 34–36] indicate that the device used in news consumption directly influences knowledge and understanding of news content, with the computer recording more positive values than the smartphone. The dimension of the elements can also be attributed to the visual focus, in which the more significant elements have greater visibility [37]. However, it is necessary to know how to size these elements so the reader can understand other variables, namely the shape [33].

2.5 Composition

Composition is the union of all the design elements within the layout and how they relate and communicate. Therefore, it should convey a sense of balance, coherence, and aesthetics. In this composition, four relevant aspects should be taken into account [38]: space, which allows the reader to obtain a greater understanding of the contents, a good positioning of the typography, from the titles to the subtitles; the use of colours to draw the reader's attention and divide up space; and the creation of different sizes and widths, which encourages a feeling of rest for the reader.

In the news context, there are similarities in the design used [21], namely dark text on a white background, the use of colours such as blue or red, and, mainly, the use of modular grids [32].

3 Subjective Factors: The Impact on Participation

Credibility, trust, perception, and aesthetic judgment are individual factors that directly affect participation precisely because they are subjective and vary from individual to individual. For example, the credibility of the brand is perceived on a personal basis. It depends on the trust of the contents, each person's perception of the news brand, and their aesthetic judgment of the platforms can lead to greater or lesser participation. As [17] points out, "journalistic quality is not always judged by its procedural or discursive characteristics (...) it is natural that it is affected by the reader's previous experience".

Credibility is an "attribute built through a relationship between the producer/issuer of information and the receiver. It is a quality attributed to the speaker, institution, or discourse from the moment it proves credible" [39]. One design principle that influences credibility and trust is Advertising [18], the higher the number of advertisements, the less credible the platform tends to be. Therefore, "it is important to understand and design websites so that they have credibility" [18, 40].

In the journalistic field, higher levels of trust lead to greater participation by the public, thus making the notion of trust one of the central pillars of the field. Mainly because "one factor that allows journalism to have a social function within a democracy is the trust that the public chooses to place in communication products" [17].

Readers' perception and aesthetic judgment of a particular news brand become even more challenging to assess and quantify. Nevertheless, the data show that the visual aesthetics of websites is crucial to attracting and retaining the public, not least because this judgment is made in less than 500 ms [13] and is made based on four categories—the beauty of the platform, the illustrations and text, the composition, and the structure.

Although these individual factors are subjective, they must be considered because they can influence public participation. Consequently, there is a connection here with the visual design adopted by news platforms.

4 Methodological Approach, Analysis, and Results

This study was conducted from October 2021 to November 2022. The results presented in this manuscript result from a detailed methodological proposal previously published [41]. Two methodological instruments were used to understand how visual communication design adopted by online news platforms can affect public participation in online journalism: a questionnaire survey and observation grids.

The questionnaire survey investigated the public's consumption habits, participation in a news context, perception, and design influence based on three Portuguese online news platforms—Público, Jornal de Notícias, and Observador. This questionnaire was available from May 13 to August 5, 2022, and had 104 responses. In addition, an analysis grid was used to compare the above-the-fold news on the front pages of the three selected media. This analysis was conducted for seven days on the three media, from the 22nd to the 28th of August 2022, where 165 news were analysed).

It is also essential to clarify why news platforms such as Público, Jornal de Notícias, and Observador were selected. According to the Digital News Report 2021, in the online brands sector, Jornal de Notícias is the second most consumed medium (23.6%), and Público is in third place (21.9%). In the sector of digitally native brands, Observador is the third most consumed brand (25.4%), and, comparing print and digital circulation, Público has values that surpass print [42]. In addition, Público is also the most awarded Portuguese newspaper with the European Newspaper Award, an award that highlights news brands distinguished by their print and digital graphics. It should also be noted that, although there are other brands with more favourable positions compared to the categories presented here, these were not selected either because some of the participation indicators are not visible or exist on the platform. Therefore, given the exposed data and the need to analyse the indicators referring to participation, Público, Jornal de Notícias, and Observador were the media that best fulfilled the requirements.

In the context of this study, whose main objective is to understand how visual communication and the design adopted by online news platforms can affect public participation in cyberjournalism, the discussion and analysis of data are based on the design and perception of respondents regarding design. Although we are aware of the importance of other factors that influence participation, such as motivations, the content of news and comments, and even the ideological positioning of newspapers, political and ideological issues, however, the focus of this study falls exclusively on the design and the perception of respondents towards design.

4.1 Public Perception

Through the questionnaire survey (available from May 13 to August 5, 2022, with a total of 104 responses), it was understood which parameters were most valued by the public on the platforms (Table 1): the credibility and trust that the brand transmits (97.1%) and the proper functioning of the platform (97.1%), the brand itself (86.5%) and the visual aesthetics (79.8%). Even though visual aesthetics are the least valued, it should be noted that the adopted design influences the platform's credibility, trust, and proper functioning.

When asked about preference in terms of information (Table 2) and design (Table 3), the results vary: in terms of information, Observador stands out (39.41%), Jornal de Notícias (46.15%) and Público (36.53%); in terms of design, Público (42.30%), Observador (35.57%) and Jornal de Notícias (41.34%) stand out.

Regarding trust (Table 4), the most trustworthy (Public 94.21%) is also preferred in design, which could indicate that design influences the public's trust [43].

Based on the results obtained, it is understood that the public's perception of news platforms could be decisive participation. Of these aspects, the relevance of parameters such as the brand itself (Table 1), the credibility it conveys (Table 1) and the way in which they perceive not only the media information (Table 2) but also the adopted design stands out (Table 3).

Table 1 Questionnaire survey: perception of the importance of platform parameters (question 3.4)

	Extremely important, very important, important (%)
The news brand itself	86.5
The credibility and trust that the brand conveys	97.1
The visual aesthetics of online platforms	79.8
The proper functioning of the news platform	97.1

Table 2 Questionnaire survey: preference for informative content (question 3.1)

	Most preferred (%)	Neutral (%)	Least preferred (%)
Público	35.57	27.9	**36.53**
Jornal de Notícias	25	**46.15**	28.84
Observador	**39.41**	25.96	34.61

Table 3 Questionnaire survey: design preference (question 3.3)

	Most preferred (%)	Neutral (%)	Least preferred (%)
Público	**42.30**	35.57	22.11
Jornal de Notícias	23.07	35.57	**41.34**
Observador	33.65	**35.57**	30.76

Table 4 Questionnaire survey: level of trust rating in brands (question 5.1)

	Extremely confident, moderately confident, confident (%)
Público	94.21
Jornal de Notícias	90.37
Observador	85.56

Table 5 Questionnaire survey: use of technological devices in news consumption (question 2.2)

	Very often, often, occasionally (%)
Smartphone	94.22
Computer	69.21
Tablet	18.25

Table 6 Survey by questionnaire: types of participation and frequency (question 4.1)

	Very often, often, occasionally (%)
Share the news on digital social networks	40.36
Comment on news content	19.21
Read the news	97.1

4.2 Participation and Behaviours

The questionnaire survey also shows that the most used device for news consumption (Table 5) is the smartphone (94.22%), computers (69.21%), and tablets (18.25%). The most common type of participation (Table 6) is reading the news (97.1%), debating (87.48%), sharing (40.36%), and commenting (19.21%).

Observador has the most significant number of debates among the public. It is also important to refer that the discussion is moderated and controlled due to hate speech and needs a login to the platform. In Jornal de Notícias, the comments are made through access to Facebook, and there is no moderation. In this case, the speeches seem to be more hostile.

4.3 Visual Design: Public Preference

Concerning the colours, the respondents have shown a preference (Fig. 1.) for the colour palette of Observador (Option 3) (43.3%), Jornal de Notícias (Option 2) (29.8%), and Público (Option 1) (26.9%). In this result, the brands mostly use blue, red, and yellow, confirming the public's preference for blue.

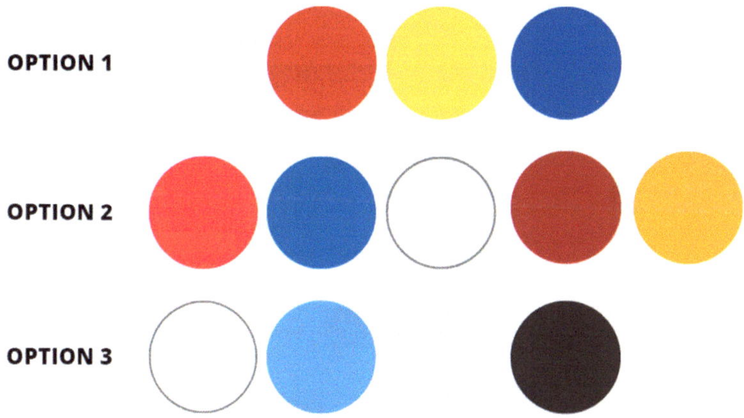

Fig. 1 Respondents' preference regarding the colour palettes of online newspapers

Furthermore, regarding navigation bars (Fig. 2), the respondents prefer bars that show all sections visible (option 1), facilitating navigation on the platform and searching for information.

Finally, when questioned about their preference for the homepages (Fig. 3), our participants preferred Observador (51%), Público (34.6%), and Jornal de Notícias (10.6%), respectively.

As will be seen below, this page considered the most aesthetically pleasing is the one with the highest levels of participation. Therefore, there is also a relationship with the design principles of Alignment, Balance, Unity, and Harmony, Space that refer to a sense of coherence, aesthetics, and balance.

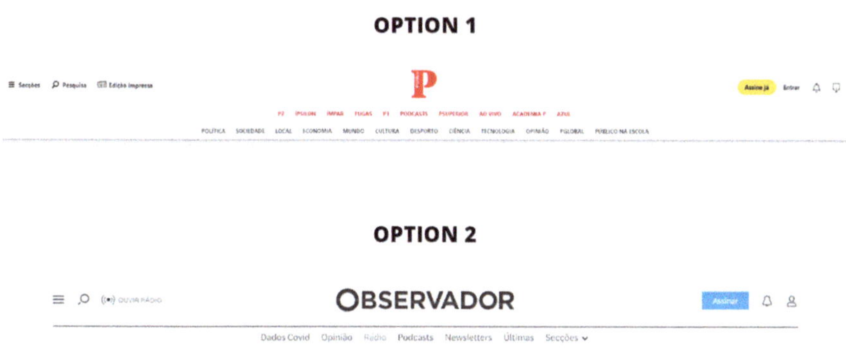

Fig. 2 Respondents' preference for navigation bars

Fig. 3 Preference of respondents regarding the front pages of online newspapers

4.4 *Participation by Newspaper and Relationship with the Design*

As for the analysis of the three selected newspapers is carried out on the three online news platforms, namely the above-the-fold news (total of 165 news). This analysis observed aspects such as the number of comments, presence of debate, the position of the news in the grid, presence of highlight, type of highlight, presence of image, size of the image, and the number of advertisements.

Through the analysis of the news of the three platforms, it appears that the medium with the most significant participation (Table 7) is the Observador, with 782 comments (average of 17 comments per news), followed by Público with 272 comments (average of 5.13 comments per news) and Jornal de Notícias with 182 comments (average of 2.75 comments per news).

Table 7 Participation by platform

	Público	Jornal de Notícias	Observador
Total number of news	53	66	46
Total number of comments	272	182	782
News average	7,57	9,42	6,57
Average participation	5,13	2,75	17

The results also show that the platform that publishes more news (Jornal de Notícias) has the lowest participation levels, and the one that publishes less news (Observador) has the highest participation levels. Such findings seem to demonstrate that a greater news flow does not correspond to greater participation; quite the contrary. This result could be related not only to the visual design but to the fact that Observador keeps non-daily news on the front page and, as mentioned before, to other issues not surveyed in the study and should be verified in further studies. Although, a more significant number of news can make the newspaper visually denser and reduce participation. A denser visual stain indicates too much news, too much information to absorb, and various colours. On the other hand, a lighter visual stain could mean less news, information easier to process, reduced range of colours. Therefore, lighter visual spots could present better participation results.

4.5 *Impact of Images and Dimensions*

Regarding images and dimensions, the results seem to indicate a value in including images (Table 8) on journalistic front pages and that this presence could increase participation. For example, the platforms with the highest levels of participation (Público, Jornal de Notícias) use the most images. To determine the size of the images, as we did not have access to the actual size, a 12-column grid was used to carry out this calculation.

Observing the data obtained, in the case of Público, there is 13 news without an image and 54 comments and 40 news with an image and 218 comments. In the case of Jornal de Notícias, there are 12 news without images with 37 comments and 54 with images with 145 comments.

However, one of the platforms (Observador) demonstrates the opposite; it uses fewer images, and participation levels are higher in this case. In this case, there is 25 news without an image and 669 comments and 21 with an image and 113 comments. As mentioned, Observador is the only medium that presents non-daily news on the front pages, which could explain these results. Furthermore, in this newspaper (Observador), the strategy used is to include relevant news in prominent

Table 8 News without and with image and participation

	Público		Jornal de Notícias		Observador	
	Without	With	Without	With	Without	With
Number of news	13	40	12	54	25	21
Number of comments	54	218	37	145	669	113
News average	1,86	5,71	1,71	7,71	3,57	3
Average participation	4,15	5,45	3,08	2,69	26,76	5,38

Table 9 Public participation through the size of the news image

	Público			Jornal de Notícias			Observador		
	9 col	4 col	3 col	9 col	3 col	$\leqq 2$ col	6 col	3 col	1 col
Number of news	6	3	28	7	19	28	7	10	3
Number of comments	11	1	199	18	49	65	12	62	39
Minimum variation in the number of comments	0	0	0	0	0	0	0	0	0
Maximum variation in the number of comments	6	1	49	10	13	17	6	22	38

positions and vary this position according to the day it is repeated on the home page. One day, a given piece of news may show an image and, the next day, find itself in a different position without an image. This could also explain the fact that there is a higher average of news without an image and participation (26.76). For these reasons and given the results of the other two platforms (Público and Jornal de Notícias), incorporating images could increase participation.

For the impact of the size of the images on participation (Table 9), it appears that the size of the images used has repercussions in the area under analysis. In the three cases, the news with smaller image sizes is those with higher levels of participation, and those with larger image sizes, which consequently are in prominent positions, have lower participation. Although the size of the image/news may indicate the importance of the theme, not least because the theme of this news was similar in the three newspapers, in terms of participation, high levels are not verified in the larger ones. Greater participation tends to occur in the news considered less important by the platforms, that is, those that occupy a smaller dimension.

4.6 *Impact of Colours Used in News Background*

Some platforms used colours strategically placed as the background of specific news. For example, Público and Jornal de Notícias use this strategy, and the results (Table 10) show that this highlighting technique can positively impact participation.

Table 10 News with colour highlight in the background and participation

	Público	Jornal de Notícias	Observador
Number of news with colour in the background	3	14	–
Number of comments	53	27	–
Average participation	17,67	1,93	–

Although this issue is visible on two platforms, it is also understandable that participation levels are higher when the platforms use these colours strategically and punctually, which allows for capturing the public's attention, especially when the spaces contain enough information. However, when there seems to be an excess of prominence in the news, participation is high, but not so high, losing the effect of capturing the public's attention in such a good way. The Observador is an exceptional case as it does not use colours to highlight the background of the news. Público is the most favourable case due to the reduced news with this strategy and the high number of comments.

4.7 *Influence of the Position of News on the Grid*

The impact of the position of the news in the grid could also confirm its influence on participation (Table 11). The three platforms use modular grids, allowing quick layout changes. Table 14 shows each platform and the lines above the fold (two for Público and Observador and four for Jornal de Notícias); it is also possible to see the position of the new in the grid (L for left, C for centre, and R for right).

This analysis has shown that the highest levels of participation are registered in the last visible line, that is, in the position immediately above the fold. The highest participation in Público is on the 2nd line, the 13 news on the left with 78 comments, and the 13 on the right with 102 comments are highlighted. In Jornal de Notícias, on the 4th line, the 12 news on the left with 39 comments, and the seven news on the right with 34 comments stand out. In this case, it also has a larger font size than the others, attracting the reader's eye. Again, Observador is an exception; the most significant participation occurs in the 1st line on the right. These results can be explained because this platform has the least visible lines (ranging between one and two lines), mainly due to the space advertising occupies.

Table 11 Position of news on the grid and participation

	Público						Jornal de Notícias									Observador	
Line	1st			2nd			1st		2nd		3th		4th			1st	2nd
	L	C	R	L	C	R	L	R	L	R	L	R	L	R	L	R	L
Number of news	11	1	14	13	1	13	10	7	7	7	8	8	12	7	7	31	8
Number of comments	20	1	66	78	0	102	24	10	16	14	15	13	39	34	12	740	30

4.8 The Impact of News Highlights by Section

News platforms, to highlight some of the contents, use specific labels among these highlights, terminologies such as: exclusive, being updated, live, up to the minute, investigation, or report, and different colours (Fig. 4). Table 12 presents the number of news with and without the highlight, the number of comments in each strand, the average news, and average participation.

Based on the results, the news highlighted by the sections mentioned above also presents greater participation than the news that does not have this highlight. As the numbers indicate, Público offers 31 news with this highlight (average of 4.43 news per day) and 197 comments (average of 6.35 comments per news). In comparison, the 22 news that does not have these highlights (an average of 3.14 news per day) have 75 comments (an average of 3.41 comments per news). The Observer follows the same line, and of the 16 news featured, this section highlight (an average of 2.29 news per day) comprises a total of 522 comments (an average of 34.50 comments per news). In comparison, the 30 news without this highlight (an average of 4.29 news per day) have 260 comments (an average of 8.67 comments per news).

Fig. 4 Examples of news highlighted by section in the three online newspapers

Table 12 News highlighted by section and participation

	Público		Jornal de Notícias		Observador	
	With	Without	With	Without	With	Without
Number of news	31	22	21	45	16	30
Number of comments	197	75	65	117	522	260
News average	4,43	3,14	3	6,43	2,29	4,29
Average participation	6,35	3,41	3,10	2,60	34,50	8,67

The exception is Jornal de Notícias since it does not follow the same logic. As can be seen, the most significant participation occurs in the 45 news that does not have this highlight (average of 6.43 news per day) and that have a total of 117 comments (average of 2.60 comments per news). On the other hand, the 21 news with this highlight (an average of 3 news per day) have 65 comments (an average of 3.10 comments per news). However, these data can also be justified on Jornal de Notícias is the medium that uses more colours to highlight the news by section (blue, yellow, black, and white), which can make the visual stain denser. Therefore, as they resort to this option frequently, the effect of drawing attention to the public ends up losing its effect.

4.9 Ease of Reading

During the literature review, namely in the approach to design principles and visual elements, the relevance of typography in understanding information content was recognised, consequently influencing public participation. The simpler and more effective the typography adopted, the more possibilities for the public to understand the contents and participate.

Concerning indicators relating to ease of reading (Table 13), and given the results obtained, it can be considered that Observador stands out for being easier to read, in which it resorts to a variation in the typeface; the typography is simpler in the sense of being more rounded and does not have a serif. On the other hand, Público and Jornal de Notícias resort to more drawn letters marked with serif. Furthermore, the variation in the typeface in Observador also facilitates reading in the sense of the information hierarchy since this variation can go up to four values. In contrast, the variation reaches six values in the others (Público, Jornal de Notícias). Therefore, as the principle of Hierarchy indicates, the ideal is three levels of Hierarchy, and therefore, the Observador can be considered to be in a more favourable position.

4.10 Advertising

The literature review has shown that excessive advertising on websites should be avoided, mainly because people evaluate the credibility and trust of platforms based on the density of existing advertising. It is also known that the more reliable the platform, the more motivated the public will be to participate. The less reliable the platform, the less motivation there is to participate. To identify the impact of advertising on credibility, the amount of publicity, the type of publicity, and the total participation was analysed in the web version of homepages (Table 14).

Observador, the platform with the highest participation (782 comments), has the highest number of advertisements (14). Therefore, contrary to what was seen in the

Table 13 Indicators relating to readability in Público, Jornal de Notícias, and Observador

	Público	Jornal de Notícias	Observador
Font and size	Publico, bold, 38 px	Noticias extra, 38 px	Halyard text, 30 px
	Publico medium, 23 px	Noticias extra, 28 px	Halyard text, 22 px
	Publico medium, 22 px	Noticias extra, 24 px	Halyard text, 22 px
	Publico medium, 19 px	Noticias text, 20 px	Halyard text, 18 px
	Publico text, 18 px	Noticias sans, 16 px	–
	Publico medium, 14 px	Noticias text, 16 px	–
	–	Noticias text,15 px	–
Alignment	Left	Left	Left
Space between the lines (size/ leading)	38/42 px	38/38 px	30/34 px
	18/27 px	28/28 px	22/26 px
	22/26 px	24/24 px	–
	19/23 px	20/21 px	
	14/18 px	16/18 px	–
Observations	Serif letters	Serif letters	Sans serif letters
Total participation	272	182	782

Table 14 Advertising present on platforms

	Público	Jornal de Notícias	Observador
Advertising number	6	12	14
Type of advertising	Newspaper subscription (1)	Football, environment, company, technology, local power, insurance, travel (8)	Discount on newspaper subscription, newspaper subscription, real estate, technology, eco-observador (5)
Total share	272	182	782
Average participation	5,13	2,75	17

literature review, there are no indications that a more significant number of advertisements can affect credibility and trust. In this case, the platform with more advertisements also has higher participation. However, this situation may be related to the fact that the advertisements in Observador are in-house. Hence there is no negative impact.

5 Final Considerations

The main objective of this study was to understand the impact of the design adopted by online news platforms on public participation in cyberjournalism—it appears that through the literature review carried out or through empirical analysis, design has a significant impact on audience participation. It can significantly increase or decrease participation levels.

Therefore, the results obtained in this investigation allow us to list a series of significant contributions in terms of the design adopted, with a view to increasing participation: (i) the importance of design to foster the credibility of platforms and trust in the public, crucial to higher levels of participation; (ii) the fact that the device used in news consumption affects knowledge and understanding of the news, with emphasis on more negative levels on the smartphone than on the computer, which means that there must be consideration on the part of the news brands for this consequence; (iii) the respondents' preference for lighter and simpler colour palettes, especially blue; (iv) the respondents' preference for navigation bars that present all sections visible, which facilitates navigation and information search; (v) our results shows that a smaller flow of news provoke higher levels of participation, that is, the smaller the amount of news above the fold, the greater the participation; (vi) a greater flow of news, according to the data, provoke lower levels of participation, that is, the greater the amount of news above the fold, the smaller the participation; (vii) a lighter visual stain, according to our data, also demonstrate positive repercussions on participation, capturing the reader's attention; (viii) a denser visual stain, based on our results, demonstrate a negative impact on participation, in which the attention-grabbing effect ends up being lost or diminished; (ix) there is value in placing images on the platforms' homepages, as a way to increase participation and keep the reader involved with the contents; (xi) even though larger images may indicate the importance of the theme, the public tends to participate in those that occupy smaller dimensions on the platform; (xii) the use of colors in the background of the news, in a strategic and punctual way, indicate an increase of the participation, but when used in excess it can harm leading to the reduction; (xiii) the position of the news in the grid also has an impact, with greater participation occurring in the last visible lines of the grid, that is, in the position above the fold; (xiv) the highlighting of news by section, that is, with labels that capture the public's attention, seems to indicate a positive impact on participation, but when used in excess the results are not so favourable; (xv) ease of reading affects public participation, namely understanding and knowledge, and in this case greater ease of reading is detected in cases where the letters are simpler and more rounded; (xvi) the number of advertisements on the platforms under analysis does not seem to affect participation, the greater the number of advertisements, the greater the participation, but it must be borne in mind that most of these advertisements are proprietary or internal.

For the above reasons, it is considered that the data obtained are of great relevance, mainly for three different areas: for the area of Communication Design because the results contribute to the adoption of more conscious and profound techniques at the level of visual design interfaces; for the public, since what is at stake is the knowledge they derive from news content; and, finally, for the news producers themselves, given that these contributions make it possible to encourage quality consumption and informed participation in increasing digital literacy through the adoption of effective techniques towards more conscious forms of interaction.

Acknowledgements This work was supported by the Portuguese Foundation for Science and Technology (FCT) in the framework of the Strategic Funding UIDB/05380/2020.

References

1. Bastos H (2012) A diluição do jornalismo no ciberjornalismo. Estudos Em Jornalismo e Mídia 9(2):284–298. https://doi.org/10.5007/1984-6924.2012v9n2p284
2. Correia JC (2004) Comunicação e Cidadania: Os media e a fragmentação do espaço público nas sociedades pluralistas. Livros Horizonte
3. Traquina N (2000) Quem vigia o "Quarto Poder"? Anais Do IX Encontro Anual Da Compós 10. http://www.compos.org.br/data/biblioteca_1431.pdf
4. Kovach B, Rosenstiel T (2014) The elements of journalism: what news people should know and the public should expect, 3rd edn. Three Rivers Press. https://play.google.com/books/reader?id=SxyOaH25-NoC&hl=pt_PT
5. Lévy P (1999) Cibercultura. Le Livros. https://mundonativodigital.files.wordpress.com/2016/03/cibercultura-pierre-levy.pdf
6. Castells M (2003) A Internet e a Sociedade em Rede
7. Rost A (2014) Interatividade: Defenições, estudos e tendências. In: Webjornalismo: 7 característicoas que marcam a diferença. Livros LabCom
8. Silva MT da, Brites MJ (2021) Audiências e jornalismo, uma relação difícil? In: De que falamos quando dizemos jornalismo? Livros LabCom. https://play.google.com/books/reader?id=jrJFEAAAQBAJ&pg=GBS.PA1&hl=pt-PT&printsec=frontcover
9. Brites MJ, Torres M (2017) The Portuguese news industry's perspectives and roles on the making of active citizens: readers' skills to comment on the news. 1:137–152
10. Kelly M (2015) Visual communication design as a form of public pedagogy. Aust J Adult Learn 55(3):390–407
11. Nakilcioğlu İH (2013) The effects of font type choosing on visual perception and visual communication. Online J Art Des 1(3):35–54
12. Tilak G (2020) Usage of visual communication design on consumer behaviour. Gedrag Organisatie Rev 33(02):963–971. https://doi.org/10.37896/gor33.02/103
13. Tractinsky N (2013) Visual aesthetics. In: The encyclopedia of human-computer interaction, 2nd edn. The interaction design foundation. https://www.interaction-design.org/literature/book/the-encyclopedia-of-human-computer-interaction-2nd-ed/visual-aesthetics
14. Wang Z (2018) Application of interaction design on visual communication design. ICEMIT, 247–250. https://doi.org/10.25236/icemit.2018.052
15. Dunlap JC, Lowenthal PR (2016) Getting graphic about infographics: design lessons learned from popular infographics. J Visual Literacy 35(1):42–59. https://doi.org/10.1080/1051144X.2016.1205832
16. Kane L (2018) Designing effective infographics. https://www.nngroup.com/articles/designing-effective-infographics/
17. Santos G, Grossi AM (2018) Jornalismo e credibilidade: uma percepção do público. Ámbitos. Revista Internacional de Comunicación 42:40–54. https://doi.org/10.12795/ambitos.2018.i42.03

18. Sousa AP de (2017) A interface na e-Health: proposta de princípios de design para a credibilidade e a confiança. https://ria.ua.pt/handle/10773/21059
19. Cyr D (2013) Emotion and website design. In: The encyclopedia of human-computer interaction, 2nd edn. The Interaction Design Foundation. https://www.interaction-design.org/literature/book/the-encyclopedia-of-human-computer-interaction-2nd-ed/emotion-and-website-design
20. Gonçalves J, Silva M, Pereira S (2017) Medir o agenda-setting nos comentários dos leitores às eleições legislativas de 2015
21. Palomo B (2014) International criteria in editorial web design. In: Shaping the news online: a comparative research on international quality media, p 348. Livros LabCom. http://www.labcom-ifp.ubi.pt/ficheiros/20141114-201403_atous_shappingnewsonline.pdf
22. Torres M, Figueiras R, Brites MJ, Amaral I, Maropo L, Santos SC, Jerónimo P, Santo PE, Pacheco L (2017) Audiências e cross-media: estudo de padrões de consumo de notícias em Portugal 1:177–199
23. Pektas PH (2012) A typographic analysis of turkish newspapers' websites. J Acad Market Mysticism Online 5(2):1–9
24. Patrick MF (2022) The UX designer's guide to typography. https://www.interaction-design.org/literature/article/the-ux-designer-s-guide-to-typography
25. MacDonald LW (1999) Using color effectively in computer graphics. IEEE Comput Graphics Appl 19(4):20–35. https://doi.org/10.1109/38.773961
26. Sasidharan S (2010) The impact of color and product congruency on user trust in B2C E-commerce. Banking 2
27. Dooley RP, Harkins LE (1970) Functional and attention-getting effects of color on graphic communications. Percept Mot Skills 31(3):851–854. https://doi.org/10.2466/pms.1970.31.3.851
28. Ahmetaj Q (2014) The influence of colours to the people mood, pp 1–10. http://psychology.about.com/od/sensationandperception/a/colorpsych.htm
29. Wright A (2010) Color psychology (the "Colour Affects" system). Color Psychol 05: 1–10. http://www.infoplease.com/spot/colors1.html
30. Snell S (2008) Newspaper website design: trends and examples. https://www.smashingmagazine.com/2008/11/newspaper-website-design-trends-and-examples/
31. Knox JS (2009) Punctuating the home page: image as language in an online newspaper. Discourse Commun 3(2):145–172. https://doi.org/10.1177/1750481309102450
32. Samara T (2007) Design elements: a graphic style manual. Rockport
33. Cooper A, Reimann R, Cronin D, Noessel C (2014) About face: the essentials of interaction design, 4th edn. John Wiley & Sons, Inc. https://www.digitalnewsreport.org/survey/2020/portugal-2020/
34. Molyneux L (2017) Mobile news consumption: a habit of snacking. Digit J 6(5):634–650. https://doi.org/10.1080/21670811.2017.1334567
35. Moran K (2020) How people read online: new and old findings. https://www.nngroup.com/articles/how-people-read-online/
36. Nielsen J (2011) Mobile content is twice as difficult. https://www.nngroup.com/articles/mobile-content-is-twice-as-difficult-2011/
37. Lowdermilk T (2013) User-centered design: a developer's guide to building user-friendly applications, 1st edn. O'Reilly Media, Inc
38. Tondreau B (2009) Layout essentials: 100 design principles for using grids. In: Technical communication, vol 56, issue 3. Rockport. http://search.proquest.com/docview/221007736?accountid=15533
39. Serra JP (2007) Manual de Teoria da Comunicação.
40. Sousa AP de, Almeida AM (2016) Habits and behaviors of e-health users: a study on the influence of the interface in the perception of trust and credibility. Proc Comput Sci 100:602–610. https://doi.org/10.1016/j.procs.2016.09.201
41. Alves S, Sousa AP de (2022) Cyberjournalism: the influence of design on public participation. In: Martins N, Brandão D (eds) Advances in design and digital communication III, vol 27. Springer Series in Design and Innovation. http://www.springer.com/series/16270

42. OberCom (2021) Digital news report—Portugal 2021. Publicações OberCom, 112. https://www.digitalnewsreport.org/survey/2020/portugal-2020/
43. Sousa AP de, Almeida AMP (2016) Trust and credibility perception in e-health. Encyclopedia of E-Health and Telemedicine, pp 1105–1122. https://doi.org/10.4018/978-1-4666-9978-6.ch086

Evaluation of the Design and Usability of the Digital Repositories of the Universities of Porto and Minho (PT), Yale (USA) and Melbourne (AU)

Ânia Loureiro, Nuno Martins, Leonardo Pereira, Daniel Brandão, and Daniel Raposo

Abstract This study is part of the Wisdom Transfer (WT) research project, which aims to contribute to the scientific inscription of the individual legacies, at risk of disappearance, of retired or retiring professors, who marked the beginning of design education in northern Portugal. In this sense, the intention was to evaluate the current panorama of the digital repositories of higher education academic institutions, in order to assess the creation's need of a specific digital repository for the Portuguese art and design universe. In Portugal, there is currently a total of 60 digital repositories of academic institutions, with none of these being exclusively dedicated to art and design education. Thus, in order to assess the need for the creation of a specific digital repository for the artistic universe, this study evaluated the design and usability of four digital repositories of reference universities (two national and two international), namely: the RepositóriUM, from Univeridade do Minho and the Repositório Aberto, from Universidade do Porto; the Cross Collection Discovery repository, from Yale University, USA; and the RMIT Research Repository, from the Royal Melbourne Institute of Technology University, Australia. The results showed that the analysed repositories aggregate a very high volume of information from many different scientific areas. This plurality and the immensity of available content make the search

Â. Loureiro
School of Design, Polytechnic Institute of Cavado and Ave, Barcelos, Portugal

N. Martins (✉)
Research Institute for Design, Media and Culture, School of Design, Polytechnic Institute of Cavado and Ave, Barcelos, Portugal
e-mail: nmartins@ipca.pt

L. Pereira
University of Aveiro, DigiMedia, DeCA, Aveiro, Portugal

D. Brandão
CECS, Institute of Social Sciences, University of Minho, Braga, Portugal

D. Raposo
Polytechnic Institute of Castelo Branco, CIAUD, Castelo Branco, Portugal

N. Martins et al. (eds.), *Perspectives on Design and Digital Communication IV*, Springer Series in Design and Innovation 33, https://doi.org/10.1007/978-3-031-41770-2_4

process difficult, and it is also difficult to meet the specific needs for sharing content from certain fields of expertise, particularly art and design.

Keywords Digital repositories · UI & UX design · Interface design · Communication design

1 Introduction

This study arises in the context of the research project called "Wisdom Transfer: Contributions to the scientific inscription of individual legacies in university and scientific retirement contexts in arts and design" (Ref. POCI-01-0145- FEDER-029038). Wisdom Transfer (WT) aimed to establish the basis for communicating and activating contributions to society from the knowledge and experiences of retired art and design academics in the northern part of Portugal [9, 11].

Currently, there is a serious problem related with the fact that the historical heritage of art and design courses in Portuguese universities is at risk of disappearing, due to the lack of an organized structure that guarantees its preservation. This wealth of knowledge and experiences is at risk of being totally lost if it is presently not recovered and validated, considering that the mentioned academic generation is aging and retiring [9].

According to [19], universities have globally a fundamental purpose regarding the production and dissemination of knowledge and scientific information from their teachers and students. With the evolution of access to information and the growing need to organize and disseminate scientific work, universities are currently under pressure. It is necessary to create alternatives to the storage systems that are currently used, taking into account that information and the respective dissemination platforms are increasingly digital [19].

The answer to this problem may lie in Information and Communication Technologies (ICT), namely digital repositories, incorporated in the university context and encompassing the collection, archiving and sharing of its content.

So, in order to assess the need for the creation of a specific digital solution for the universe of art and design, this study consisted on the identification and analysis of four reference digital repositories (two national and two international) within the university context.

During the process of identification and selection of the evaluated platforms, it was uncovered, through the OpenDOAR platform (a directory of academic repositories with free and open access, where it is possible to access statistical data and academic resources of repositories from all over the world) that, as of November 2021, the United States of America were the country with the largest number of academic repositories, with a total of 913 repositories; in Portugal, there is a total of 60 repositories.

In the recent years of the Portuguese context, the trend has been growing: in 2016 there were 48 academic repositories, and in 2021 the number rose to 60 [4]. Another found positive point was the growth in the number of scientific publications, which was influenced by the increased "(…) effort in the modernization of the scientific and higher education system in Portugal, allowing publications to multiply several times" [3]. In 1981, there was a scientific production in Portugal, including papers published in international scientific journals, of 3.1 documents per 100 thousand inhabitants, and, by 2018, this number had significantly increased to 233.6 documents per 100 thousand inhabitants [13].

It should also be noted that the repository software package adopted by Universities to create their platforms is DSpace, a free and open source software. In Portugal, DSpace is adopted by about 85% of the academic institutions while worldwide it drops to about 39%.

In some sense these results explain why the structure and organization of the information available in the different academic repositories is very close, especially at national level [12].

2 Platforms' Selection for Analysis and Adopted Methodologies

According to [6], during a study, it is necessary to analyse the same questions in different reference examples and compare them with each other, making sure that the selection of these examples is consistent with the research problem. For these reasons, and in order to carry out a more complete and contextual study, four digital repository platforms of four university institutions were selected, two of the platforms being Portuguese and the other two international [20].

The selected Portuguese platforms were the following: RepositóriUM, from Universidade do Minho (UM); and Repositório Aberto, from Universidade do Porto (UP). As for the selected international platforms, these were the following: the Cross Collection Discovery repository, from Yale University, United States of America; and the RMIT Research Repository, from the Royal Melbourne Institute of Technology University, Australia.

Regarding Portuguese repositories, the RepositóriUM of UM was chosen for analysis, due to the fact that it is not only one of the largest Portuguese universities, but also for being the first institutional repository to appear publicly in Portugal, and for being the first Portuguese university to implement and translate the open access software DSpace. It should be remembered that 85% of national institutional platforms use this software, including the largest Portuguese universities. The selection of the Repositório Aberto of UP, was due it being the most widely used repository in Portugal and because it's from the university that, at the scientific level, has most contributions, with the largest number of documentation available amongst Portuguese repositories.

The Cross Collection Discovery repository at Yale University was selected, not only because Yale University is, as of November 2021, in the top 10 universities in the United States of America, but also because it is the North American university with the most documentation regarding the arts field [14]. Another important factor for its selection was the fact that some of the stored art collections are hundreds of years old, so it was important to understand the methods used for the organization, recording and dissemination of such old content.

Considering that platforms from Portugal and the United States of America were chosen for analysis, the 2020 ranking of the best Art and Design universities in the world from the QS World University Ranking (a portfolio of rankings that assesses the performance of higher education institutions globally) was examined in order to identify the first university listed that was not from the European or American continent. It turned out that Australia's Royal Melbourne Institute of Technology University was ranked 11th, so this university's digital repository, the RMIT Research Repository, was selected [14]. RMIT University is also considered the best university of Art and Design in Australia, and its research portfolio is mainly concentrated in Design fields, which reinforces this selection option.

3 Platforms' Analysis

3.1 RepositóriUM—UM

RepositóriUM (Fig. 1), belonging to UM, was created with the aim of "(…) storing, preserving, disseminating and providing access to the intellectual production of Minho's University in digital format" [16]. It is possible to find several types of documentation, such as: journals, doctoral theses, master's dissertations, among other types of documents.

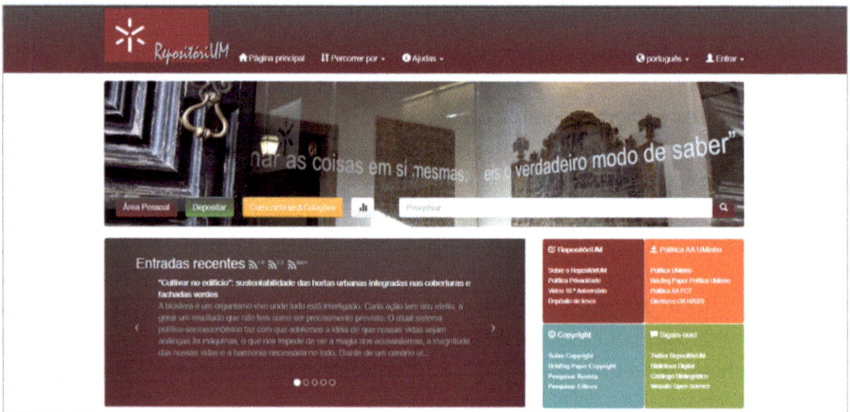

Fig. 1 RepositóriUM, Universidade do Minho, Portugal. *Source* RepositóriUM [16]

This repository was created in November 2003 and was the first institutional repository to appear publicly in Portugal. Its external motto was a "(…) challenge launched by the Portuguese Government, through the Mission Unit for Innovation and Knowledge (UMIC), to Portuguese universities to present ideas and projects in the framework of a modernization initiative that the Government intended to launch" [3].

UM was also, besides being a pioneer in this issue, the main driver of the Open Access movement and, in general, of Portuguese repositories. It was this institution that implemented a first version of the DSpace software platform to create repositories in Portuguese, becoming one of the first universities in Europe to translate and implement this open access software [3]. Later, in 2006, it also managed to get the "(…) Council of Rectors of Portuguese Universities (CRUP) to recommend to national universities the creation of institutional repositories, also recommending that they defined institutional policies for deposit in the repositories" [3].

Analysing the content of the mentioned repository, one sees that there are two menus: a search menu with text field and topics for filtering (Fig. 3); and a Communities and Collections menu.

The informational structure entitled "Communities and Collections" is a structure by which the DSpace software is well known, which consists of dividing the platforms' content around communities "(…) which may correspond to administrative entities such as schools, departments, laboratories and research centers. Within each community, there can be an unlimited number of sub-communities and an unlimited number of collections. Each collection can contain an unlimited number of items" [15]. The Communities and Collections menu of RepositóriUM has as categories the following ones: schools; departments; centers; and university courses, which makes it quite extensive and makes it difficult to perform specific searches, since it is divided into 18 communities and approximately 70 sub-communities, which are divided, themselves, by even more topics.

Looking at the search panel, it is visible that while there is an attempt to simplify the search process through the text field, this panel has a very confusing menu regarding the additional search filters, which are in total more than 71,000 filters. These filters are divided into pages, with the information inside small text boxes, that require the user to select the "Next" or "Previous" button in order to search for the desired filter. However, the filters are ordered by the amount of documentation available and are divided into so many pages that, in practice, they make this section unusable. As an example of how difficult it is to perform a search, we tried searching for the topic "Digital Preservation" in the "Subject" section of the filters. The result was that to be able to select the "Digital Preservation" filter it was necessary to go through about 139 pages.

Through the repository's statistical information, one can observe a large increase in the total number of downloads and consultations of the files deposited in the platform: in 2006, there was a total of 405,469 downloads, rising in 2020 to more than 2 million. The country that most contributed to the mentioned rise of these numbers was Portugal, followed by Brazil and the United States of America. The repository is available in Portuguese, English, French and Spanish.

3.2 Repositório Aberto—UP

Repositório Aberto belongs to UP, which aggregates two institutional repositories: the Repositório Aberto (Fig. 2), which is related to the preservation and free access to the UP academic community's intellectual production; and the Repositório Temático, which may require authentication and stores information resources produced in specific fields or for specific audiences [15]. It should be noted that the latter Repositório Temático has documentation that is mostly related to the University's services—minutes, summaries, recruitment processes—and not exactly scientific documentation [3]. For this reason, and due to the fact that its main function is not the scientific community, this case study only included UP's Repositório Aberto (RAUP).

RAUP became public on the 3rd November 2007, with a clear mission of "(...) enhancing the visibility and impact of the scientific production of its academic community, as well as preserving the intellectual memory of the University" [15]. With this repository, UP also intends to allow the dissemination of research results that are developed in its institution, accelerating the pace of knowledge transfer.

This university was the fourth one in the country to make a repository available and, although the option of developing a completely new platform from scratch was considered, they opted to adapt the already existing DSpace software [3]. At the beginning of 2008, their repository had about a thousand documents, but as of November 2021, they had a total of more than 95 thousand documents. These documents cover all major fields of knowledge and are responsible for 20% of the Portuguese scientific articles indexed annually in the Web of Science, making UP the largest producer of science in Portugal [3].

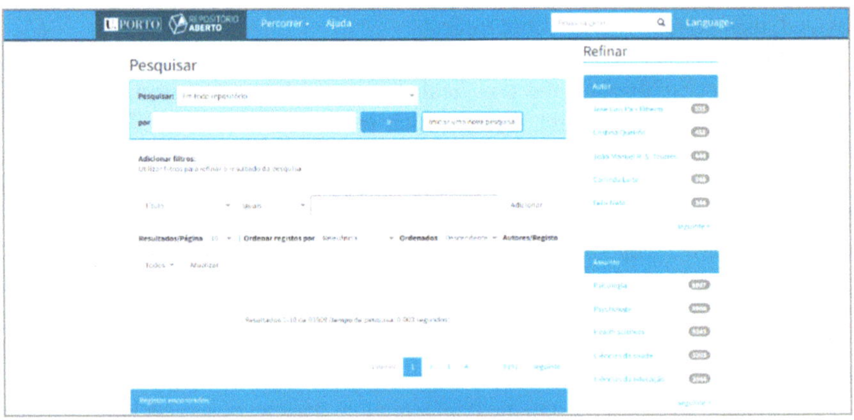

Fig. 2 Repositório Aberto, Universidade do Porto, Portugal. Page regarding the search menu with a text field and filtering topics. *Source* Repositório Aberto.UP [15]

Fig. 3 Cross Collection Discovery, Yale University, EUA. *Source* Cross Collection Discovery [5]

Like the UM's repository, UP's Repositório Aberto has a Communities and Collections menu, with 19 communities in total and an average of 29 collections per community, which, in fact, only represent document types (dissertations, reports, theses, articles, books, among others). An immediate problem with this menu is that they put all the document typologies as submenus, or collections, of all the communities. This results in a too long and confusing menu, and there are empty submenus, without any document, because the communities do not have all the document typologies. These empty submenus represent 35% of all submenus, which in turn means that 35% of the information available in the menu is not needed.

Just as with RepositóriUM, RAUP also has the possibility of performing a search with filtering and text field, and the topics for filtering are quite extensive (Fig. 2). The main search filters are four: author; subject; document type; and publication date, which are divided into 22,686 pages, with 113,430 filters in total. With these numbers we conclude, again, the uselessness of placing such extensive filtering options, with the text field for search being the most practical format. Regarding languages, the platform only offers the options Portuguese and English.

3.3 *Cross Collection Discovery, USA*

In the United States of America, the initiatives regarding the creation and organization of repositories emerged in 1991, through a project called ArXiv [3]. This project consisted in enabling researchers in the field of physics to make their research work available on a central digital platform, so that other researchers, regardless of their location, could access those documents. However, and although similar projects opened the doors for the creation of these platforms, it was only from the beginning of the XXI century that repositories were given their real and due importance in the United States of America [3].

Cross Collection Discovery (CCD) (Fig. 3) is a digital repository of Yale University, located in New Haven, Connecticut. Yale University was founded in 1701 and is the third oldest institution of higher education in the United States of America. Its activities are divided among 14 different schools and it also has 15 libraries [5].

The CCD digital repository provides a way to search the collections of art, history, videos, photos, books, audios, events, and maps that are part of the university's institutional identity and contribution. This repository also aims to virtually unite the collections of all the different schools and libraries on campus, and makes an ongoing disclosure of its cultural and intellectual, digital and non-digital assets, allowing the discovery of content held by different units on its campus, these being mostly art-related or related with the university's Museum of Natural History [5].

This platform's users can search all the collections in a single public access website and, if necessary, they can also be redirected to the unit where the content is kept in order to obtain more information. Regarding the amount of documentation, this platform has almost 1 million resources available online and refers to another 1.5 million resources not available online.

A very positive point of the platform is the fact that it is available in 26 different languages, including Portuguese, which allows easy access to a larger number of users. This fact can be justified by the percentage of foreign students that the university has, about 21%, from 123 different countries.

Regarding negative points, the platform is quite confusing and does not meet usability standards, such as: applying consistent and easy-to-understand styles, text hierarchy, colours and typography; avoiding redundant content; reducing long texts, leaving clear and objective information; and avoiding a high number of steps to access information [2]. To be able to view a document, it is necessary to open several pages where there is no visual coherence or typographic hierarchy, which leads to a poor understanding of navigation and how to open the files you want.

Regarding menus, this platform has only one search menu with a text field and filters. The search filter categories are 10 in total, and are divided into 186 filters, quite a large number for users trying to find a specific topic to refine their search.

3.4 RMIT Research Repository, Australia

The RMIT Research Repository (Fig. 4) is the digital repository of the Royal Melbourne Institute of Technology (RMIT), University of Melbourne, Australia. This institutional repository has as its mission: to acquire, organize, preserve and disseminate records that provide evidence of activities undertaken at RMIT [17].

Regarding its content, the repository has a total of more than 65 thousand documents, mostly scientific papers, which represent 61% of the total documentation available. There are also other types of documentation, such as meeting documents, journals, theses, reports, books, among others. These documents are also available

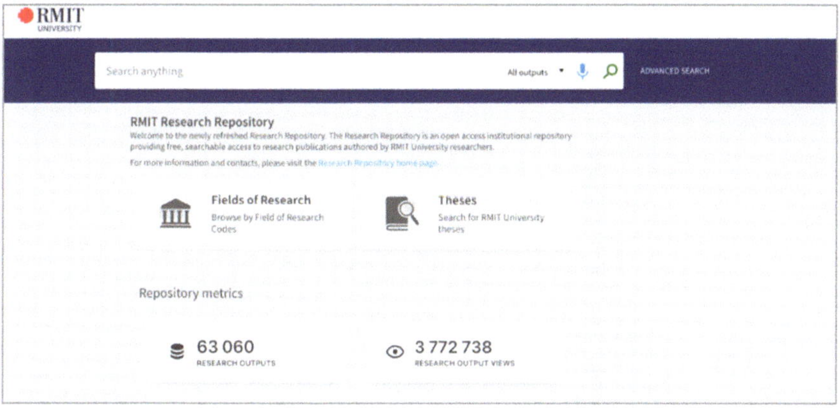

Fig. 4 RMIT Research Repository, Royal Melbourne Institute of Technology, Australia. *Source* RMIT University [17]

for learning, teaching, marketing and even promotional purposes. However, doctoral theses are only available in open access from the publication date of 2006. Theses prior to 2006 are only made available with the author's consent.

Overall, this platform is quite practical and has simplified menus that are easy to understand, unlike the examples previously analysed. It is possible to perform a search through two menus: a Search Academic Units menu; and a search menu with text field and filters. This search menu is organized by seven filter categories: Availability; Resource Type; Research Unit; Author/Creator; Date; Research Code; and Language. The filter categories are divided into 97 filters in total, and are quite easy to find (since they can be opened in their entirety and are not divided by pages). Note, also, that all documents have very detailed information about content, document type, and metrics, such as the number of views and references on other platforms.

3.5 Comparative Analysis

Table 1 identifies the main features and functionalities of each of the analysed repositories.

With the exception of RMIT Research Repository, all platforms reveal an extensive and confusing set of menus. In both RepositóriUM and the UP's Repositório Aberto, the filtering topics are unusable due to their length and complexity. It is concluded that these repositories clearly fail to meet the following usability standards: avoid a high number of steps to enable access to information; and avoid information overload, leaving information clear and objective [2].

In both of the Portuguese repositories, the Communities and Collections menus are also not practical, especially when compared to the direct search possibility of the name of the record you want. The Communities and Collections menus, once opened,

Table 1 Comparison and presentation of the research results of the analysed repositories

Repositories	Software	Documentation	Menus and submenus	Filters	Login	FAQ'S	Contacts	Multilingue	Responsiveness	Usability
RepositóriUM	DSpace	83,012	18	10	✓	✓	✓	✓	✓	✗
			70	71,496						
Repositório Aberto da Universidade do Porto	DSpace	95,127	19	8	✗	Only mentioned on the homepage		✓	✓	✗
			503	113,430						
Cross Collection Discovery	Not specified	2,336,495	–	10	✓	✗	✓	✓	✓	✗
			–	186						
RMIT Research Repository	Fez	65,381	3	7	✗	✗	✗	✗	✓	✓
			–	97						

Source Self-formulation (2021)

have other communities or sub-communities that are split again and are sometimes empty. This becomes a "(…) clear example of the difficulties and constraints placed on the user regarding the understanding of the internal organizational structure of the repository" [4], due to its high extension and complexity for search purposes.

With the exception of RMIT Research Repository, in all the analysed repositories there is an excess of categorization. There is a big dispersion and compartmentalization of content, because of the existence of too many menus, submenus or filters, these sometimes even being repeated or empty, without any document inside them.

There is also a considerable difference in the classification of documentary typologies, which vary from institution to institution, and there is no common model or standard being used. Each institution chooses to "(…) organize knowledge by the designations it believes best represent the science produced in the institution" [4]. However, this aspect may become confusing for users, since, when performing their searches in the various platforms, they will have to constantly adapt to new terms and new menus.

Given the results obtained in this analysis, it was concluded that RMIT Research Repository respects a greater number of usability standards, namely: the avoidance of redundant content; making available only items of interest and with well-defined objectives; and the avoidance of a high number of steps to enable access to information [2]. However, this repository fails in certain points, such as: it does not have a contact page; and it does not have a question and answer page. The usability standards for digital repositories, mentioned in the study by Camargo and Vidotti [2], indicate that a platform should provide a contact page and an information section, with answers to questions that may arise among users. The RMIT Research Repository fails on these points, which results in a reduction in the understanding of the platform and, consequently, a decrease in the likelihood that users will revisit the repository because they have no way of finding answers to questions that may arise [7].

Regarding the analysed platforms' structure and visual environment, it is generally considered that the structures of RepositóriUM, UP's Repositório Aberto and Cross Collection Discovery are unappealing and disorganized—an aspect that can be a disincentive and a major obstacle to the true mission of the platforms for sharing and depositing knowledge.

Given the results of this analysis, we believe that it would be beneficial to establish clear distinctions in the platforms' menus, standardizing the classification of documents, removing informational overload and avoiding the existence of empty menus. This way, it will be possible to facilitate the search for information and enable a more assertive comparison of bibliometric parameters between institutions [4].

4 Usability Tests

In order to obtain a more rigorous and technical evaluation of the quality of the interfaces of the four selected platforms, it was considered important to conduct usability tests. These tests allow us to identify problems in user interaction with an interface and to collect the necessary data to correct potential problems, therefore, improving the interface. It is a method that is based not only on data collected through tests with real users, but also on the observation of those users as they interact with the platform being tested [10]. The results will also make it possible to prevent possible errors and identify the best solutions in the creation of a specific digital solution for the fields of art and design.

Regarding the ideal number of users needed to perform the tests, there are different perspectives [18, 20]. For this project, we chose to follow the methodology of [10], which argues that five is the ideal number of users to perform the tests [1, 8].

In order to start the tests, a set of tasks was created for navigation in the platforms and five users were selected to perform these tasks. All users are master's students: three students are from the Design area; one student from the Computer Engineering area; and one student from the Digital Marketing Management area. The selection criterion was their areas of study, with the goal of identifying problems in the design and understanding and navigation of the under assessment platforms, through different perspectives—not only from a designer's perspective, but also as a website programmer and as a digital manager. The users are all already working in their respective fields of expertise and are divided as follows:

- Users 1, 2 and 3—Design Students;
- User 4—Computer Engineering Student;
- User 5—Digital Marketing Management Student.

For the tests, five interaction tasks with the digital repositories were defined, in order to understand how easy it was to perform the tasks, and to observe the users' behaviour when performing them. It should be noted that, in case the user cannot perform a task, he can go back and skip that task. The tasks defined for the usability tests were as follows:

1. loging in to the digital repository platform, by searching the name of the respective platform in the Google search field;
2. Identifying the location of the search bar and search for "digital repository" or "digital repository", depending on whether the platform is national or international;
3. Filtering the search results by "thesis" or "dissertation" for national platforms or "thesis" or "dissertation" for international platforms;
4. Opening the page of a record of the user's choice from the list of search results;
5. Downloading the record, or opening it for viewing.

After performing the tasks, a short satisfaction questionnaire was applied about the users' experience when performing the tasks, in order to analyse the reasons for possible obstacles encountered, or difficulties in completing the tasks (Table 2).

Table 2 Conducted questionnaire as part of the usability testing

Questionnaire		I agree	Undecided/ Without opinion	I disagree
Comprehension	The information presented in the platform is easy to be comprehended			
	It was easy to find the information that I wanted			
	The menus are easy to navigate			
	I would need help in order to use this platform			
Interaction	I was able to complete my tasks without problems			
	The platform has an appealing visual look			
	In general I am satisfied with the platform			

Source Self-formulation (2022)

After conducting the usability tests, the phase of collecting and analysing the data obtained in the respective tests began. Table 3 was first created to evaluate the success of the tasks, where three colours are distinguished to evaluate whether or not there was success: green represents *the successful completion of the task*; yellow, *the task completion, but with difficulty*; and red, *the non-completion of the task.*

The results of the tests on the success in completing the tasks show a serious problem with the North American Repository Cross Collection Discovery: four users could not complete the task successfully and one user completed it with difficulty, due to the fact that the platform gave an error when searching. We were able to confirm that this problem has persisted for several days.

Regarding UM's RepositóriUM, two users had difficulties in performing task number three of filtering the search, while the remaining three users could not perform this task and proceeded to the next step. This indicates that this platform's filtering system isn't working properly, because it forces users to navigate through quite extensive search results.

The UP's Repositório Aberto was the only analysed repository in the usability tests in which all users were able to complete all tasks successfully. In the case of the RMIT Research Repository platform, two of the users found it difficult to complete task number three of filtering the search, reporting that it was necessary to press a specific button to add the filter, after selecting that filter, which was not perceptible in their navigation. User number two also had a hard time completing task number five, reporting that the platform's interface does not have a good typographic hierarchy or visual coherence.

In order to obtain a more complete analysis of these usability tests, we evaluated the user experience satisfaction questionnaires, which were answered individually and immediately after the end of the task completion experience. The results obtained from the analysis of the questionnaires can be seen from Tables 4, 5 and 6, where the

Table 3 Assessment of user success in completing usability test tasks

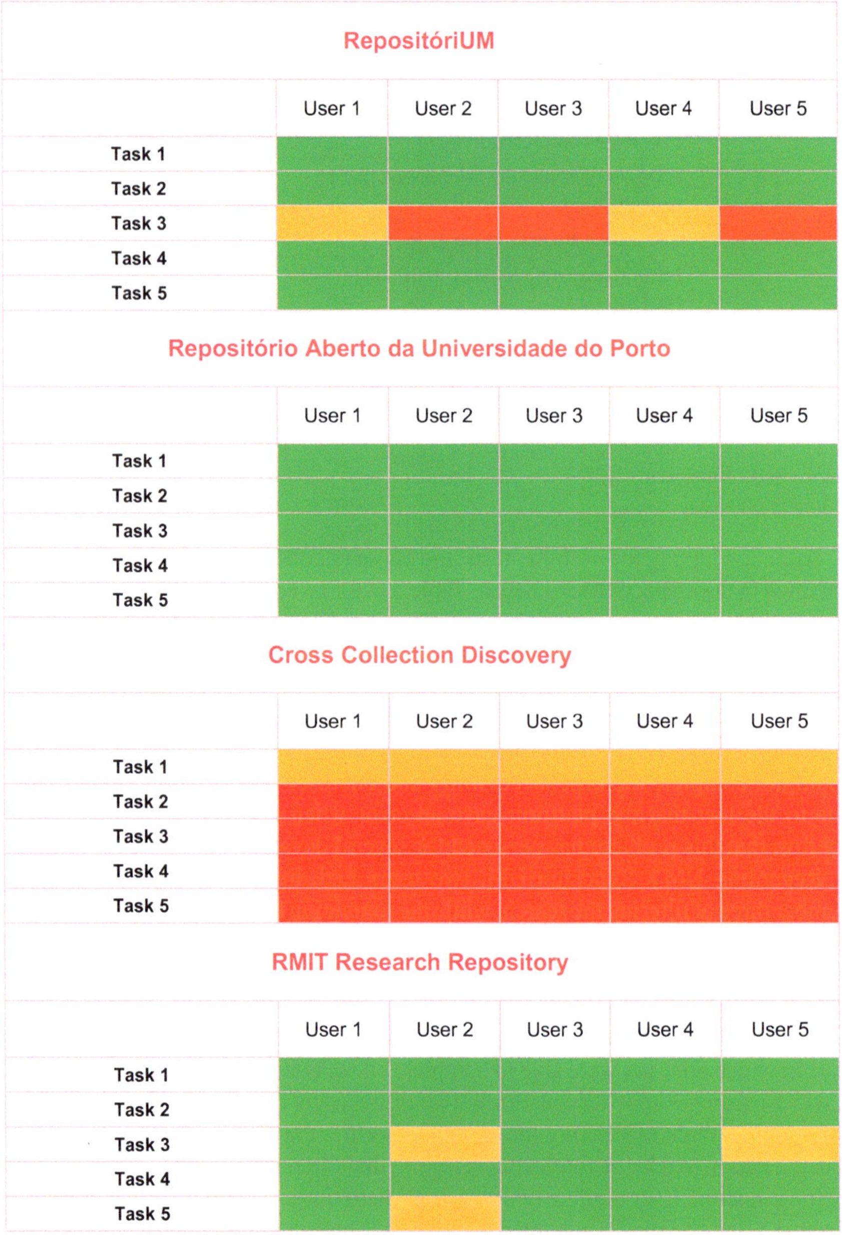

Caption: Concluded the task successfully; Concluded the task but with difficulty; Did not conclude the task successfully

Source Self-formulation (2022)

number of users who agreed, disagreed or had no opinion about each of the topics in the questionnaire, among the five users who performed the tests, was placed. The results of the questionnaires were divided by each of the analysed platforms and a satisfaction graph was created (Graph 1), which shows the percentages of satisfaction for each of the platforms, obtained through the number of users who agreed or disagreed with the questions of the questionnaire of the respective platform.

It was possible to conclude that the repository that provided the best browsing experience to users was UP's Repositório Aberto, with 88.57% satisfaction rate, followed by RMIT Research Repository, with 68.57% satisfaction. The repository with the highest level of dissatisfaction was UM's RepositóriUM, with an 85.7% dissatisfaction rate.

The repository with the best visual appearance, from the users' perspective, was the UP's Repositório Aberto, as four out of five users answered that the platform has a pleasant visual appearance (Table 7). The repository rated as having a less pleasing visual appearance was UM's RepositóriUM, with all five users disagreeing on the topic: "The platform has a pleasing visual appearance" (Table 6). Several users also added that this UM's repository needs a redesign in all aspects, because they considered the navigation and the information presented on the platform difficult to understand due to the design.

The results obtained in these usability tests were convergent with the first qualitative analysis performed to each of the platforms, reinforcing the argument that design and usability contribute decisively to the success of a digital solution.

Table 4 Results from satisfaction questionnaires regarding UM's RepositóriUM

UM's RepositóriUM				
		I agree	Undecided/ Without opinion	I disagree
Comprehension	The information presented in the platform is easy to be understood	–	1	4
	It was easy to find the information I wanted	–	–	5
	The menus are easy to be navigated	–	1	4
	I would need help to use this platform	3	2	–
Interaction	I was able to complete my tasks without problems	–	–	5
	The platform has an appealing visual look	–	–	5
	In general, I am satisfied with the platform	–	1	4

Source Self-formulation (2022)

Table 5 Results from satisfaction questionnaires regarding UP's Repositório Aberto

UP's Repositório Aberto				
		I agree	Undecided/ Without opinion	I disagree
Comprehension	The information presented in the platform is easy to be understood	4	1	–
	It was easy to find the information I wanted	5	–	–
	The menus are easy to be navigated	4	1	–
	I would need help to use this platform	–	–	5
Interaction	I was able to complete my tasks without problems	5	–	
	The platform has an appealing visual look	4	–	1
	In general, I am satisfied with the platform	4	1	

Source Self-formulation (2022)

Table 6 Results from satisfaction questionnaires regarding Cross Collection Discovery

Cross Collection Discovery				
		I agree	Undecided/ Without opinion	I disagree
Comprehension	The information presented in the platform is easy to be understood	–	–	5
	It was easy to find the information I wanted	–	–	5
	The menus are easy to be navigated	–	–	5
	I would need help to use this platform	2	3	–
Interaction	I was able to complete my tasks without problems	–	–	5
	The platform has an appealing visual look	–	4	1
	In general, I am satisfied with the platform	–	–	5

Source Self-formulation (2022)

5 Conclusions and Future Research

Through this study it was possible to confirm that academic repositories have a high demand and that digital media, by enabling a fast, easy and universal access, should be a central focus for the archiving and availability of academic and scientific content to citizens.

It was also proved that the design of the four analysed platforms should be improved, especially the UM's repository.

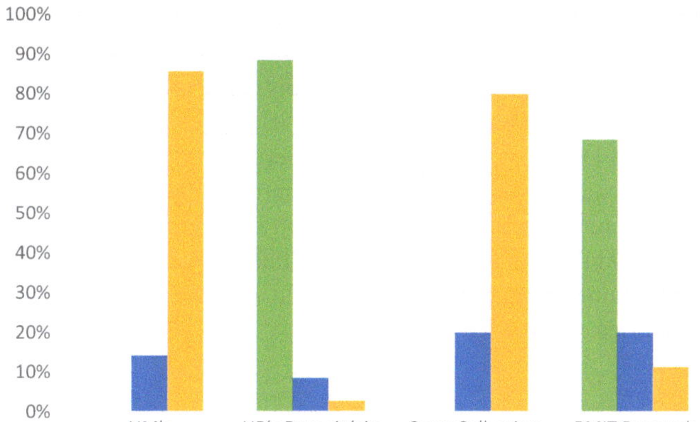

	UM's RepositóriUM	UP´s Repositório Aberto	Cross Collection Discovery	RMIT Research Repository
Satisfyed	0	88.57	0	68.57
Undecided/Without opinion	14.3	8.57	20	20
Unsatisfyed	85.7	2.86	80	11.43

Graph 1 User satisfaction assessment with each of the platforms. *Source* Self-formulation (2022)

Table 7 Results from satisfaction questionnaires regarding RMIT Research Repository

RMIT Research Repository				
		I agree	Undecided/ Without opinion	I disagree
Comprehension	The information presented in the platform is easy to be understood	4	1	–
	It was easy to find the information I wanted	3	–	2
	The menus are easy to be navigated	4	1	–
	I would need help to use this platform	–	–	5
Interaction	I was able to complete my tasks without problems	3	1	1
	The platform has an appealing visual look	2	2	1
	In general, I am satisfied with the platform	3	2	–

Source Self-formulation (2022)

In general terms, it was found that making available a large volume of information, divided into different scientific fields and types of content is a complex issue, where the best solutions for an easy and effective search are not always found. And the more different areas of knowledge the repository aggregates, the more difficult it is to find a solution that can answer to the different specificities of each field. In addition, when issues associated with usability are not properly considered, platforms become confusing and impair the participation of the academic community in accessing and sharing knowledge.

So, given the problem identified in this study about how hard it is to organize such a large volume of information; and given the specific needs related to the themes of Art and Design, namely sharing different types of works and works of the most varied artistic expressions, the argument about the need and relevance of studying in the future an online platform solution specific to the universe of Art and Design gains strength. It is also believed that this hypothesis can also be applied to other fields of knowledge, which have specific contents that the current generic academic repositories cannot answer.

References

1. Bevan N, Barnum C, Cockton G, Nielsen J, Spool J, Wixon D (2003) The "Magic Number 5": is it enough for web testing? Extended Abstracts on Human Factors in Computing Systems 3:698–699. https://doi.org/10.1145/765891.765936
2. Camargo L, Vidotti S (2009) Arquitetura da informação para repositórios científicos digitais. Implantação e gestão de repositórios institucionais: políticas, memória, livre acesso e preservação. In: Sayão L, Toutain LB, Rosa FG, Marcondes CH (eds) Implantação e gestão de repositórios institucionais: políticas, memória, livre acesso e preservação, pp 55–82.
3. Costa J (2015) A comunicação científica das ciências sociais e das humanidades nos repositórios institucionais das universidades públicas portuguesas: estudo comparado (Doctoral dissertation, Universidade Fernando Pessoa). http://hdl.handle.net/10284/5332
4. Costa JL, Freitas JAG (2016) Repositórios institucionais: constrangimentos no processo de disponibilização de conteúdos científicos. Cadernos BAD (1):69–81. https://doi.org/10.48798/cadernosbad.1516
5. Cross Collection Discovery (2021) Discover yale digital content. https://discover.odai.yale.edu/ydc/
6. Ghauri P (2004) Designing and conducting case studies in international business research. Handbook of Qualitative Research Methods for International Business 5:29–40. https://doi.org/10.4337/9781781954331.00019
7. Gibson C (2006) Website engineering in the real world. ACM-SE: Proceedings of the 44st Annual Southeast Regional Conference 44:573–578. https://doi.org/10.1145/1185448.1185573
8. Leavitt M, Shneiderman B (2006) Research-based web design & usability guidelines. https://commons.erau.edu/publication/1028
9. Lima C, Alvelos H, Barreto S, Penedos-Santiago E, Santos R, Martins N (2022) Case studies of a trans-generational pedagogy of art and design. Int J Des Soc 16(1):113–129. https://doi.org/10.18848/2325-1328/CGP/v16i01/113-129
10. Nielsen J, Landauer T (1993) A mathematical model of the finding of usability problems. In: Proceedings of the conference on human factors in computing systems, pp 206–213. https://doi.org/10.1145/169059.169166
11. Penedos-Santiago E, Martins N, Barreto S, Lima C (2020) The rise of communication design in Portugal: an overview of the higher education teaching methodologies. In: Advances in human factors in training, education, and learning sciences. AHFE 2020. Advances in Intelligent Systems and Computing, vol 1211. Springer, Cham. pp 86–91. https://doi.org/10.1007/978-3-030-50896-8_13
12. OpenDOAR (2021) Global directory of open access repositories—Statistics. https://v2.sherpa.ac.uk/view/repository_visualisations/1.html

13. PORDATA (2020) Base de Dados de Portugal Contemporâneo - Publicações científicas por 100 mil habitantes. Portugal. https://www.pordata.pt/Portugal/Publica%c3%a7%c3%b5es+cient%c3%adficas+por+100+mil+habitantes-1196
14. QS Top Universities (2020) QS world university rankings. https://www.topuniversities.com
15. Repositório Aberto.UP (2021) Repositório Aberto da Universidade do Porto. https://repositorio-aberto.up.pt
16. RepositóriUM (2021) Sobre o RepositóriUM Universidade do Minho. https://repositorium.sdum.uminho.pt/about/about.htm
17. RMIT University (2021) RMIT archives collection. https://www.rmit.edu.au/about/our-heritage/rmit-archives-collection#digitalresources
18. Silva J, Martins N, Raposo D, Brandão D, Neves J, Almendra R (2022) The role of visual curation as a contribution to the development of interfaces between academia and business. In: Advances in design and digital communication: proceedings of the 5th international conference on design and digital communication, Digicom 2021, November 4–6, 2021, Barcelos, Portugal. Springer Series in Design and Innovation, vol 19. Springer, Cham. pp 365–372. https://doi.org/10.1007/978-3-030-89735-2_30
19. Sobral R, Santos C (2017) Repositórios institucionais digitais de informação científica: implementação com o software DSpace como solução técnica. Prisma.com, 35, pp 152–184. https://doi.org/10.21747/16463153/35con2
20. Veríssimo F, Raposo D, Barradas V, Neves J, Silva J, Martins N, Brandão DA (2022) Checklist proposal to evaluate the quality of university websites. In: Advances in design and digital communication: proceedings of the 5th international conference on design and digital communication, Digicom 2021, November 4–6, 2021, Barcelos, Portugal. Springer Series in Design and Innovation, vol 19. Springer, Cham. pp 134–145. https://doi.org/10.1007/978-3-030-89735-2_12

Pedagogy, Society and Design Practice

Designing Playful Artefacts for People Living with Dementia: A Methodological Approach with Undergraduate Students

Cláudia Lima

Abstract This paper reports on a set of methodologies and strategies applied in an academic environment with undergraduate students, in the context of a semester-long project aimed at creating playful artefacts for people living with dementia. This project was carried out under the *REMIND—Design for People with Dementia* research and resulted from a partnership between Lusofona University and Memória de Mim Daycare Center, an Alzheimer Portugal service. Its main goal was to design artefacts based on biographical and cultural characteristics of the Center's users. In addition, it aimed to introduce concepts of social design and inclusive design in academic learning through practice in real contexts. The students had regular direct contact with health professionals from the Center, developed extensive research on dementia, main symptoms, and forms of manifestation, and gathered a set of biographical and cultural information that allowed the design of artefacts potentially more oriented to the interests and characteristics of the Center's users. These artefacts were created according to a set of premises, namely, accessibility, flexibility, and inclusiveness in the forms of use; focus on the users' experience, knowledge, and interests; cognitive stimulation; encouragement of communication/conversation among users. In this paper, the methodologies used, positive aspects and constraints of the work process, and results obtained are reported, aiming at the replicability and continuity of learning practices integrated into social and community contexts.

Keywords People living with dementia · Cognitive stimulation · Playful artefacts · Inclusive design · Social design · Alzheimer Portugal

---

C. Lima (✉)
ID+/Unexpected Media Lab—Faculty of Fine Arts, University of Porto, Av. de Rodrigues de Freitas 265, 4049–021 Porto, Portugal
e-mail: claudiaraquellima@gmail.com

Lusofona University, Rua Augusto Rosa nº 24, 4000-098 Porto, Portugal

N. Martins et al. (eds.), *Perspectives on Design and Digital Communication IV*, Springer Series in Design and Innovation 33, https://doi.org/10.1007/978-3-031-41770-2_5

1 Introduction

It is estimated that in Portugal the number of dementia cases exceeded 193,000 in 2018 and that this will exceed 340,000 by 2050. Portugal is the 4th OECD country with the highest prevalence of dementia. The high rate of cases and the estimated increase in the coming years is due, among other reasons, to the growing population aged over 70 years and the high number of people aged over 85 years, age groups that will more than double between 2018 and 2050 [1].

Dementia covers various neurocognitive changes and associated symptoms. According to Alzheimer's Disease International [2] there are more than 100 forms of dementia, with Alzheimer's disease being the most common, accounting for 50–70% of cases. Other types of dementia include vascular dementia, dementia with Lewy bodies, frontotemporal dementia, Huntington's disease, mixed dementia.

Symptoms of dementia may include "memory loss; difficulty performing familiar tasks; problems with language; disorientation in time and place; poor or decreased judgement; problems with concentration, planning or organizing; misplacing things; changes in mood or behavior; trouble with images or spatial relationships; withdrawal from work or social activities" [3].

The symptoms of dementia are progressive and vary from person to person. Alzheimer's Society [4] defines 3 stages of dementia based on the progression of symptoms, as mild (or early) stage, moderate (or middle) stage and severe (or late) stage.

At mild stage the person is still autonomous and common symptoms include forgetting things, losing or misplacing objects, starting to have difficulty finding the right words for a situation and having poor judgement in the planning or decision making. At moderate stage, the person already needs help with daily tasks, shows greater confusion and memory loss, judgment tends to worsen, disorientation increases, both in space and time, and has difficulty remembering personal information. At the severe stage, the person needs continued support (becomes dependent), is no longer able to communicate fluently or hold a conversation (although he/she can speak) and physical abilities are severely compromised—difficulty to eat and swallow food, loses sphincter control, difficulty in locomotion with increased risk of falls [4–6].

Several studies have underlined the importance of non-pharmacological interventions alongside pharmacological ones, such as Reality Orientation, Validation Therapy, Reminiscence Therapy and other cognitive stimulating activities. Reminiscence therapies and activities focused on the person's life story have integrated interventions in people living with dementia at different stages [5–7]. And Reality Orientation was the basis of Cognitive Stimulation Therapy, a program recognized "for its benefits for cognitive performance, language function and quality of life" [8, p. 78]. This program is based on group interventions including approaches such as reality orientation, reminiscence and validation therapy and activities focused on people and their personhood, seeking to maximize their potential and strengthen interpersonal relationships. Several studies have shown positive results in improving cognitive function, wellbeing and independence of the person living with dementia [9–13].

Moreover, research in this area also highlight the importance of activities focusing on biographical and cultural aspects of the person [9, 12, 14], a principle advocated by Kitwood, when he published *Dementia Reconsidered: the Person Comes First*, in 1997, and widely supported by several researchers.[1]

Vernooij-Dassen [18] states that meaningful activities for people living with dementia—activities with a positive impact—are those that provide them with pleasure and promote their involvement, giving them a sense of connection and belonging, of autonomy and self-identity. Concurrently, van Rijn et al. [19] note that leisure activities can contribute to fulfilling needs of people living with dementia such as the expression of thoughts and feelings, sense of belonging, engagement and involvement, and their self-esteem, while providing comfort and reducing "boredom, agitation, and isolation" (p. 74). And Anderiesen et al. [20] point out the therapeutic value of playful experiences through games for older people, as they can contribute to "slow down the deterioration of, or might even improve, memory, handeye coordination, reaction times, and self-esteem", as well as stimulate greater physical and social activity (p. 156). The authors argue that games elicit several types of playful experiences such as relaxation, exploration, expression, fantasy, fellowship, humor, challenge, competition (among many others) having an added potential to explore reminiscences when designed for people living with Alzheimer's Disease. They conclude that reminiscence, relaxation, and sensation are play experiences suitable to all stages of the disease strongly recommending the inclusion of these experiences in games for these audiences.

Despite the importance of playful activities with people living with dementia, both van Rijn et al. [19] and Anderiesen et al. [20] stress the scarcity of games suitable to the characteristics and interests of this public as well as to their remaining abilities, namely their cognitive, perceptual, physical and interaction needs.

In Portugal, there are limited cognitive stimulus materials available in the market that reflect biographical and cultural aspects or even in Portuguese language [12, 21]. Playful activities as games suitable for people living with dementia are also scarce. As a result, healthcare professionals often create, in an amateur fashion, their own artefacts to support cognitive stimulation activities and provide playful experiences, based on the characteristics and culture of the people living with dementia with whom they work [21].

Based on this perception, in 2021–2022 a pedagogical project was developed with undergraduate students from the Lusofona University, Porto, with a view to creating materials for cognitive stimulation activities carried out at the Memória de Mim Daycare Center, a specialized Center for people living with dementia, located in Matosinhos [21, 22]. This was done through a partnership with Alzheimer Portugal, North Delegation, and involved health professionals from the Center. The results—cognitive stimulation artefacts—were positive and responded to the proposal made

[1] Examples include works by Costa [15], Lindsay et al. [16], Rodgers [17], Tsekleve and Keady [6], and Wallace et al. [7].

by the Technical Director of the Center and some of them are currently being used in the activities. The most appreciated artefacts were two traditional games based on Portuguese culture and on biographical components of the people living with dementia who attend the Center: a Goose Game[2] and a Bingo Game. Through playful activities, these games aimed to encourage interaction and communication among participants, sharing of life stories and memories and, above all, the wellbeing of people living with dementia.

However, the project had limitations. The students did not have access to the Center's users due to the restrictions resulting from the pandemic. Mediation was done by health professionals, as well as usability testing at a final stage. The absence of direct contact with the target audience and usability testing in intermediate stages had repercussions on less functional aspects in certain cases. For example, a few illustrations developed in the artefacts were too complex for users with more impaired cognitive abilities. The project also had financial limitations and there was no possibility to create prototypes of all the designed artefacts [21].

Based on this experience, a second pedagogical project was held in the academic year 2022–2023. This project involved students from the 3rd year of the Communication Design BA, at Lusofona University, Porto, and was carried out within the scope of the discipline Design Lab. This paper reports methodologies applied and main results. Although the project has included signage and playful artefacts, the focus will be on the latter.

This study was carried out within the framework of *REMIND—Design for People with Dementia: cultural, contextual, and biographical components in the production of artefacts for cognitive stimulation activities*—, a project that aims to identify contributions of design in the creation of artefacts to complement and improve the effectiveness of cognitive stimulation activities for people living with dementia in Portugal. As aforementioned, this paper will focus mainly on methodological approaches within class for the development of playful artefacts and subsequent results.

2 Methodological Approach

2.1 Briefing and Research

This pedagogical project was held within the scope of Design Lab, a semester-long course of the 3rd year that ran from September 2022 to January 2023. Nine students participated, and the work was developed during the 15 sessions of the discipline (3 h each).

The first session of Design Lab was dedicated to the presentation of the project, methodologies and results obtained in the previous year. This was followed by an

[2] In Portugal known as Jogo da Glória.

Fig. 1 Seminar on Dementia & Design that brought together a representative of the Alzheimer Portugal association; researchers specialized in designing for and with people living with dementia; and alumni who developed similar projects in the academic year 2021–22. Photography: Cláudia Lima

extracurricular seminar on Design & Dementia focused on contributions of Design to the health, well-being, and quality of life of people living with dementia. This seminar included presentations by a representative of the Alzheimer Portugal Association; Rita Maldonado, researcher and designer specialized in dementia; Cláudia Lima, designer, and author of the REMIND Project; Anabela Thomé and Mafalda Marinho, both alumni of the Lusofona University Communication Design BA who designed the Bingo Game and the Goose Game adapted to people living with dementia in the prior academic year (see Fig. 1).

This seminar was essential for a first approach to the knowledge of dementia, of the different types and forms of manifestation as well as daily experiences of people living with dementia; to learn about the mission and role of Alzheimer Portugal; and to learn about various contributions and design projects for and with people living with dementia. It was also essential to understand the nature of the REMIND project, in the framework of which the students' projects would be developed, and for a contact with two projects previously carried out under the same lines, including research and work processes applied by the students and results obtained.

A second session of the Design Lab course, also dedicated to project presentation, included the Technical Director of the Memória de Mim Daycare Center, who reinforced the mission and goals of Alzheimer Portugal and of the Center in particular, the characteristics of the people who are attending it, activities they carry out and dynamics of the Center's daily life. In this session several needs of the Center and its users and possible contributions of Design were identified, part of them related to daily activities and non-pharmacological therapies, such as Reality Orientation or Cognitive Stimulation.

Considering the importance of participatory methodologies in projects oriented towards people living with dementia [17, 23], it was proposed that the students had direct contact with the Center's users for a greater knowledge of their characteristics and personhood and possible collaborative work. This direct contact was not authorized since the presence of the students could become disruptive to the daily dynamics of the Center, thereby becoming a stress factor for its users. The students were, however, allowed to visit the Center at a time before the arrival of the users and divided in two groups to interfere as little as possible with the work of the Center's professionals.

This visit took place in mid-October, early in the morning, on two different days (1 for each group). In each of these visits there were 4–5 students, the teacher, and a health professional. The latter led the visit, showing the spaces of the Center and explaining the type of uses and activities carried out in each one of these. The health professional also talked about the daily dynamics of the Center's activities, namely the type of interactions developed with users, activities and tasks performed. Several artefacts used in the activities and therapies carried out with people living with dementia were shown, including cognitive and sensory stimulation artefacts, recreational games, a prototype of a "book of life" to be filled by people living with dementia and their caregivers, a board for Reality Orientation Therapy among other objects.

Although students had been given a lot of information about dementia and its manifestations through documentation, presentations from health professionals and research, it was felt that this would not be sufficient for a holistic understanding of the issue. This perception also came from the results obtained in the previous year's project. Hence, an extracurricular workshop was organized in partnership with Servilusa, entitled *Dementias—Expressive Therapies and Art: "In your skin"* and conducted by Ana Costa, post-graduate in Neuropsychology and Dementia from the University of Barcelona and author of the *hOpeningDementia*, a project that addresses dementia through psychology, design, and participatory methodologies.[3] This workshop allowed students to experience everyday symptoms and difficulties felt by people living with dementia using accessories that limited their abilities: diving goggles compromised vision; headphones with loud music compromised hearing; rubber bands on the fingers compromised hand mobility; tight gown also hindered the body mobility and cloth rags stuffed into shoes impaired the locomotion. Facing these difficulties students had to perform a set of tasks similar to those performed by people living with dementia in daycare centers (see Fig. 2).

As reported by the students, during this experience they felt moments of confusion—"it was a lot of information at the same time, because Ana explained everything in a row and with the music playing simultaneously it was even more difficult"; they had difficulty to understand the instructions given, remember the tasks and move their fingers; they felt discomfort with their clothes and shoes. Some of them were a bit disturbed by the experience.

Ana Costa explained that she compromised the students' senses, trying to simulate some of the sensations and difficulties experienced by people living with dementia. For example, she explained that as we age, there is a tendency for the foot to widen and there is a need to widen shoes, but this does not always happen. The same with clothing. Sometimes a person is not comfortable but does not have the ability to verbalize it and stays that way all day. Often, the person also does not understand how to communicate poor seeing or mishearing. If the person does not communicate

---

[3] About the *hOpeningDementia* project see Costa [15].

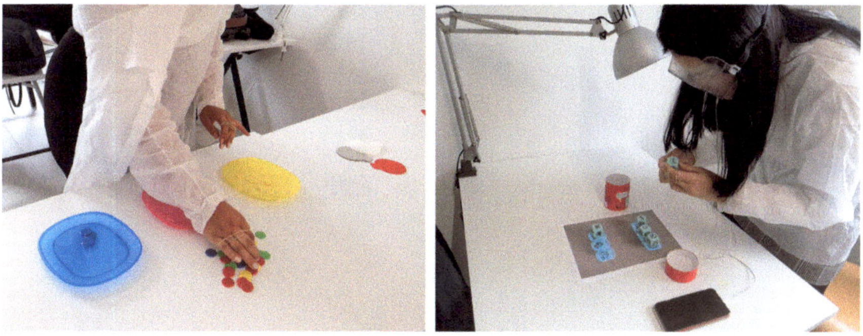

Fig. 2 Images captured during the workshop *Dementias—Expressive Therapies and Art: "In your skin"*, October 2022. Photography: Cláudia Lima

the vision or hearing problem, he/she may perform a given task poorly or not respond at all.

After the workshop, aiming a better knowledge of the Center's users (and since it was impossible to meet them personally), a survey was designed by the students focusing on their biographical and cultural characteristics, including place where they live, prior occupation, their interests and preferences, music they like to listen to, activities they most like to do in the Center, among others. This survey was sent to the health professionals of the Center and later answered.

2.2 *Project Development*

Based on research findings and survey answers, a mind map was made on one of the walls of the classroom using post-its of various colors, each color being assigned to one of the following themes: dementia & symptoms; life experiences [in dementia]; design contributions; biographical features (of the Center's users); Portuguese culture (see Fig. 3). Although there was one session dedicated to building this mind map, information (post-its) was added and consulted regularly until the end of the project.

The fact that the students were in permanent visual contact with the mind map allowed them to constantly recall characteristics of people living with dementia and of the Center's users in particular, possible functionalities and features to be included in the artefacts, design aspects to be taken into account (such as chromatic contrast, no infantilization of graphics, larger font sizes, adaptation to each person's abilities) and aspects of Portuguese culture more significant for the users' generation.

During the development of the project, in the Design Lab class, three interim oral presentations were held (one at the beginning of October, one at the end of that month and one at the end of November) with a view to sharing, analyzing, and discussing the different stages of the project. These presentations required the students to reflect and structure the work done so far and to develop skills in oral communication,

Fig. 3 Mind map on dementia & symptoms; life experiences [in dementia]; design contributions; biographical features; Portuguese culture. 2022–2023

an important aspect since the projects would be presented orally to the Technical Director of the Center in January. The information gathered during the research phase (including seminar, visit to the Center, meetings with health professionals, user surveys, and documentation reviewed), the experiences in the workshop *Dementias—Expressive Therapies and Art: "In your skin"*, and the information posted on the mind map on the classroom wall was frequently referred by students to justify conceptual and functional options for the projects under development.

These interim presentations provided moments for discussion of ideas and projects among students and between students and faculty, with issues such as feasibility, functionality, and materials to be used thoroughly analyzed and debated.

Periodical contact was established with health professionals of the Center and in early December prototypes of the projects under development were presented to the Technical Director for analysis and discussion of aspects related to the adequacy of formats and materials used, the dynamics of the playful activities designed and how the artefacts would work. Corrections were made at this stage related to material aspects and chromatic selection, but there were no conceptual changes or changes related to the functionality of the artefacts.

At the end of January, a presentation of completed projects was made to the Technical Director of the Center and prototypes of the designed artefacts were provided, some of these ready to be used at the Center.

At the conclusion of the projects, students were asked to complete an anonymous survey about this pedagogical project, including questions about positive and negative aspects of the project; what could be improved in the teaching and working process; new learnings provided; and what they would do differently if they repeated the project. This survey aimed, on the one hand, to listen to students about the impact of the project both in Design learnings and in social terms (awareness of dementia issues) and, on the other hand, to improve pedagogical methodologies as it is planned to continue this practice in the next academic year.

3 Results

3.1 *Contextualization of the Memória de Mim Daycare Center and Its Users: People Living with Dementia*

In *Demência—Desdramatizada e Colorida*, Costa [15] notes that the groups attending daycare centers or other spaces dedicated to people living with dementia are quite heterogeneous in terms of life experiences, socio-demographic characteristics, personal interests, needs and expectations. The absence of direct contact with the Memória de Mim Daycare Center users made it impossible to get to know them better, but the visit to the Center's facilities, the subsequent survey on the users and regular contact with health professionals allowed us to learn about biographical aspects becoming evident the diversity of population referred by Costa.

Back then, the Center served 12 users, 4 men and 8 women, aged between 70 and 87, all living in the district of Porto. Prior occupation was different for each one of them and personal interests varied. Common interests included music, mostly traditional Portuguese music and fados, although occasionally other types such as English, French, or Italian music were mentioned; going for walks; and housekeeping, since the majority of the women did domestic chores all their lives. Occasionally, many other interests were mentioned, such as gardening, movies, animals, crossword puzzles, or painting, emphasizing the heterogeneity of the users.

Preferred activities held at the Center included cognitive stimulation activities (the most referred), listening to music, going for walks, helping with the Center's chores (such as setting the table, folding laundry, or wipe down the tables and sweep the floor), logical thinking activities, matching games, domino, and bingo. Many other activities were mentioned occasionally such as cooking, gardening, creative activities, dancing, or chatting. We were informed that the users usually adhere to the activities performed at the Center, but in certain situations they may not do so well if they have limitations that hinder the performance of the task.

Regarding the Center's facilities and its uses, there is a large room at the entrance where users spend part of their time and do most of the activities: it's a space with a dining area and a leisure and rest area with several armchairs and a sofa. Two boards are fixed on the walls of this room, a blackboard in the dining area where health professionals write the daily menu in chalk, and a wooden board painted in green in the leisure area for Reality Orientation activities with removable pieces fixed with Velcro to facilitate the manipulation by users. The Center also has a Snoezlen room and the library room, two less frequented spaces and generally used by users who want to be quieter, away from the group.

Activities frequently performed at the Center include bingo; color matching; memory game; sequences of daily activities; scrabble—building words based on a question; activities with more sensory objects—working fine motor skills; fine arts activities such as painting; manicure, among others. The importance of using artefacts that work as conversation starters was also emphasized.

In the initial contacts with the Technical Director of the Center and with the health professionals, several possibilities of Design intervention were mentioned in the scope of the activities carried out by them and the dynamics of the Center itself. These included artefacts for cognitive and playful activities like textured dominoes, games that stimulate handling and matching games; conversation starters, since interaction between users tends to be reduced given the difficulty they have in starting conversations; redesign or complement the menu with pictures; Life Story Book to be completed with each user; lockers personalization.

The design of artefacts for the encouragement of communication and social interaction between users has been often emphasized. Indeed, several authors highlight the importance of interpersonal communication and social interaction, as the richness of human contact and interactions, along with person-centered care, contribute to the quality of life and well-being of people living with dementia [5, 6].

Other aspects were highlighted both by health professionals and in the literature review, namely: reduce excessive visual and audio stimuli that may cause restlessness, stress, and anxiety; focus on the users' experience and knowledge, their personal interests; consider different levels of difficulty for greater inclusion—flexibility; use language that is accessible and appropriate to users; do not infantilize objects and activities; stimulate autonomy, participation, sharing of opinions and decision-making.

Based on the findings, the students developed projects under the following premises: focus on themes related to biographical and cultural features of the Center's users; create conversation starters and encourage dialogue between users; provide pleasurable activities; provide complementary artefacts to activities performed at the Center; contribute to the dynamics of the Center's daily life.

Although several artefacts have been made to be used at the Center, the following section will focus on the playful ones.

3.2 *Cognitive Stimulation Activities Through Playful Artefacts for People Living with Dementia*

One of the playful artefacts developed by a student is a domino game that can be played by several users (see Fig. 4). The game consists of 28 pieces made of varnished wood and glued materials with compositions of different colors (contrasting) and different textures. Textures include one similar to pearls, so appreciated by women, another made of cork, a raw material characteristic of Portugal, some are smooth, others rougher.

The domino was designed considering aspects such as flexibility and inclusiveness with the possibility to be played by several participants according to traditional rules or used individually for activities of sequences of colors and textures. Additionally, the different textures applied to the pieces work as sensory stimulation. This flexibility

Fig. 4 Domino by João Pedro Martins (2023). Photography: Cláudia Lima

in how the game can be played allows it to be used by people living with different stages of dementia.

Two students developed a card game based on matching images of objects from the users' daily life, presenting cards with the visual representation of these objects in the past and their representation in current times (see Fig. 5). The selection of objects was drawn from information gathered from the Center's health professionals. The representations were made through simple and contrasting illustrations, complemented by a subtitle with the name of the object. A background of identical color was used for each pair of images, acting as an additional clue for the performance of the activity. The images, in an approximate format of 6 × 9 cm, were printed, glued on Kline, and plasticized to facilitate handling by people living with dementia and cleaning of the pieces after use.

This game is designed for people living with dementia in the early stages. It is believed that the combination of visual representations of everyday objects in their older appearance and in their current form may act as a trigger for conversations between players.

Three students made two matching games inspired by Portuguese culture: one based on popular sayings, one on cultural symbols. In the first case, pairs of cards were developed with the beginning and the end of sayings written in a *sans serif* bold font (for easier reading) on a white background, with the text of each saying in a different color (each pair of cards) to facilitate matching (see Fig. 6). Popular sayings are very present in Portuguese culture, particularly in these older generations, as stated by the health professionals at the Center. It is believed that a game based on

Fig. 5 Card game by Sarah Nogueira and Danielly Correa (2023). Photography: Cláudia Lima

popular sayings may thus become engaging for users and stimulate communication among them, since this activity can be performed both individually and in groups.

In the second case, sets of 4 cards were created with illustrations of cultural symbols such as the Barcelos rooster, the Portuguese guitar, or the carnation, associated with the April 25th revolution, a landmark for the users' generation (see Fig. 7). The illustrations, of simplified shapes, are presented on white backgrounds for greater contrast and visibility. The printing of 4 identical cards for each image enables the possibility of different ways of playing: with these sets of cards, it is possible to play Go Fish, memory games, and matching games. These cards can also work as conversation starters by referring to such iconic symbols of Portuguese culture. Hence, this artefact becomes flexible in the way it can be used and more inclusive and adapted to everyone's abilities by offering various levels of difficulty.

Both games—popular sayings cards and cultural symbols cards—were digitally printed on 15 × 11 cm cards and plasticized, facilitating their handling and cleaning.

4 Final Considerations

The artefacts were presented and delivered to the Technical Director of the Center and will be subject to usability tests in the presence of the Design Lab teacher. This phase will be essential to evaluate the suitability of the artefacts, namely functional

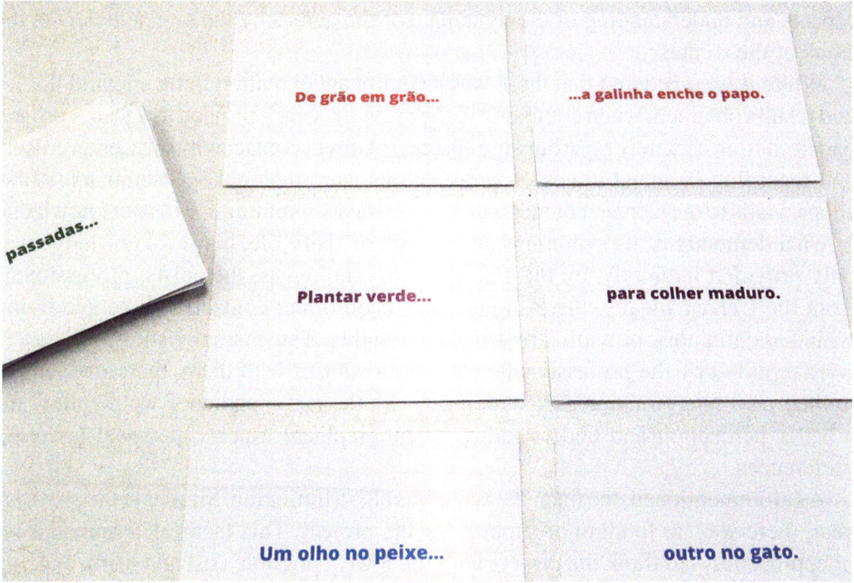

Fig. 6 Popular saying cards by Gustavo Santos, Hugo Carvalho and Luís Góis (2023). Photography: Cláudia Lima

Fig. 7 Matching cards by Gustavo Santos, Hugo Carvalho and Luís Góis (2023). Photography: Cláudia Lima

aspects and understanding of the dynamics of the games by the target audience, the users of the Center.

While it is considered that there was a significant evolution in the methodologies and results obtained, when compared to the previous year's project, this year's project had limitations, namely regarding the absence of direct contact with the Center's users and regarding financial issues. Several actions were undertaken—seminars, workshops, visits to the Center, contact with professionals—aiming at a deeper knowledge of what dementia is, its symptoms and impact on daily life, but this knowledge was only provided indirectly, by third parties—psychologists, therapists, professionals from the Center. Ideally, there should have been direct contact with people living with dementia, thus providing first-person insights. The characteristics of the users were reported by the professionals who are in contact with them, therefore subject to their own interpretation. Direct contact with the target audience would allow for a better perception and understanding of biographical aspects, personal interests, preferences.

As aforementioned, the financial issue was also a limitation. Similar to the previous year, there was no funding or support for the project. This factor determined a set of approaches—to think the project with the lowest possible cost and using existing materials such as leftover Kline or fabrics.

The prototypes of the games presented in this paper were all produced by students and faculty, but it was not possible to do it with the quality initially planned. This was the most emphasized issue in the surveys made to the students after the conclusion of the project. As negative aspects of the project, several students mentioned the financial investment in the production of the prototypes (by students and teacher), and to the question about what they would do differently if they repeated the project, one of the students even wrote "maybe the models could have better quality, which was limited by the cost and time [available]". In this survey, the lack of direct contact with users was also highlighted as a negative aspect.

As the survey consisted exclusively of open-ended questions, the answers given by the students were very diverse, although certain aspects related to learning and methodology, production, results, and the experience itself stood out for the frequency with which they were referred, such as those mentioned above.

On the positive side, more than half of the students stated the importance of the workshop for understanding what it is like to live with dementia. This was also mentioned several times in the students' oral discourse when presenting their ideas. Other positive aspects identified were the methodology applied to the project, the content provided, and the constant contact with the client.

To the question about what could be improved in the teaching and working process, 2 students reinforced the budget issue, but most of them answered that nothing could be improved. One of the students developed this answer by stating "I think that the way the whole process of teaching and working was approached was quite appropriate, I find the development of practical and real projects more engaging".

Regarding new learnings in design, the answers were very diverse, covering aspects such as learning to budget and to make models, or aspects more related to visual communication (typography, color, visual hierarchies), but the answer that stood out most often was related to the concept of social design and inclusive design.

In terms of new learnings beyond the design domain, most students reported the understanding of dementia and how it affects people's lives. One student pointed out that he had learned "a lot about dementia, a learning that I know can be useful for life", then stated that "it was an engaging and challenging job". Another student wrote "I enjoyed being involved in this topic and I became quite interested in learning more". Aspects such as crafts, caring for others, and thinking in alternative ways were also mentioned.

In conclusion, while, on the one hand, the answers to the survey highlight the limitations of the project, such as the financial constraints and the lack of direct contact with the Center's users, on the other hand, they also highlight the importance of the methodologies applied and the subsequent learning in design—in social design and in inclusive design, in particular—and beyond design, providing knowledge about dementia, its symptoms and how it affects the daily lives of people living with this diagnosis. Thus, the learning acquired in the Design Lab course went far beyond the design domain, reinforcing humanitarian values and raising students' awareness about dementia, a growing social issue in our country and worldwide.

Acknowledgments This work is financed by national funds through the FCT – Fundação para a Ciên-cia e a Tecnologia, I.P., under the scope of the project UIDB/04057/2020.

References

1. OECD (2021) Health at a glance 2021: OECD Indicators. OECD Publishing, Paris. https://doi.org/10.1787/ae3016b9-en. Last Accessed 01 March 2022
2. Alzheimer's Disease International: Alzheimer's Disease. https://www.alzint.org/about/dementia-facts-figures/types-of-dementia/alzheimers-disease/. Last Accessed 10 Aug 2022
3. Webster C (2021) What is dementia, why make a diagnosis and what are the current roadblocks? In: World Alzheimer Report 2021, Journey to a Diagnosis of Dementia. https://www.alzint.org/resource/world-alzheimer-report-2021/
4. Alzheimer's Society: what do we mean by signs and stages of dementia? (2021). https://www.alzheimers.org.uk/about-dementia/symptoms-and-diagnosis/how-dementia-progresses/progression-stages-dementia. Last Accessed 2023
5. Kitwood T (2019) Dementia reconsidered, revisited; the person still comes first. Open University Press
6. Tsekleve E, Keady J (2021) Design for people living with Dementia: interactions and innovations (Design for Social Responsibility). Routledge, London and New York
7. Wallace J, Wright PC, McCarthy J, Green PD, Thomas J, Olivier P (2013) A design-led inquiry into personhood in Dementia. In: Proceedings of the SIGCHI conference on human factors in computing systems (CHI '13). Association for Computing Machinery, New York, NY, USA, pp 2617–2626. https://doi.org/10.1145/2470654.2481363
8. Brooker D (2019) Personhood maintained: commentary by Dawn Broker. In: Kitwood T (ed) Dementia reconsidered, revisited; the person still comes first. Open University Press, pp 78–82
9. Câmara J, Nunes MVS (2021) Impacto de um programa de estimulação cognitiva na comunicação e cognição: Um estudo com pessoas com demência institucionalizadas. In Joaquim Pinheiro (coord.) Olhares sobre o envelhecimento, vol I. Estudos interdisciplinares, pp 273–285

10. Kim K, Han JW, So Y, Seo J, Kim YJ, Park JH, Lee SB, Lee JJ, Jeong HG, Kim TH, Kim KW (2017) Cognitive stimulation as a therapeutic modality for Dementia: a meta-analysis. Psychiatry Investig 14(5):626–639. https://doi.org/10.4306/pi.2017.14.5.626
11. McDermott O, Charlesworthb G, Hogervorstc E, Stoner C, Moniz-Cook E, Spector A, Orrell M (2019) Psychosocial interventions for people with dementia: a synthesis of systematic reviews. Aging Ment Health 23(4):393–403. https://doi.org/10.1080/13607863.2017.1423031
12. Pereira GA, Sousa I, Nunes MVS (2020) Cultural Adaptation of Cognitive Stimulation Therapy (CST) for Portuguese people with Dementia. Clin Gerontol 1–12. https://doi.org/10.1080/07317115.2020.1821857
13. Spector A, Fisher E (2022) Cognitive stimulation. In: World Alzheimer Report 2022 life after diagnosis: navigating treatment, care and support, pp 245–247
14. Aguirre E, Spector A, Orrell M (2014) Guidelines for adapting cognitive stimulation therapy to other cultures. Clin Interv Aging 9:1003–1007. https://doi.org/10.2147/CIA.S61849
15. Costa A (2019) Demência—Desdramatizada e colorida. Prime Books
16. Lindsay S, Brittain K, Jackson D, Ladha C, Ladha K, Olivier P (2012) Empathy, participatory design and people with dementia. In: CHI 2012: 30th ACM conference on human factors in computing systems, pp 521–530. https://doi.org/10.1145/2207676.2207749
17. Rodgers PA (2018) Co-designing with people living with dementia. CoDesign 14(3):188–202. https://doi.org/10.1080/15710882.2017.1282527
18. Vernooij-Dassen M (2007) Meaningful activities for people with dementia. Aging Ment Health 11(4):359–360. https://doi.org/10.1080/13607860701498443
19. van Rijn H, van Hoof J, Stappers PJ (2010) Designing leisure products for people with Dementia: developing "the Chitchatters" Game. Am J Alzheimer's Disease Other Dementias 25(1). https://doi.org/10.1177/1533317509333039
20. Anderiesen H, Scherder E, Goossens G, Visch V, Eggermont L (2015) Play experiences for people with Alzheimer's disease. Int J Des 9(2):155–165
21. Lima C, Penedos-Santiago E (2023) Designing for people with Dementia: a Portuguese case study. In: Houben M, Brankaert R, Hendriks N, Wilkinson A, Morrissey K (eds) Dementia Lab 2022: the residue of design. D-Lab 2022. Design for inclusion, vol 3. Springer, Cham. https://doi.org/10.1007/978-3-031-14466-0_9
22. Lima C (2023) Designing for people with Dementia in academic contexts. In: Martins N, Brandão D, Paiva F (eds) Perspectives on design and digital communication III. Springer Series in Design and Innovation, vol 24. Cham. https://doi.org/10.1007/978-3-031-06809-6_12
23. Hendriks N, Huybrechts L, Wilkinson A, Slegers K (2014) Challenges in doing participatory design with people with dementia. In: Proceedings of the 13th participatory design conference: short papers, industry cases, workshop descriptions, Doctoral Consortium papers, and Keynote abstracts—Volume 2 (PDC '14). Association for Computing Machinery, New York, NY, USA, pp 33–36. https://doi.org/10.1145/2662155.266219

Design Culture Perception: Diagnosis of Knowledge and Vision of Traditional and Innovative Industries in Portugal

Hugo Palmares, Miguel Terroso, and Emília Costa

Abstract This chapter focuses on how design culture is perceived in an industrial context, through the diagnosis tool Design Culture Ladder (DCL) in which the deliverable outcomes are provide both qualitative and quantitative data base. In a first phase, research based on the literature review (1) compared studies on the indices/paradigms of culture, culture evolution, design culture, and organizational/corporate culture, contextualized to the industry, (2) the analysis of indicators extracted from the paradigms, and (3) the triangulation of the dimensions found; in a second phase, empirical study applied in (4) interviews and (5) surveys—to 18 companies in traditional and innovative sectors, respectively from the footwear industry and the software industry in Portugal—along with a multidisciplinary framework of the companies' employees; in a third phase, analysis of found data, such as (6) sub-categories, (7) frequency and metric analysis, (8) categories, and (9) main concepts; and in a last phase, (10) the analysis and comparision of quantitative data between the analysed industries. The results/findings of this chapter show that this diagnosis tool can be applied and be transversal to different industries in Portugal, provide comparative results and validate the questionnaire script of the DCL model.

Keywords Design Culture Ladder · Diagnosis tool · Qualitative and quantitative research · Footwear industry · Software industry

H. Palmares (✉) · E. Costa
ID+ Research Institute in Design, Media and Culture, Faculty of Fine Arts of the University of Porto, 4049-021 Porto, Portugal
e-mail: hpalmares@ipca.pt

E. Costa
e-mail: ecosta@fba.up.pt

M. Terroso
ID+ Research Institute in Design, Media and Culture, School of Design, IPCA Campus, 4750-810 Barcelos, Portugal
e-mail: mterroso@ipca.pt

N. Martins et al. (eds.), *Perspectives on Design and Digital Communication IV*, Springer Series in Design and Innovation 33, https://doi.org/10.1007/978-3-031-41770-2_6

1 Introduction

The chapter aims to demonstrate the perception that professionals in the footwear and software sectors in Portugal have regarding the design culture that is conscientious in their organizations and in their respective sectors. Dealing with a design culture in an organization means incorporating a complex matrix of anthropogenic activities, perceptions and manifestations, rather than remaining fixed and monocultural ([23], p. 3). This study is research through design, in partnership with a multidisciplinary team (e.g., designers, design directors, marketing directors, engineers, and CEO's), from companies in the footwear and software sectors, thus promoting knowledge sharing between industry and academia. An empirical observation of researchers—9 interviews (to 8 footwear companies) and 10 surveys (to 10 software start-ups)—carried out, and through a systematic literature review, it is argued that the perception of professionals in the industry regarding design culture in organizations identifies the need for change (concerning organization), and for innovation (related to processes and products).

In the business context, change and flexibility are factors associated with organizational efficiency. In this sense, for an organization to be able to evolve in its design culture, it must promote change and take risks with that change. Therefore, it is necessary to deal with risk if innovation and excellence are to be achieved. It will be demanding to migrate to innovation culture if there is a perception that the design can only be used to solve problems with known products or technology issues ([9], p. 105).

According to authors, there have been several approaches to measure how organisations perceive design and the depth of their design resources, from both industry and academia ([48], p. 535). Performing a diagnosis and providing feedback, so the change agent understands the employees' perception of the proposed change through the collection and analysis of appropriate data. Personal interviews, surveys, observations, analysis of previously collected data, and other techniques can be used to collect data to diagnose insights and generate progress assessment ([18], p. 9).

Based on the research hypothesis that the presence of a design culture tool can be an interface for innovation and evolution in organizations, capable of contributing to overcoming traditionalist management modes by different sectors of Portuguese industry. Accordingly, to improve design practice, management, education and its results, design research aims to formulate and validate models and theories about design phenomena with all their dimensions—e.g., people, products, knowledge/methods/tools, organization, microeconomics, and macroeconomics ([4], p. 2).

The present chapter is part of a larger research that involves the process of co-creating a *toolkit*, as an interface for diagnosis, prescription, verification, and validation of the tool in organizations. This *toolkit* will be able to emancipate companies so that they can evolve the design culture, contributing to a multidisciplinary business management as a vehicle for innovative processes, products and/or organizations. Thus, the starting point tool developed in the present chapter—in the context of the footwear and software sectors—in the future could be applied in other industries in Portugal, and with the potential to be tested also in an international context.

2 Industry

2.1 *Overview of Design Perspective into Industry*

Allong time the relationship between design and industry became closer, more balanced but also more dependent on each other, favouring a linkage of new competencies associated with a systemic approach ([10], p. 1318). However, the relatively low value put on design issues and design management by the industry may lead to a poor culture for design ([22], p. 5).

It was in the 1990s that industrial design and the emerging design applications went through the user-centered perspective. The fundamental notion was that everyone had some level of skill, which might serve as inspiration for design ([30], p. 18). Thus, bringing in external inputs, it will no longer be necessary to place the blame solely on the perspective of designers, who have long seen themselves as speakers for people in the industry.

Several anthropologists were hired by major companies in the 1990s (e.g., Apple, and Intel), and another inspiration was fieldwork done in design firms (e.g., IDEO, and Fitch). To manage cultural issues in industry, design ethnography integrates with industrial design ([30], p. 69). In previous decades, Sony was the Apple and IDEO in its day, profitable, innovative, efficient, and it treated its employees well. At the beginning of a company, however, one can still aim and hope for something bigger, and vaguer, as highlighted by Nike co-founder ([28], p. 281).

However, the industry's unique and resolute way of *doing things* the way it always has, as well as its deep resistance to change, make it difficult for people to evolve an inside culture—e.g., mentalities, conventional methods, and business practices ([25], p. 2).

2.2 *Traditional and Innovative Industries in Portugal*

The manufacturing industry in Portugal, according to 2021 data from the Bank of Portugal,[1] consists of 43,013 companies, responsible for a turnover of almost 102 million euros. More than 71% were micro companies, 22% small and 5.6% medium companies.

Through the report promoted by the Agency for Competitiveness and Innovation (IAPMEI), the number of *PME Excellence* companies increased exponentially, a total

[1] Available at https://bpstat.bportugal.pt/conteudos/publicacoes/1348.

of 3,881 companies representing the various sectors of activity in Portugal. Regarding sectoral distribution, industry is the second most representative activity, with 984 companies selected, representing 25.4% of all sectors in the Portuguese economy. The clusters with the highest concentration of *PME Excellence 2021* companies, which were analysed study cases in the present article, are in the districts of Porto (20.7%), Braga (11%), and Aveiro (10%) [21].

The metallurgy, textiles and footwear sectors helped reinforce the weight of exports in the national gross domestic product, which is expected to increase to 49% in 2022. Since the most attractive sectors to work in Portugal are information technology (IT)—e.g., software, and applications—consulting, health, tourism, accommodation, and leisure, results from the *Randstad Employer Brand Research* study [26].

For this chapter we focused on two different industries, the footwear industry seen as a traditional sector—despite having companies that present innovative strategies and processes—and the software industry, seen as a booming sector associated with innovation. Both sectors in Portugal work predominantly with international companies and clients, with an export aspect of their products and services.

Footwear Sector. In Portugal is one of the national sectors with the most international recognition and dissemination. Being an industry with a tradition of knowing and doing well with quality, knowledge that had been passed down from generations, and which positioned this sector—followed by the Italian industry—as the most expensive footwear producer on the European continent. Portuguese companies have been working over the last few decades as producers requested by international brands, thus leaving national companies, for the most part, to bet on their own brand.

The footwear industry in Portugal is sometimes understood as a traditional sector, although we can diagnose and distinguish companies that use traditional processes and others that use innovative processes (see Fig. 1).

These processes are the reflection of a vision on the management perspective and its employees, and this vision can drive and evolve the business, encouraging new working practices. In an empirical observation, it was verified that the footwear sector, on the one hand, mostly presents companies with traditional values from a managerial point of view, on the other hand, it presents, on a smaller scale, companies with a

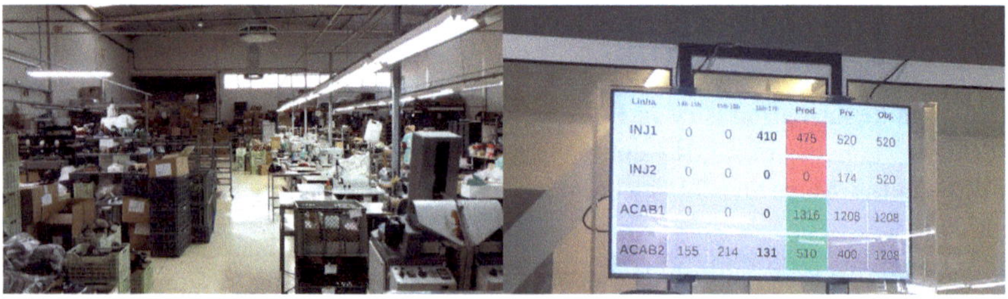

Fig. 1 Traditional layout and processes (left), and digitization processes (right), at footwear industry

vision for innovation and for the change of processes, products and/or organizations. Fostering a culture of innovation is critical to success, and that's why it's often necessary to shank a few before your swing smooths out, isn't about perfection ([27], p. 297).

A sector made up of 1,500 companies and responsible for around 40,000 jobs, the footwear sector exports more than 95% of its production to 172 countries on five continents. Exports from the Portuguese footwear and leather goods sector surpassed 2.35 million euros in 2022, setting a new record and increasing by 22.2%, compared to 2021. CEO's and retailers expect the price of footwear to record a strong rise next year, pointing to values between 5% and 20%, according to data from the *Business Conditions Survey* of surveys of specialists from more than 40 countries, within the scope of *World Footwear* publication,[2] which the Portuguese Footwear, Components and Leather Goods Manufacturers' Association (APICCAPS) publishes twice a year.

Exporting clusters frequently involve more natural resources or labour in a middle-income economy like Portugal ([40], p. 87). Cluster analysis, in addition to classifying subjects using measuring instruments into groups that are as similar as possible to each other, and as different as possible in relation to other groups, has also been used to detect personal constructs and group perceptions on socio-educational issues ([13], p. 311). In the geographical location of the footwear sector clusters, around 90% of this industry is located in the north of the country, 7.6% in the center and around 2.5% in the metropolitan area of Lisbon. The footwear industry in Portugal is divided into two clusters in the north of the country, one specializing in classic footwear located in the district of Aveiro (S. João da Madeira, Santa Maria da Feira, Oliveira de Azeméis); the other, located in the Porto and Braga districts (Vila Nova de Gaia, Vila do Conde, Felgueiras, Guimarães), specializing in casual and sports footwear. A third pole with less representativeness is located further south, in the Benedita area.

In an innovation plan, according to a APICCAPS' publication,[3] investments are expected to be made in the next two years in terms of digitalization and sustainability—140 million investments in total—will create the conditions for Portugal to be, increasingly, an international reference in the development of excellent footwear, with a focus on Industry 4.0.

Software Sector. Industry focused on innovation (e.g., data storage devices, computer architecture, software, and data communication to speed up the process of storing, processing, and distributing information) in such a way as to enable the creation of new industries and restructure existing ones, and thus transforming society's lifestyle.

The current world as we know it would be impossible to exist without the software sector, as national infrastructures and services are controlled by computer systems, and most electrical products are made up of a computer and software that controls

[2] Available at https://www.apiccaps.pt/publications/world-footwear/113.html.

[3] On this matter see: https://www.apiccaps.pt/news/calcado-portugues-investe-140-milhoes-de-euros-e-prepara-industria-do-futuro/7359.html.

Fig. 2 Software solutions (left), and digitization processes for Industry 4.0 (right)

that same computer. In general, manufacturing and industry are fully computerized, as is the financial system.

In recent years there has been an increase in the turnover of this sector due to greater dynamism and a growing number of companies in the sector of computer programming services in Portugal. IT jobs (e.g., business analyst, software development, helpdesk, project manager, java developer, full stack developer, and web developer) are among the most competitive, according to the *Great Place to Work* 2022 editorial.[4]

IT companies in Portugal are sometimes subsidiaries of multinationals, sometimes they are start-ups that annually seek presence at the tech conference *Web Summit*[5] in Lisbon, Portugal. Currently, it can be said that most of the IT companies in Portugal are concentrated in the hardware and software segments (see Fig. 2), with the services segment being one of those that is expanding.

According to the consultancy *IDC Portugal* report,[6] in September of 2022, the Portuguese IT market reached a value of 5 million euros, which represents a growth of 3.9%, compared to 2021. The Portuguese industry will have to accelerate the digital transformation even more, around 50% of the investment in software will be made in the *as a service model*, with cloud technology already dominant.

The IT sector, with almost 90% of professionals who regularly work remotely, clearly stands out from other more traditional sectors. Since the current trend that is most verified is hybrid work, as pointed out by the National Institute of Statistics (INE).[7]

Currently, the concept of Industry 4.0 contemplates a wide variety of different technologies, from traditional IT to operational technologies (OT). The industry's digital transformation efforts are divided into these main fronts: operational planning procedures of production lines, concentrating on the optimization and scheduling of

[4] Available at https://www.greatplacetowork.pt/conteudos-menu/relatorio/melhores-lugares-de-tecnologia-para-trabalhar-em-portugal.

[5] On this matter see: https://websummit.com/.

[6] Available at https://static.computerworld.com.pt/media/2022/10/Estudo-da-Economia-e-da-Sociedade-Digital-2022-ACEPI-IDC-PT-Vers%C3%A3o-Completa.pdf.

[7] On this matter see: https://www.ine.pt/.

production orders, have received a lot of attention from this sector as part of its cultural (r)evolution.

3 Diagnosis Tool

3.1 Design Ladder' *Evolution*

The starting point to build a design culture model was used as a reference the original model of the *Danish Design Ladder* (DDL), developed by the Danish Design Centre, published by Kretzschmar ([31], p. 28). This model still appears today as a reference for worldwide companies to map the firms in terms of design maturity level—even created nearly two decades ago. The main conclusion of using the DDL model, is that the data from the interviewed Portuguese companies, suggest that the company's perception and actual use of the design in their business differs, indicating some lack of understanding and skill in its use ([44], p. 35).

The framework of the DDL model in this research served as a departure for a reformulation and contextualization of a ladder model, based on the design culture of organizations, as a diagnosis tool from the context of Portuguese industry. This research purposes a diagnosis tool for companies to map and indicate recommendations in terms of the level of design culture maturity in their organizations. Based on a ladder structure and by evolutionary way as the DDL model—in which for the new model *Design Culture Ladder* (DCL)—the levels represent the evolution of the design culture and where companies can see where to improve themselves and their organization—through a questionnaire divided by levels (i.e., *dimensions*), in which the perception of each of participants has on those themes is evaluated ([38], p. 529).

3.2 Design Culture Ladder*' Framework*

The framework of the DCL model intends facilitate the organizations to identify what the design maturity they have and where they need to improve to complete entirety each level. The DCL model exposes the determining factors for the maturity of the design culture to emerge, and which in turn, the framework of this ladder model—in an increasing and evolving way—require the domain of the following *dimensions*: Technical Skills, Business Skills, Social Factors and Cultural Factors (see Fig. 3).

The *triangulation* of concepts here consisted of combining two or more points of view, theoretical approaches, and data collection methods in research so that we can obtain as a final result a more reliable picture of reality or a more complete understanding of the phenomena to be analysed ([13], p. 239). Thus, to the theoretical framework, the concepts of the authors Anders Kretzschmar (*Danish Design*

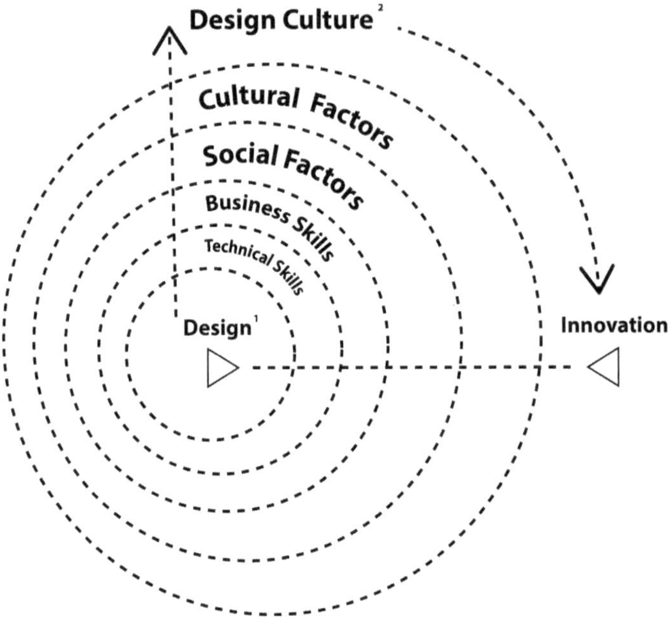

Fig. 3 Design culture introducing diagram

Ladder), Frank W. Geels (*Multi-Level Perspective*), and Sam Bucolo (*Design Led Innovation*), were triangulated.

The factors associated with design culture, are correlated to the perspective of the human being—individual axis—and their relationship with the community to which they belong—collective axis. According to Geels' analysis ([17], p. 1257), in a *multi-level perspective*, on the axis relationship between *individual—family—community—society*, referenced to the context of this research on the design culture in the industry, relating here respectively to the *technical—business—social—cultural* levels. According to Bucolo' *design led matrix*, for each stage there use a set of questions and tools, which allows to have a structured conversation with staff across the organisations, providing value to and how they could change a needed mindset ([9], p. 142).

For all of four levels, a guide of 35 questions were prepared, and the interviews followed a structured and open-ended based questions—for quantitative and qualitative analysis—with formulated questions (a questionnaire script) and is given to each interviewee or respondent in the order listed in the script. This structure facilitates data comparisons and is useful when used by other interviewers ([4], p. 271).

3.3 *Theoretical Framework for Questionnaire Script*

The structure of this research was segmented by *indices* (i.e., paradigms, concepts extracted from the hypothesis), by *indicators* (extracted from the state of the art from the previous indices), culminating in the correlation with the *dimensions* (previously found and explained in this chapter), as can be analysed in Table 1.

Indices. In this phase, the paradigms (i.e., *indices* that are concepts extracted from the research hypothesis) of culture are clarified and explored with focus on the business and industrial context, where they are divided by: design culture, culture evolution, organizational culture, and culture definition. These research paradigms can be defined as articulated sets of postulates, known values, common theories and rules that are accepted by authors (i.e., elements of a scientific community) at a given historical moment.

Thus, the paradigms in this research fulfil the function of unifying concepts, points of view, belonging to a common identity with theoretical and methodological issues, and the legitimation among the authors, given that a given paradigm points to criteria of validity and interpretation [12, 37].

Indicators. Extracted from the authors' common theories, the *indicators* are represented by a list of concepts of a precise theoretical and methodological framework, and consequently, a sharing of experiences and an agreement regarding the nature of research and the conception of knowledge [4, 37]. Among others were found matrix of human activities; values, knowledge, and vision; cultural reform; collective culture; innovation ecosystem; workplace culture.

4 Methodology

4.1 *Research Methods*

As this study is about a cognitive nature, through a systematic, flexible, and objective inquiry process, it is characterized both by its multiplicity (of paradigms and models) and by its contextual dependence (industry). Through the application of quantitative and qualitative techniques, using inductive and deductive methods, and other procedures that promote a dichotomous discussion of the epistemological and methodological components of this investigation ([13], p. 7).

This chapter is characterized by being an exploratory investigation of a provisional nature insofar as it is carried out to obtain a first knowledge of the situation of the present context and of previously demonstrated multiplicity. This contextual relationship between academia and industry, can generate new reflexive practices and highlight values that can produce significant change in both, resulting in significant perceptions of individuals and society ([10], p. 1331).

Table 1 Indices, studies, indicators, dimensions

Indices	Studies	Indicators	Dimensions
Design culture	Papanek [39]	– Matrix of life (design culture)	Cultural factors
	Julier [23]	– Complex matrix (human activities, perceptions, articulations)	
	Jolly et al. [22]	– Values, knowledges, and behaviours (surround design act)	
	Viniegra [46]	– Evolution of results (design exercises)	
	Deserti and Rizzo [14]	– System of knowledge, competences, and skills (operate, mediate, coordinate) – Cultural reform (design as guide)	
	Buchanan [8]	– Close relationship (design, engineering, marketing)	
	Brown and Martin [7]	– Promoter of innovation (design helping stakeholders in a process and in the system)	
	Manzini [34]	– Knowledge, values, and vision – Meaningful context (new meanings are conceived, developed, produced)	
Culture evolution	Hall [19]	– Communication (as culture)	Social factors
	Manzini [33]	– Mindset review (models, instruments)	
	Filson and Lewis [15]	– Workplace culture (culture change, difficulty of dealing)	
	Carlos Viniegra [46]	– Human culture (dynamic, evolutionary, complex system) – Exchanged (information, knowledge, art, and technology)	

(continued)

Table 1 (continued)

Indices	Studies	Indicators	Dimensions
Organizational or corporative culture	Santos et al. [43]	– Collective mental capacity (create, innovate, systems, processes, products, and services)	Organizational skills
	Gilley et al. [18]	– Culture change (regularly reminded purpose, and goals)	
	Mozota [35], Kootstra [29]	– Attributes, and methods (internal to the organization)	
	Turner [45], Norman and Verganti [36]	– Environment, and conditions (for innovation, radical or disruptive)	
	Junginger [24]	– Use of design (organizations knows when, where, and how)	
	Liem and Sanders [32]	– Co-creation (design tools and methods by perspective)	
	Badding and Leigh [2]	– Organizational change (flexibility, adaptability, effective leadership, and strategic thinking)	
	Hernández et al. [20]	– Organizational innovation (external factors, stakeholders, with internal design process)	
	Freire [16]	– Values, beliefs, expectations, and practices (unique and common sense of identity)	
	Barauna [3]	– Innovation ecosystem (culture of anticipation of change by organizations)	
Culture	Hall [19]	– Behaviour (patterns, attitudes)	Technical skills
	Rittel and Webber [42]	– Cultural diversity (sector, cluster)	

(continued)

Table 1 (continued)

Indices	Studies	Indicators	Dimensions
	Bonsiepe [5]	– Culture identity (daily artifacts)	
	Manzini [33]	– Group motivations (practices, culture)	
	Jolly et al. [22]	– Values, beliefs, and behaviours (common)	

4.2 Variables Sampling

The questionnaire script was previously tested in three pilot interviews (interviewees were designers with experience in the footwear industry, accessories or leather goods, and textile industry); and was applied in 9 interviews with professionals in the footwear industry in Portugal. Of the 30 invitations sent to the researchers' networking and recommendations by respondents, 30% of the guests agreed to participate in this study.

Qualitative variables were identified here, of polytomic category either at the level of the interviewee's positions—e.g., designers, creative directors, marketing directors, engineers, and CEO's—either at the level of qualifications of each individual—of the 89% of those interviewed have a higher education degree (Bachelor's degree), and of these, 50% have a postgraduation (Master's degree or MBA)—in the areas of Management, Economics, Engineering, Marketing, Design and Arts.

The object of analysis chosen for this empirical study is a collective of mature firms in terms of design and innovation, called design driven firms, and the type of design and innovation teams consists of group from different backgrounds situated at the original DDL between level 2 and the top of level. Were chosen footwear companies from different clusters (e.g., Aveiro, Braga, and Porto), polytomous category variable, to avoid formalised groupings and similar cases to each other than to those in other clusters [4, 11].

Participants were asked to sign a consent form, were explained of the risks and benefits of participating in this academic investigation. The continuous quantitative variable was identified, in which the duration of each interview varied between 40 and 90 min. The interviews were recorded via video conferencing platform Zoom with the recording function, transcribed, and analysed, both individually and at the level of comparison.

Regarding the applied surveys into software industry, 10 start-ups in software engineering were answered and sent by email—each with 8–10 employees with a higher education degree—incubated at the Faculty of Engineering of the University of Porto (FEUP), and inserted in the Project Management Laboratory (LGP). Each of these software companies, between May and July of 2022, worked exclusively for a reference company, both nationally and internationally, such as Deloitte, Vestas, Stellantis, Talkdesk, or Fashable.

A criterial sampling was identified, in which industry segments and their clusters were selected for this study—according to a pre-defined criterion—companies and employees that communicate innovation and promote a disruptive culture; as also a *snowball sampling* ([13], pp. 95–97), in which target members were identified and in which they suggested others, having been a useful technique to reach hard-to-reach participants. In this chapter, care was taken in selecting the sample rather than its size.

5 Results

5.1 Qualitative Analysis of Diagnosis Tool

The questionability of the results is imposed because more than the study of large samples, in case studies the interest is in the subjects who act in situations, since the meanings they share are *meanings-in-action* ([37], p. 28). At a conceptual level, the object of study in the investigation is not behaviours, but intentions and situations, i.e., it is about investigating *ideas, discovering meanings* in *individual actions* and *social interactions* from the perspective of the actors involved in the process ([13], p. 28).

At a methodological level, the investigation of a qualitative nature is based on the inductive method, in this sense the theory of this research is of an interpretative type, i.e., it does not precede the data but arises from these same data, in a constant and dynamic relationship with practice, without precise normative intentions ([13], p. 30).

Sub-categories. It is understood by outputs extracted from the field study/interviews, in the encounter of *patterns* occurred by the analysed perception of the interviewees. Interviews are a powerful data collection technique, generating new information that implies a *reconceptualization of the topics under study* ([13], p. 141), and following this study the raised hypothesis is submitted to empirical confrontation under strict experimental control.

The sub-categories extracted from the qualitative analysis of the interviews were found by the segmented script in each of the four *dimensions*: cultural factors (e.g., value, specialization, creative process, disruptive thinking, inspiration to innovate, failure to innovate, among others); social factors (e.g., investment in human resources, literacy, culture and identity); organizational skills (e.g., collaborative work, deepening relationships, spirit of harmony, didactic work, among others); and technical skills (e.g., categorization and segmentation, technical specializations, commitment to internal procedures, among others); as can be analysed in Table 2.

Frequency. This phase of the study ends with the occurrence of qualitative data, the *metric* and quantification of patterns and statistical analysis of texts, overcoming the

Table 2 Dimensions, sub-categories, frequency, categories

Dimensions	Sub-categories	Frequency	Categories
Cultural factors	– Value (creation and intellectual)	9	
	– Specialization	4	
	– Disruptive thinking (agencies)	6	
	– Creative process	9	
	– Process update (creative/productive)	**21**	– Disruptive culture
	– Differentiation (design, technology, process)	**18**	– Innovation
	– Strategy (sales-commercial-design chain)	8	
	– Inspiration to innovate	8	
	– Err to innovative	8	
	– Content production	5	
	– Brand vision	**11**	– Differentiation
	– Aesthetic vision	1	
	– Functionality (design)	5	
Social factors	– Investment (human resources)	6	
	– Culture and identity	6	
	– Literacy	4	
Organizational and/or corporate skills	– Collaborative work (partners/brands)	**13**	– Collaboration
	– Deepen relationships	4	
	– Activism spirit	7	
	– Work on relationships (internal/external)	**10**	– Inclusion
	– Harmony spirit	6	
	– Didactic	4	
Technical skills	– Tradition-Industry	**11**	– Cluster integration
	– Specifications (technical, registered)	5	
	– Categorization and segmentation	2	
	– Sustainability (design)	5	
	– Tolerance (external procedures)	**20**	– Sectorial culture
	– Commitment (internal procedures)	7	

qualitative/quantitative dichotomy and the use of both qualitative and quantitative research methods [4, 41].

The mentions or references most used by the group of interviewees were: the updating of creative and/or productive processes (cultural dimension); tolerance in external procedures (technical dimension); differentiation in design, technology, process (cultural dimension); collaborative work with partners or brands (organizational dimension); the brand vision (cultural dimension); the relationship between tradition and the industry itself (technical dimension); and work on internal and external relations to the organization (organizational dimension).

Categories. In this phase, after data collection *the categories emerge from them* ([47], p. 217), the search for thought patterns, words, phrases, i.e., in this study regularities in the measured data that justify a *categorization* (theorization of dimensions), *coding* (selection of sub-categories), and *data reduction* (analysis of categories) [6, 47].

Therefore, the categories analysed here are rubrics or classes that bring together a group of elements due to common characteristics, and that must have, among others, the following qualities: homogeneity, objectivity, fidelity, and mutual exclusion. The following categories *emerged* from the most frequent patterns: disruptive culture, innovation, differentiation, collaboration, inclusion, cluster integration, and sectorial culture. At this stage, the categories serve to unify the study both in conceptual and methodological aspects, in a process of summarizing the data and their interpretation.

5.2 Quantitative Analysis of Diagnosis Tool

The questionnaire script/model was the basis for both qualitative and quantitative analysis, where *Likert* scale was used for quantitative approach. A scale with a type of psychometric response, where respondents specify their level of agreement with each question. For this purpose, it was necessary to use a measuring instrument on which the *informative quality of the data obtained in the investigation* will depend ([13], p. 110). The scale was divided into 5 possible answers: (a) total agreement, (b) agreement, (c) neither agree nor disagree, (d) disagreement, (e) total disagreement.

The objective was to measure opinions, perceptions, and values, of a continuous nature—of interval variables—according to the *Likert* scale measure used in the questionnaire for quantitative analysis of the data, instead of which it was complemented with a request for justification and/or example of each question for the qualitative analysis.

Quantitative analysis of surveys. In this statistical analysis of the data resulting from the measurement of ordinal scale variables, it represents qualitative properties that can be ranked. When collecting this data by survey—based on the interview guide tool, where the questionnaire included the same questions as the interview, except for the request for justification—care was taken in terms of designing the layout and general appearance of the form to be easily accessible ([13], p. 108).

Footwear sector. In this phase of quantitative data analysis, it is pertinent to compare the two scale instruments previously analysed, DDL and DCL, to perceive the proximity of the resulting values. It can be seen from the outset that from the DCL model there is more information and deepening—of the design culture domain by level and globally—and of less redundant values and label typology. This DDL classification was easy to classify due to the possibility of interviewing and getting to know the companies better, than if it were through surveys without direct contact with employees. In Table 3, only the means in percentages and in *Likert* scale are presented, however there are percentages for each dimension of the DCL model for greater return of information provided to the respondents (but for this article it was decided that this information was not present in the table, for clarity relevance).

Software sector. Compared to the interviews, the surveys does not provide the richness of details of an interview or face-to-face observation. However, it should be noted that similar variables in the same cluster can result in similar data, more precisely comparing the average result between the software and the footwear sectors, in numerical value the global average is similar, as can be seen in Table 4.

Table 3 Comparation of design models from quantitative data at footwear industry

Variables correlation	Danish Design Ladder (2003)	Design Culture Ladder (2023)
Manufacturing and brand companies (100+ employees)		
Company **#1** (marketeer) Cluster: Braga	level **4** (design as strategy)	**4.3** (scale 1/5) **86.7%** (average)
Company **#1** (creative director) Cluster: Braga	level **4** (design as strategy)	**3.9** (scale 1/5) **81.6%** (average)
Company **#2** (CEO) Cluster: Porto	level **3** (design as process)	**4.2** (scale 1/5) **85.3%** (average)
Brand companies (2–10 internal employees; 50+ outsourcing)		
Company **#3** (creative director) Cluster: Braga	level **3** (design as process)	**4** (scale 1/5) **81.7%** (average)
Company **#4** (CEO) Cluster: Aveiro	level **3** (design as process)	**4.1** (scale 1/5) **83.4%** (average)
Company **#5** (creative director) Cluster: Aveiro	level **4** (design as strategy)	**3.7** (scale 1/5) **76.1%** (average)
Company **#6** (CEO) Cluster: Porto	level **3** (design as process)	**3.9** (scale 1/5) **80.2%** (average)
Company **#7** (creative director) Cluster: Braga	level **4** (design as strategy)	**4.2** (scale 1/5) **84%** (average)
Design and Consultant Companies (2–10 employees; 50 + outsourcing)		
Company **#8** (creative director) Cluster: Braga	level **2** (design as styling)	**3.3** (scale 1/5) **66.9%** (average)
Global Average	**3.3** (scale 1/4) **82.5%** (average)	**3.9** (scale 1/5) **80.6%** (average)

Table 4 Questionnaires at software industry

Variables correlation	Main client' sector	Design Culture Ladder (2023)
Start-up **#1** (implementation manager)	Client: Deloitte (audit and assurance)	**4.2** (scale 1/5) **84.4%** (average)
Start-up **#2** (software engineer)	Client: zerozero (football database)	**3.6** (scale 1/5) **73.4%** (average)
Start-up **#3** (software engineer)	Client: Vestas (wind energy turbines)	**3.8** (scale 1/5) **78.6%** (average)
Start-up **#4** (software engineer)	Client: DEUS.AI (data and artificial intelligence)	**3.6** (scale 1/5) **71.2%** (average)
Start-up **#5** (software engineer)	Client: Stellantis (automaker and mobility)	**3.9** (scale 1/5) **80.7%** (average)
Start-up **#6** (project manager)	Client: Jumpseller (e-commerce platforms)	**3.7** (scale 1/5) **74.8%** (average)
Start-up **#7** (CEO and quality manager)	Client: CustEasy (information services)	**4.4** (scale 1/5) **89%** (average)
Start-up **#8** (CEO)	Client: Talkdesk (cloud contact center)	**3.6** (scale 1/5) **73.5%** (average)
Start-up **#9** (developer)	Client: Fashable (artificial intelligence, fashion)	**3.4** (scale 1/5) **66.4%** (average)
Start-up **#10** (CEO)	Client: Erasmus Programme (academic mobility support)	**3.6** (scale 1/5) **71.1%** (average)
Global Average (design culture domain)		**3.7** (scale 1/5) **76.3%** (average)

Comparative analysis and triangulation of sectors. Comparing the 9 interviews and the 10 surveys carried out to two different industries, with 35 responses according to the *Likert* scale, it's possible to visualize spots of dominant color, blue more dominant in the answers and red less dominant (see Fig. 4). The analised model in this visual form serve to better communicate to the interviewee or respondent their perception of the domain of design culture that represents their own company.

The visual analysis of the diagnosis tool can be done as an indicator for the interviewee or respondents to realize where the steps to be filled are; thus, focus on these gaps to work and improve in the future, according to the existence of the red or orange colours in the corresponding bars.

In reading the final average obtained, it's possible to perceive that there is a tendency for less dominance—with more responses of disagreement and less of agreement—in cultural and social factors than in technical and organizational skills. It can also be highlighted the fact that there is a substantial difference in the comparative data of the two sectors in relation to technical skills, and for the software sector there is a perception of less mastery than for the footwear sector. It should be noted that among the sectors there is a very similar perception in relation to the cultural dimension, because it is the level with the lowest questioned domain compared to the other levels. If previously analysed average numerical value is similar, here the real

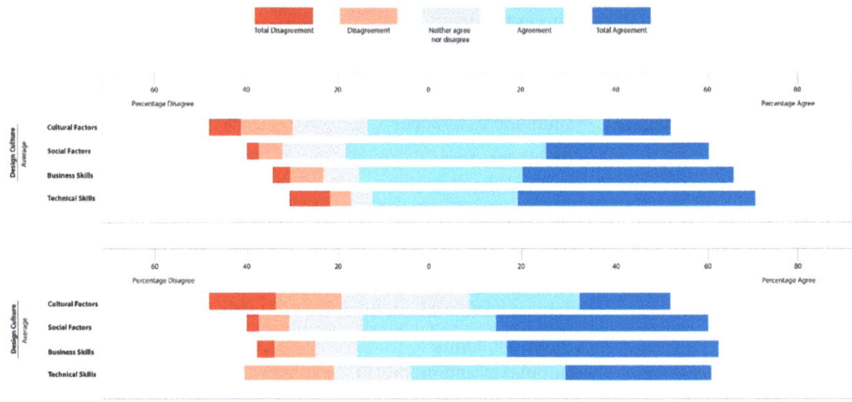

Fig. 4 Design culture average from footwear (top) and software (bottom) industries, applied by *Likert* scale

differences between the analysed sectors and the difference by level can be visually verified in the Fig. 4.

6 Conclusions

In this chapter it was crucial to strengthen the evidence using the triangulation of sources and methods ([4], p. 108) to gather the evidence about the design culture into companies, comparing traditional and innovative industries. The chapter helped to answer and to understand how the questionnaire script/model in design culture could be applied and be transversal to two different industries, one so-called traditional such as footwear and the other so-called innovative such as software.

This chapter is not without limitations, the small sample has low businesses and geographic cases diversity, thus generalization of results has this caveat. Concerning generalization of results for the sample used of companies could be more representative for the quantitative analysis, was the available sample at the time of this chapter, however more than the study of large samples interested in understanding the case studies in depth, in qualitative research.

Nevertheless, the strengths of this research are the fact the questionnaire script was based on a theoretical framework and was assessed for empirical study validity. Therefore, the intention of a future study in other industries and contexts is to replicate this methodology to understand, in the broadest sense of the term, not only the results of the scientific method, but the process itself ([13], p. 25).

If the validity of this research reflects empirically true facts drawn from interviews and surveys, it should be noted that validity is the complement of reliability ([1], p. 12), and in which the data that were obtained are independent of context. According to the results obtained in this study should be emphasised that there are differences in terms of design culture perception between the two industries analysed here, with greater importance for the use and awareness of design mindset in the footwear industry—as key aesthetic and functional components—than in the software industry.

Further developments for the tool presented and tested in the current chapter, in the context of the footwear and software sectors, are in the future research phase, such as building and testing a *toolkit* to applied to other industries, both nationally and internationally, either for its functionality or for its validation. There is also a intention to spread this tool in other formats through industry associations to reach a significant number of cases, as a contribution from academia to industry.

Acknowledgements Research funded by Foundation for Science and Technology (FCT)—Grant: 2020.08088.BD, hosted by the ID+ Research Institute for Design, Media and Culture. The authors thanks to Adriana Mano, João Alberto Azevedo, Eduardo Bonini, Gil Gonçalves, Jorge Pinto, Marita Setas Ferro, Nelson Amorim, Orlando Andrade, Pedro Castro, Pedro Maçana, Raul Moreira Vidal, Vítor Costa.

References

1. Anderson G, Arsenault N (1999) Fundamentals of educational research. Falmer Press Teachers Library, London
2. Badding S, Leigh K (2015) Frontiers of thought: the integration of design thinking & human resource development. In: AHRD international conference in the Americas. St. Louis, Missouri, pp 1–38. https://www.researchgate.net/publication/299655606
3. Barauna D (2021) Ecossistemas de inovação como ambientes de aprendizagem e cultura de design: uma proposição de reflexão pelo Design Estratégico. Design em Questão 1(1):18–30. https://design.revistas.ufcg.edu.br/index.php/designemquestao/article/view/31
4. Blessing LTM, Chakrabarti A (2009) DRM, a design research methodology. Springer-Verlag London Limited. https://doi.org/10.1007/978-1-84882-587-1
5. Bonsiepe G (1992 [1975]) Teoria e prática do design industrial: elementos para um manual crítico [Industrial design theory and practice: elements for a critical manual]. Centro Português de Design, Lisboa
6. Bravo MP, Eisman LB (1998) Investigación educativa (3ª Ed.). Ediciones Alfar, Sevilha
7. Brown T, Martin RL (2015) Design for action: change management. Harvard Business Review, Cambridge. https://hbr.org/2015/09/design-for-action
8. Buchanan R (2015) Worlds in the making: design, management, and the reform of organizational culture. She Ji: The Journal of Design, Economics, and Innovation 1(1):5–21. https://doi.org/10.1016/j.sheji.2015.09.003
9. Bucolo S (2016) Are we there yet?: Insights on how to lead by design. BIS Publishers, Amsterdam
10. Camacho B, Alexandre R (2019) Design education. University–industry collaboration, a case study. Design J 22(sup1):1317–1332. https://doi.org/10.1080/14606925.2019.1594958
11. Cirella S, Murphy S (2022) Exploring intermediary practices of collaboration in university–industry innovation: a practice theory approach. Creat Innov Manag 31(2):358–375. https://doi.org/10.1111/caim.12491
12. Coutinho CP (2005) Percursos da investigação em tecnologia educativa em Portugal: uma abordagem temática e metodológica a publicações científicas (1985–2000). Universidade do Minho, Braga, CIED
13. Coutinho CP (2011) Metodologia de Investigação em Ciências Sociais e Humanas: teoria e prática (2ª Ed.). Edições Almedina, Coimbra

14. Deserti A, Rizzo F (2014) Design and the cultures of enterprises. Des Issues 30(1):36–56. https://doi.org/10.1162/DESI_a_00247
15. Filson A, Lewis A (2000) Cultural issues in implementing changes to new product development process in small to medium sized enterprise (SME). J Eng Des 11(2):149–157. https://doi.org/10.1080/09544820050034240
16. Freire A (2020) Estratégia - Criação de Valor Sustentável em Negócios Tradicionais e Digitais. Bertrand Editora, Lisboa
17. Geels FW (2002) Technological transitions as evolutionary reconfiguration processes: a multi-level perspective and a case-study. Res Policy 31(8–9):1257–1274. https://doi.org/10.1016/S0048-7333(02)00062-8
18. Gilley JW, Quatro SA, Hoekstra E, Whittle DD, Maycunich A (2001) The manager as change agent: a practical guide for high performance individuals and organizations. Perseus, Cambridge
19. Hall E (1959) The silent language. Doubleday and Co. Inc., New York
20. Hernández RJ, Cooper R, Tether B, Murphy E (2018) Design, the language of innovation: a review of the design studies literature. She Ji: The Journal of Design, Economics, and Innovation 4(3):249–274. https://doi.org/10.1016/j.sheji.2018.06.001
21. IAPMEI (2022) Excelência'21. Especial. Impresa Publishing - Dossiês Especiais, Lisboa
22. Jolly L, Radcliffe D, Nycyk M, Andersen J (2005) An evolving model of design culture. In: International conference on engineering design Iced 05. Melbourne, pp 1–8. https://www.academia.edu/25355824
23. Julier G (2000) The culture of design. Sage, London
24. Junginger S (2009) Design in the organization: parts and wholes. Des Res J 2(9):23–29. https://www.academia.edu/213937
25. Kajewski S, Tilley P, Crawford J, Remmers T, Chen S-E, Lenard D, Brewer G, Gameson R, Kolomy R, Martins R, Sher W, Weippert A, Caldwell G, Haug M (2001) Industry culture: a need for change. Polit Sci
26. Kantar (2021) Randstad Employer Brand Research (REBR) 2021. London
27. Kelley T, Littman J (2001) The art of innovation: lessons in creativity from IDEO, America's leading design firm. Currency, New York
28. Knight P (2016) Shoe dog. A memoir by the creator of Nike. Simon & Schuster, London
29. Kootstra G (2009) The incorporation of design management in today's business practices. An analysis of design management practices in Europe. In: The Hague and Rotterdam, Design Management Europe (DME) consortium
30. Koskinen I, Zimmerman J, Binder T, RedstrÖm J, Wensveen S (2011) Design research through practice: from the lab, field, and showroom. Morgan Kaufmann. https://doi.org/10.1109/TPC.2013.2274109
31. Kretzschmar A (2003) The economic effects of design. National Agency for Enterprise and Housing, Copenhagen
32. Liem A, Sanders EB-N (2011) The impact of human-centred design workshops in strategic design projects. In: Human centered design—Second international conference, pp 110–119. Orlando. https://doi.org/10.1007/978-3-642-21753-1_13
33. Manzini E (1993 [1989]) A Matéria da Invenção [The Material of Invention]. Centro Português de Design, Lisboa
34. Manzini E (2016) Design culture and dialogic design. Des Issues 32(1):52–59. https://doi.org/10.1162/DESI_a_00364
35. Mozota BB (2003) Design and competitive edge: a model for design management excellence in European SMEs. Des Manag J Acad Rev 2:88–103. https://doi.org/10.1111/j.1948-7177.2002.tb00014.x
36. Norman DA, Verganti R (2014) Incremental and radical innovation: design research versus technology and meaning change. Des Issues 30(1):78–96. https://doi.org/10.1162/DESI_a_00250
37. Pacheco JA (1993) O pensamento e a ação do professor em formação [PhD's thesis]. Instituto de Educação e Psicologia da Universidade do Minho, Braga

38. Palmares H, Terroso M, Costa E (2023) Design culture ladder: creation of a diagnosis tool to evolve industry. In: Martins N, Brandão DC (eds) Advances in design and digital communication III. Springer Nature Switzerland, pp 523–535. https://doi.org/10.1007/978-3-031-20364-0_45
39. Papanek V (1972) Design for the real world: human ecology and social change. Pantheon Books, New York
40. Porter ME (1998) Clusters and the new economics of competition. Harvard Business Review 76(6)
41. Peréz Serrano G (1998) Investigación qualitativa. Retos e interrogantes (2ª Ed.), vol 1. Editorial La Muralla, Madrid
42. Rittel HWJ, Webber MM (1973) Dilemmas in a general theory of planning. Policy Sci 4(2):155–169. http://www.jstor.org/stable/4531523
43. Santos AR, Pacheco FF, Pereira HJ, Bastos Jr, PA (2001) Gestão do conhecimento: uma experiência para o sucesso empresarial. Editora Universitária Champagnat, Curitiba, pp 11–48. https://www.academia.edu/39505702
44. Silva RM (2014) Design in business: an on-line survey to Portuguese innovation-driven firms about design maturity [master's dissertation]. School of Economics & Management, Lisboa
45. Turner R (2004) Herramientas de gestión del diseño y liderazgo empresarial para maximizar su valor. In: Temas de Diseno En La Europa de Hoy. BEDA, Barcelona, pp 14–16. https://www.academia.edu/16672698
46. Viniegra C (2011) Cultural evolution, design and philosophy: for the change of era. Int J Des Nat Ecodyn 6(3):171–212. https://doi.org/10.2495/DNE-V6-N3-171-212
47. Wiersma W (1995) Research methods in education: an introduction, 6ª. Allyn an Bacon, Boston
48. Yeo Y, Lee JJ (2018) Mapping design capability of public service organisations: a tool for collaborative reflection. In: Proceedings of the service design and innovation conference (ServDes'18). Linköping: Linköping University Electronic Press. Milano, pp 18–20

Design Thinking Methodology and Text-To-Image Artificial Intelligence: A Case Study in the Context of Furniture Design Education

Carlos Sena Caires, Gerald Estadieu, and Sandra Olga Ng Ka Man

Abstract The design thinking methodology is a problem-solving approach that involves empathising with end-users, (re)defining problems, brainstorming solutions creatively, and experimenting with prototypes and testing. It has been widely adopted in education to help students develop critical thinking, creativity, and problem-solving skills in design. On the other hand, text-to-image artificial intelligence is a method used to generate images from natural language descriptors (usually referred to as prompts). Design thinking methodology can teach students to think creatively and critically about real-world problems when applied in the classroom. In the context of design teaching at the University of Saint Joseph, Macao, students use the design thinking methodology to develop innovative proposals for furniture design solutions. Combining design thinking methodologies with text-to-image artificial intelligence can further enhance the learning experience by allowing students to generate visual representations of their ideas during the ideation phase. The authors developed a systematic approach to generate images for ideation on furniture design based on prompting text-to-image (PTI). The analysis related students' results who applied the design thinking methodology without using AI tools and the results generated using a standard text-to-image programme. By combining both methods, teachers can help students develop critical thinking, creativity, and problem-solving skills, while also allowing them to generate visual representations in a different paradigm and, by so, being able to communicate their ideas with the most appropriate support for them.

Keywords Design thinking · Artificial intelligence · Furniture design education · Education quality · Text-to-image prompt

C. S. Caires (✉) · G. Estadieu · S. Olga Ng Ka Man
University of Saint Joseph, Macau, Estrada Marginal da Ilha Verde, 14-17, Macau, China
e-mail: carlos.caires@usj.edu.mo

G. Estadieu
e-mail: gestadieu@usj.edu.mo

S. Olga Ng Ka Man
e-mail: sandra.ng@usj.edu.mo

N. Martins et al. (eds.), *Perspectives on Design and Digital Communication IV*, Springer Series in Design and Innovation 33, https://doi.org/10.1007/978-3-031-41770-2_7

1 Design Thinking in the Field of Education

Design thinking in education is a pedagogical approach that strongly emphasises on the development of cognitive, technical, systemic, and strategic problem-solving and critical thinking skills inside the classroom. This approach is widely discussed and referenced in the academic literature [1–4], with multiple authors describing "design thinking" as a human-centred approach that considers the behaviours, needs, and desires of individuals in problem-solving processes. This approach has gained widespread acceptance in areas including product and service design and other domains such as architecture, management, and business strategy.

The application of design thinking in education involves engaging students in a structured and systematic process that encompasses several key stages (see Fig. 1), including the identification and definition of a problem, the collection and analysis of relevant information, the generation of innovative ideas, the creation and testing of prototypes, and the evaluation and improvement of solutions. The implementation of this process is marked by a series of iterative stages, which comprise empathy, the understanding of the intended user, the formulation of concepts, and the ideation process. This is followed by the creation of prototypes and their subsequent evaluations, enabling students to engage in a form of experiential and reflective learning.

The inclusion of these phases in the process of design thinking in education offers students the chance to engage in hands-on, experiential learning and reflection, promoting the cultivation of creativity, innovation, and an entrepreneurial mindset, where participative learning, and involving potential users and other stakeholders

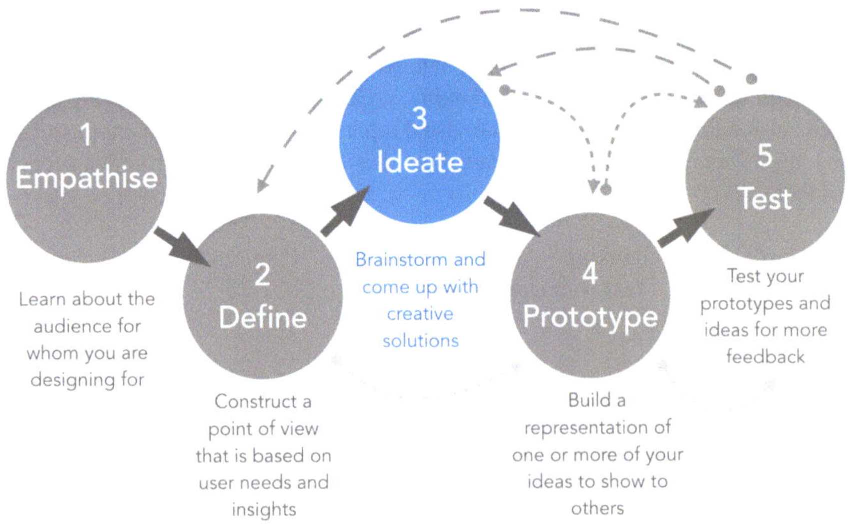

Fig. 1 The design thinking methodology iterative process. Highlighted is the area where this study focuses

is an added value to this approach. Additionally, it allows students to cultivate critical professional skills, including collaboration, effective communication, and adaptability. By integrating the principles of design thinking into their educational practices, teachers can facilitate their students a systematic and thorough approach to solving complex and context-specific issues that affect society.

The concept of design thinking has been widely adopted within higher education since the mid-twentieth century, particularly in design schools but also in the fields of arts and architecture. This widespread adoption was driven by an increased focus on the study of cognition and design methods [5], which led to the development of design research methodologies and techniques, procedures, tools, and principles of design [6]. These developments culminated in the idea of design thinking as a singular and inventive method for experimentation and problem resolution. The seminars conducted by Professor and mechanical engineer John Edward Arnold, which were rooted in the psychology of creative thinking and imagination, significantly contributed to the evolution of this field [7]. Later, between 1960 and 1970, research in design at the Royal College of Art in London and by Professor Leonard Bruce Archer helped to systematise design processes through evidence-based evaluations in industrial design [8]. The term "design thinking" emerged in 1969 in the book "The Science of the Artificial" [9] and became widely recognised when David M. Kelley, professor at Sandford University and the founder of the global design and innovation company IDEO (an American Design consultancy firm),[1] defined it as a form of creative action and a set of both mindsets and design-based activities that foster the collaboration required to solve problems in human-centred ways. Generally, design thinking refers to adopting a designer's perspective in problem-solving, which entails addressing real issues, seeking efficacious solutions, and demonstrating empathy towards the end-user by comprehending their emotions, behaviours, and desires. This approach, which is based on empathy, enables the translation of end-user insights into solutions that have the potential to enhance the quality of a product or service, and by so, the quality of life.

At the start of the twenty-first century, there was a marked increase in interest in the areas of design thinking, which spread beyond its traditional design domain to other fields, such as business and management. This has resulted in design thinking no longer being of interest only to designers and the way they approach solution and product development progressively and iteratively, but also to other areas of knowledge, such as architecture, management, and business administration. According to Elsbach and Stigliani [10], the introduction of design thinking as a procedural method for problem-solving, and the identification of the most appropriate tools for this purpose, has been applied and adopted in management and administration. Thus, the design thinking methodology is no longer limited to the use of designers and design educators.

The subject of design thinking in academia has been a matter of extensive discussion in recent years, with growing attention to its application in the educational context. The number of scientific publications exploring this area is also on the rise.

[1] https://www.ideo.com/people/david-kelley.

In the publication, "Design Thinking in Education: Perspectives, Opportunities, and Challenges," Panke et al. [11] delves into the processes and mindset necessary for collaborative problem-solving in various educational settings. Panke et al. conduct a systematic examination of the characteristics, benefits, and limitations of this methodology and aim to demonstrate that incorporating design thinking into pedagogical practices can be beneficial to students, teachers, and educational institutions alike. According to the authors, the competencies that result from using design thinking in the educational context can be encapsulated as follows: eliciting implicit experiences, fostering empathy and comprehension, mitigating cognitive biases, promoting play-based and flow-oriented learning, fostering inter- and cross-disciplinary collaboration, embracing productive failure, catalysing surprise, and innovative solutions, and fostering creative self-assurance. The authors also refer that academic research in the design field needs to evolve from exploring "What is Design Thinking?" to exploring "What is the Design of Design Thinking" in education. The distinction between "Designerly Thinking" and "Design Thinking", as described by Johansson-Sköldberg et al. [12] in their study on "Design Thinking: Past, Present and Possible Futures: Creativity and Innovation Management" provides an insightful perspective on this appropriation. The authors distinguish between "Designerly Thinking," which refers to the academic study and understanding of the professional designer's practices, skills, and competencies, and theoretical reflections on how to characterise this non-verbal competence of the designer. This approach highlights the academic construction of the designer's practice, which is reflected in the theoretical studies in this field. On the other hand, the concept of "Design Thinking" is embodied when design methodologies, techniques, and practices are applied beyond the traditional design domain by individuals who may not possess a formal design education or background. Thus, Johansson-Sköldberg et al. reinforce the idea that "Design Thinking" is a simplified form of "Designerly Thinking", the former representing a mode of thought and a design methodology incorporated within a context or discipline that is not exclusively linked to design professionals. Therefore, there are multiple ways to apply the methodology of design thinking in contexts beyond the realm of design, encouraging participants or students to think and act as designers.

Accordingly, the pedagogical approach of design thinking, in design education in particular, should encompass two interconnected approaches. Firstly, the teaching and schooling of the methodological principles of design thinking, including its origins, historical development, and practical application, within the realm of design and related fields such as art, architecture, as well as interdisciplinary areas such as business management and entrepreneurship. Secondly, the hands-on implementation of the design thinking method in concrete project-based learning experiences, where students and teachers actively participate in all product or service creation stages, from empathising, problem definition, ideation, prototyping, to testing and improvement. Through structured and repeated engagements with these different stages over the course of their academic journey, students develop ways to understand their users, gain problem-solving skills, and the ability to visualise and validate their ideas through prototyping and testing, all while building creative confidence.

2 The Role of Prompting in Text-To-Image Generation

The field of computational visual representation involving generating digital images from textual descriptors is experiencing an increasing interest among the research community. In 2018, Qiao et al. [13] referring to the process of text-to-image generation, indicated that it was still a complex task that poses significant challenges due to the large disparity between semantic analysis[2] and the computational science of visual representation, computer graphics, and image processing. The emergence of generative adversarial networks (GANs)[3] has stimulated research into these areas, particularly in generating visual representations from textual descriptors. This method, originally developed by Ian Goodfellow and as proposed by Reeds et al. [14], provides a flexible and intuitive means for conditional image generation and has demonstrated significant advancements in visual realism, diversity, and semantic alignment in recent years. In fact, the use of GANs has been widespread in the realm of image generation using base descriptors, images, and texts [15–17]. GANs consist of a generator and a discriminator, which operate competitively to produce high-quality images aligned with textual descriptions. Specifically, the generator acquires the capability to synthesise images based on textual descriptions while the discriminator assesses the authenticity of the generated images. By means of an adversarial training mechanism, the generator acquires the ability to produce images that are indistinguishable from authentic images, resulting in enhanced visual realism and diversity. Recent advancements in GANs have aimed to improve the semantic coherence between textual descriptions and the generated images. This entails developing methodologies that guarantee that the generated images reflect the intended textual descriptions accurately, resulting in more coherent and visually meaningful images.

In the domain of image generation from text, prompting denotes the act of feeding natural language descriptions or "*prompts*"[4] as input to a neural network model that is designed to generate the corresponding images. Specifically, Prompting Text-to-Image (PTI) is a compelling method in the field of artificial intelligence and computer vision for producing images based on textual descriptions. This methodology entails training a neural network model, typically a generative adversarial network (GAN), to create images that are consistent with the provided natural language descriptions. Thus, using prompts in text-to-image generation facilitates a more versatile and user-friendly approach to conditional image generation, enabling users to produce images

[2] Semantics is the analysis of meaning in natural languages. That is, it is devoted to the study of linguistic meaning, such as words and sentences.

[3] Generative adversarial networks (GANs) are innovative machine learning technology based on competitive neural networks to generate models based on training data. For example, GANs can create images that look like photographs of human faces, even though the faces don't belong to any real person.

[4] As described on the Midjourney Documentation website, a "Prompt is a short text phrase that the Midjourney Bot interprets to produce an image. The Midjourney Bot breaks down the words and phrases in a prompt into smaller pieces, called tokens, that can be compared to its training data and then used to generate an image. A well-crafted prompt can help make unique and exciting images." https://docs.midjourney.com/docs/prompts, last accessed 2023/03/04.

Fig. 2 Two visual representations of the prompt "a bus in the shape of a banana" generated using the software Midjourney bot

based on textual descriptions without relying on sophisticated image editing software or technical proficiency. Furthermore, prompting plays an important role in artificial intelligence as a technique for providing cues or hints to a computer system to guide its behaviour or decision-making process.

Prompting techniques are widely employed in various artificial intelligence applications to govern behaviour or decision-making processes. In the domain of natural language processing (NLP),[5] prompting is leveraged to provide context or constraints for text generation tasks such as language translation or text summarisation. For instance, Google Translate uses prompts to generate text translations from one language to another. In personal assistant applications like *Siri* or *Alexa*, prompting is used to facilitate users' interactions with the system via natural language commands (usually voice commands), thereby providing an intuitive and user-friendly interface. Furthermore, in computer vision, prompting is used to generate images based on natural language descriptions or prompts (usually a text composed of words and sentences), allowing users to create visual content without the need for specialised image editing software expertise or technical skills. For example, *DALL-E,* an AI model developed by OpenAI, or Midjourney.ai can generate images based on textual prompts such as "a bus in the shape of a banana" (see Fig. 2). Hence, prompting is a versatile technique that can be applied to many AI applications to provide users with more intuitive and flexible ways to interact with intelligent systems.

Considering the recent progress in visual applications for generating images from textual input, such as DALLE-E, Midjourney, and Stable Diffusion, various websites for generating prompts have emerged, including the Midjourney Prompt Helper,[6]

[5] Natural language processing (NLP) refers to the branch of computer science—and more specifically, the branch of artificial intelligence or AI—concerned with giving computers the ability to understand the text and spoken words in much the same way human beings can.

[6] https://prompt.noonshot.com/, last accessed 2023/03/02.

Stable Diffusion Prompt Builder,[7] and DALLE-E Prompt Generator,[8] among others.[9] These prompt-generating tools assist users in generating effective prompts for producing high-quality images, offering the ability to specify styles, details, artists and even generate new results based on a given image as a starting point. In October 2022, Borji [18] conducted a quantitative and qualitative comparative analysis of high-quality facial image generation using Stable Diffusion, Midjourney, and DALL-E. Based on his findings, Borji suggests that, in most cases, it is possible to distinguish between computationally generated and real faces. He proposes future work in the areas of expression detection (e.g., sadness, happiness, anger), age, and camera viewpoints to improve further the quality and realism of generated facial images. An extension of Borji's comparative analysis can be applied to other domains of visual representation, including industrial products, animals, and furniture design.

The efficacy of prompting techniques for generating high-quality images in these domains may depend on the individual's ability to describe the desired image, environment, or attributes. Therefore, further research is needed to explore effective prompting strategies and their impact on the quality and realism of generated images. For example, the Midjourney platform utilises a vast database of over 650 million internet images to create visually striking images that align with the input textual prompt. The level of detail and uniqueness of the generated images generally depends on the text's quality and descriptiveness. However, the platform offers several advanced settings and options that enable users to create variations in the images, adjusting the level of detail to their preferences. As described by Nielsen [19], users have the option to prompt images based on various parameters such as themes, artists, image resolution, rendering, and effects, as well as camera and lens formats. These parameters include settings such as "style", "chaos", "resolution", "aspect ratio", and "prompt weights", as well as the ability to pass an image as a prompt through a URL. The style parameter allows users to specify design styles or artistic genres, artist names, or lighting properties to influence the image output. Users can also adjust the level of detail or abstraction in the output, as well as customise the image size. A list of parameters can be found on the Midjourney website to help users describe what they want as a result. It is suggested to think about, the subject (i.e. person, animal, character, location, object, etc.), the medium (i.e. photo, painting, illustration, sculpture, etc.), the environment (i.e. indoors, outdoors, underwater, in the moon, etc.), the lighting (i.e. soft, ambient, overcast, neon, studio light, etc.), the mood (i.e. sedate, calm, energetic, etc.), the composition (i.e. portrait, closeup, birds-eye view, etc.) and the colour (i.e. vibrant, bright, colourful, black and white, etc.).

---

[7] https://promptomania.com/stable-diffusion-prompt-builder/, last accessed 2023/03/02.

[8] http://dalle2-prompt-generator.s3-website-us-west-2.amazonaws.com/, last accessed 2023/03/02.

[9] https://mpost.io/6-free-prompt-builders-and-helpers-that-artists-actually-use-in-2022/, last accessed 2023/03/02.

3 Applying Design Thinking Methodology in the Classroom, the Context of Furniture Design Education at the University of Saint Joseph

Throughout the history of design education, the design studio has emerged as a fundamental pedagogical method and is likely to maintain its prominent role [20]. By providing an optimal space for students to engage in their creative potential, the design studio promotes an iterative process that encourages students to examine prior examples as a source of design inspiration and to work from there on the design of their project works. Hence, and through ongoing presentation and revision of their work-in-progress with both teachers and peers, students engage in a continuous and iterative cycle of improvement. Furthermore, a central tenet of the design studio is the individualised desk critique (crit), a practice that entails regular and frequent discussion between the student and the teacher(s) concerning the ongoing design work. As a learning-by-doing environment, this practice-based methodology provides a context in which students can develop their design skills and knowledge with the guidance and support of their teacher(s).

The Furniture Design Studio is an essential module in the Bachelor of Design curriculum at the University of Saint Joseph, Macau SAR, China. It is offered during the programme's third year and carries a weight of six Macau credits or 12 ECTS (European Credits Transfer System). The course runs for the first semester of the academic year, from September to December, and comprises thirty sessions of three hours each, twice a week. The primary goal of the studio is to enable students to design and build a full-scale prototype of an original piece of furniture. Through this exercise, students gain comprehensive knowledge of the entire furniture design production and manufacturing processes, from conception to ideation and to the creation of a 1:1 scale prototype. The studio is not limited to practical learning but also provides students with a theoretical grounding in the history, genres, expressions [21, 22], and contemporary references in furniture design [23, 24].

Over the last three years, our students have been tasked with developing their Furniture Design projects under the main theme of "Useless Design" [25]. This theme revolves around designing products that align with the post-structuralist-hypermodern era's useless spirit, seeking to create compatible and consistent solutions [26, 27]. The concept challenges the false dichotomy between the notions of uselessness and usefulness in design and encourages exploration of the potential for useless design practice. This movement aims to traverse the growing chasm between the industrial and post-industrial design paradigms, ultimately discovering the value in the uselessness of a design. The idea that design must be useful and functional has long been the norm, but could a useless design practice bring new methodology and discussion to contemporary furniture product design? Is innovation possible through a useless design approach? These are the fundamental questions and challenges posed to our students at the beginning of each semester in the Furniture Design Studio.[10]

[10] It is expected that students undertaking the Furniture Design studio know how to present their projects in terms of technical drawings and following ergonomics standards. Previous modules on

Table 1 List of sub-concepts chosen by students during the school years from 2020 to 2023

Useless synonyms and sub concepts			
Broken	Expendable	Meaningless	Unfunctional
Deconstructed	Fail	Non-functional	Unusable
Deformed	Hopeless	Of no use	Unworkable
Dysfunctional	Impossible	Pointless	Valueless
Good-for-nothing	Inutile	Purposeless	Worthless

The trend towards sustainable, multifunctional, and interactive products has sparked a discourse among designers about the role of design in society. Criticism has been directed towards the overproduction of gadgets, bad design, and useless objects, as well as the excessive emphasis on multifunctionality, which can result in consumers failing to fully use the products they purchase. To address this issue and the dichotomy between functionality and utility, a new approach is proposed for the furniture design studio. This approach explores the concept of designing furniture with features of uselessness, while still maintaining functionality and usability as a requirement. In this context, students are challenged to create a usable and functional piece of furniture (i.e. a stool, a chair) that exhibits conceptual or aesthetic characteristics of uselessness. Inspiration is drawn from architect Katerina Kamprani and her series of "uncomfortable"[11] products. In addition, students are encouraged to explore related synonyms or sub-concepts (see Table 1) that complement this central theme. By introducing this thought-provoking topic, the aim is to engage students in critical thinking, stimulate their creativity, and encourage them to question traditional and historical features of furniture design.

In the academic setting at hand, students are required to choose a sub-concept to inform their furniture design project. This selection is based on thorough research aimed at comprehending the sub-concept's potential applications in the realm of design. This knowledge is then used to inform the creation of the furniture design pieces. To facilitate this process, a comprehensive survey is conducted that encompasses real-world examples from various design-related disciplines such as architecture and art. The survey is designed to explore a range of possibilities for application and acquire tangible references that can inform the design process, looking at precedents, main uses and functions, main problems to solve and main opportunities for improvement.

In the academic context of this module, the design thinking methodology is being implemented, specifically focussing on the first two stages. During the first stage, students identify the intended user archetype for their project. The user archetype may involve designing a chair set for children in a day-care centre, a family stool set for

year 1 and 2 focus on topics such as Technical Drawing, Design Principles and Theory, Ergonomics, Graphic and Visual Design (https://www.usj.edu.mo/en/course/ba-design/).

[11] Katerina Kamprani, The Uncomfortable website: https://www.theuncomfortable.com/, last accessed 2023/02/25.

the living room, or an outdoor bench for public parks in Macao, among other possibilities. This stage is crucial in establishing research objectives and identifying optimal solutions to address the target user's most pertinent desires and needs. The second stage of the methodological process involves organising all the gathered information to identify potential areas of opportunity and relevant solutions for the designated end-user. This phase includes gathering, defining, and synthesising information, from the conceptual investigation to the confirmation of the typical user. The definition phase is initiated by the primary question and challenge posed by the teachers—The challenge is to design a practical and useful piece of furniture based on the concept of uselessness. For instance, a student who chooses the sub-concept of "deconstructed" would face the significant challenge of creating a functional and usable piece of furniture that embodies the formal and aesthetic characteristics of deconstruction within the framework of uselessness. Overall, this phase requires students to integrate the information collected from their research to develop creative and practical solutions that address the needs and desires of the target user. The implementation of this stage in the methodological process ensures that the resulting furniture design is not only aesthetically pleasing but also functional and useful to the end user.

During the third stage of project development, the emphasis is on ideation, which can present challenges for students in terms of visual representation. Although they may strongly grasp a concept or idea, translating it into a graphic format may prove difficult. To address this difficulty, teachers employ various creative methods to aid students in comprehending and conceptualising potential solutions for their furniture designs. Techniques such as drawing-by-abstraction, drawing-by-addition and subtraction, and drawing-by-thought suggest ideas are used to facilitate the exploration of innovative and unique solutions. Later, the selection of the most suitable ideas and solutions is a collaborative effort between teachers and students. During the prototyping phase of furniture design, students acquire hands-on experience in manufacturing their designs using wood as the primary material (see Fig. 3 for students' final prototypes). This process occurs in two primary settings: the woodworking workshop at ZaWood[12] and the fabrication lab at USJ. Depending on the design solutions they propose, students may have the opportunity to experiment with alternative materials such as resin, plastic, or 3D-printed components. Throughout the iterative process, teachers provide support and guidance, which includes modifying designs to improve functionality and ergonomics, as well as testing different materials and types of wood. Following the construction of prototypes, both students and teachers evaluate them in the studio to identify any previously undetected errors that require correction. Even so, the comprehensive iterative process of redesign and prototyping may only be partially feasible due to time constraints and limited sessions of the module.

[12] Zawood Workshop, website https://www.zawood.com/, last accessed 2023/02/25.

Fig. 3 Final prototypes (1:1 scale) of stools and chairs designed by the students of the Bachelor of Design at USJ, from the academic year 2020–2023

4 Exploring the Use of Text-To-Image Prompting for Ideation on Furniture Design

As previously discussed, the pedagogy of design in higher education has been transformed through the integration of novel design thinking techniques [28]. While this approach has demonstrated significant benefits in enhancing students' creative capacities [29], it is crucial to acknowledge that advanced graphic recording skills are indispensable during the ideation phase to achieve a precise representation of concepts on paper [30]. To visually capture and communicate ideas, graphic records such as sketches, drawings, and diagrams can be instrumental. Multiple approaches can be adopted during this stage, including visual brainstorming (employing quick sketches of ideas), mind maps (depicting the relationships between ideas and exploring new connections), storyboards (creating visual narratives with characters), and basic graphic schemes. For instance, basic graphic schemes can be employed to visualise the interrelation between different furniture components and illustrate their functionality or how the joint parts align. Furthermore, three-dimensional models can be developed using 3D modelling software, albeit requiring a comprehensive understanding of digital resources that may not yet be familiar to students at this stage of their study.

The emergence of artificial intelligence (AI) and its incorporation of text-to-image prompting capabilities have recently brought about a significant transformation in the ideation process [31, 32]. Specifically, AI algorithms have been shown to improve efficiency and productivity, enabling the generation of more ideas within a shorter period. Using AI algorithms, ideation can be automatically generated by analysing textual data and information, freeing students, and teachers to concentrate on refining and assessing the most promising ideas. Additionally, the text-to-image feature of AI can facilitate the generation of more innovative and effective ideas by providing a larger number of options more rapidly and efficiently than traditional methods. In the context of higher education and specifically in the field of furniture design, it is worth exploring the use of text-to-image prompting resources in the classroom. These resources can assist students in visualising their ideas more efficiently and quickly through automatic generation based on provided textual information. While the technology of writing a word or sentence and quickly generating a visual representation of it is impressive, it is important to note that there is still room for improvement. Specifically, there are areas for procedural and graphical development, such as the need for faster generation of results (as each iteration currently takes around 1 min to generate four options) and greater efficiency in generating results when inserting specific text into the visual solution, for instance when asking to generate an image with words or sentences or drawing realistic hands. Also, fabrication feasibility is another concern, considering these creative generations may go beyond human technological capabilities.

To conduct our study, we followed a systematic approach that involved analysing various solutions for designing a piece of furniture. We specifically focussed on designing a stool first and later a chair. This process included a practical exercise. At

Fig. 4 Five results for the prompt "A stool" using Midjourney bot

Fig. 5 Five results for the prompt "design a stool", using Midjourney bot

first, only the word "stool" was used as a prompt to generate various visual iterations of a stool. Subsequently, more words and meanings were incorporated to enhance the generative process. The models produced were consistent with the provided instructions and largely followed a conventional form that is typically associated with a traditional "stool". Figure 4 provides an example of the relationship between the visual output and the corresponding textual meaning. The generated results exhibited a classic and almost monotonous design style that closely aligns with the traditional understanding of a stool.[13]

In a second iteration, we provided the prompt with the wording "design a stool" to the programme. The generated results are more interesting as they indicate a more geometrical and designed approach to furniture design while maintaining a formality level, as shown in Fig. 5.

To investigate the software's ability to differentiate between various modes of graphical representations, we modified the verb "design" in the prompt and replaced it with several possibilities, such as "draw_, sketch_, ideate_, shape_, depict_, originate_, make_, imagine_, _a stool". The results, as shown in Fig. 6, indicate that the generated stools remain in a conventional and traditional register (the use of wood as the main material), with the most creative being the prompt "imagine a stool."

To further explore the capabilities of the Midjourney bot, we included specific details about the piece of furniture in the prompt, such as the height, number of legs, and materials like wood, glass, and cushion. The programme was able to understand the meaning of the text and generate reliable results, as evidenced by the outcomes

[13] Stool: a seat without a back or arms, typically resting on three or four legs or on a single pedestal.

Fig. 6 Eight results generated in Midjourney modifying the first word (verb) and keeping the noun "stool" in the prompts

presented in Fig. 7. To ensure that the results obtained were not based on previous iterations, we used a new seed (i.e. –seed 0000)[14] for each new generative process. This ensured that each result had a unique origin.

Our initial experiment found that the software generates visual representations that correspond directly to the prompt. By changing the input text, we were able to generate reliable visual representations of different solutions. We also tested the software's ability to incorporate specific qualitative characteristics into the generation of images by using words such as "professional", "high quality", "realistic", "accurate representation", "detail", and "realistic materials" in the prompt. The results of this experiment can be seen in Fig. 8.

In the context of the furniture design studio of the Bachelor of Design at the University of Saint Joseph, we continue our study to investigate how results generated from prompts with complex or connotative meanings can aid in the development of novel and meaningful design solutions, ultimately aiding students in better conveying their ideas. To explore this and based on the main concept of "Useless Design", we prompted the software with words and sub-concepts meanings such as "failed",

[14] The "Seed" command allows to set a specific seed value, resulting in more consistent and interdependent results. This parameter is useful if you need to create multiple images, return to the initial idea and choose to modify it or to ensure that your results always look the same.

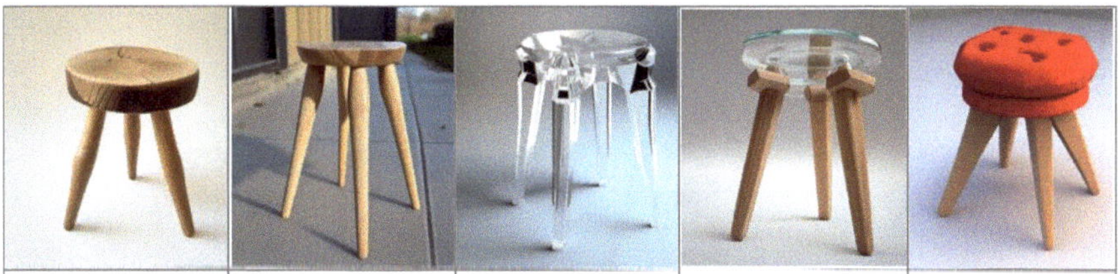

Prompt: "A stool in wood with 3 legs 38 cm height".	Prompt: "A stool in wood with 4 legs 60 cm height".	Prompt: "A stool in transparent glass with 4 legs and 40 cm height".	Prompt: "A stool wit top in transparent glass and 3 legs in wood".	Prompt: "A stool with a seat in red cushion and 4 legs in wood".

Fig. 7 Specifying design features in the prompts such as number of legs, height, and materials

"impossible", "deconstructed", and "dysfunctional" while specifying materials such as "transparent Plexiglas" or "wood". The results of these prompts can be seen in Fig. 9.

To further elaborate, we found that even when using more unconventional concepts such as "Dysfunctional", "Deformed", "Wrong Useless", and "Impossible-deconstructed", the generated results were still impressive, as shown in Fig. 10.

This generative tool also allows for the creation of new images based on a specific photograph or a previously generated solution. For instance, in this next study, we instructed the software to generate a chair with specific characteristics and the following prompt: "A deconstructed impossible design chair".

The Midjourney bot generated four solutions based on the prompt "a deconstructed impossible design chair", which can be seen in the first row of Fig. 11. Using one of the generated images, namely the third image in the first row, as the seed URL image, we asked the bot to (re)generate four new iterations with the new prompt "functional and useful four legs chair" (images of the second row). Despite starting this process from a deconstructed and impossible sub-concept, the Midjourney bot was able to generate new solutions that look functional and could be produced and used as chairs. This is promising for students who want to focus on a specific concept or idea and generate diverse and creative solutions. In the third iteration (images of the third row of Fig. 11), the new prompt included materials such as "wood and transparent glass" and "accurate size" for the dimensions. In the fourth iteration (images from the fourth row of Fig. 11), the same seed URL image was used but prompted with new plexiglass as material and added the colours red, yellow, and green.

To establish an initial testing protocol, we decided to engage our students in the traditional design thinking ideation phase, as it is important for them to understand the objectives and the constraints of this step. Students defined their keywords and started producing graphic representations of stools and chairs to express their ideas. Our

Fig. 8 Eight images generated with Midjourney specifying design render quality and materials in the prompts

experience shows us, in the past years, that students, on average, would generate about 10 graphical representations of their ideation step within a period of 2 sessions, three hours each. We then introduced the text-to-image application Midjourney as a mean to provide support for the ideation step with a different process, as a textual description is required before being able to produce any graphical representation. The proposed exercise used the same keywords from the define phase to generate images. Over a period of 1 session (3 h), it was generated 25 graphical representations per student, on which 5 were kept as interesting concepts to consider further. The selection criteria were: (1) a plausible correlation between the chosen sub-concept and the visual generated solution, (2) an aesthetic consideration as a consensus between the student and teachers and (3) a preliminary evaluation of the functionality, usability, and physicality of the object.

Fig. 9 Twelve different results generated with Midjourney specifying useless design sub-concepts in the Prompts

By the results obtained, it seems clear that using PTI can generate visual representations that correspond directly to the prompt given. By changing the input text, we were able to generate reliable visual representations of different solutions. We also tested the software's ability to incorporate specific qualitative characteristics into the generation of images. Even when using more unconventional concepts, the generated results were impressive, indicating that the technology has the potential to aid in the development of novel and meaningful design solutions. Overall, the integration of AI and text-to-image prompting resources in the field of design education has the potential to significantly enhance the generation of graphical representations during the ideation process—both on the time spent and the number of offered solutions.

	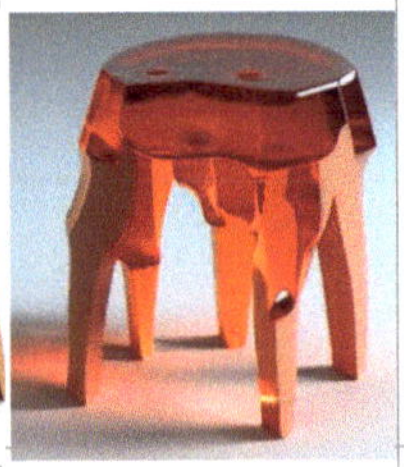		
Prompt: "Dysfunctional Design Stool, Wood, transparent-blue acrylic, professional product high quality detailed accurate representation realistic materials accurate dimensions realistic high resolution".	Prompt: "Deformed Design Stool, Wood, transparent-red acrylic, professional product high quality detailed accurate representation realistic materials accurate dimensions realistic high resolution".	Prompt: "Wrong Useless Design Stool, Wood, transparent-blue acrylic, square seat, four legs, professional product high quality detailed accurate representation realistic materials accurate dimensions realistic high resolution".	Prompt: "Impossible Deconstructed Stool, Wood, transparent-blue acrylic, square seat, four legs, professional product high quality detailed accurate representation realistic materials accurate dimensions realistic high resolution".
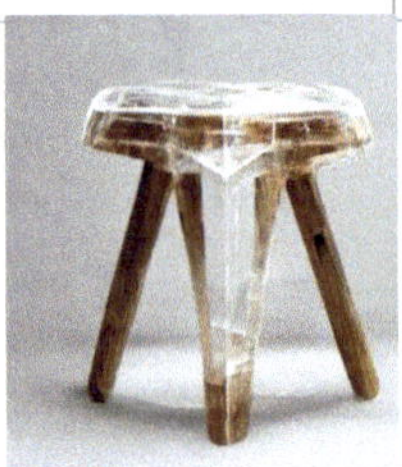			
Prompt: "Impossible Design Stool, wood and transparent acrylic, realistic materials".	Prompt: "Impossible Design Stool, wood and transparent acrylic, realistic materials".	Prompt: "Impossible geometric Stool, Wood, square seat, four legs, professional product low quality accurate representation realistic resolution".	

Fig. 10 Height different results generated with Midjourney specifying useless design sub concepts and materials in the Prompts

Fig. 11 Results generated with Midjourney using prompts for the design of a chair: a deconstructed-impossible chair and using an URL seed image as reference

We believe it can enable more efficient and productive design thinking solutions, and we intend to validate this statement in further research.

5 Conclusion

In recent decades, design thinking has emerged as a prominent human-centred and problem-solving approach within the field of education, particularly in higher education. Through the structured, systematic, and iterative design thinking process, students are provided with a platform to cultivate critical thinking, creativity, innovation, collaboration, and an entrepreneurial mindset. In furniture design, applying design thinking has proven to be a crucial pedagogical method for developing innovative and creative solutions.

The Bachelor degree programme in Design at the University of Saint Joseph, Macau, has successfully implemented design thinking methodologies within its furniture design studio. By challenging students to design and build a full-scale prototype of an original piece of furniture, such as a chair or a stool, under the theme of "useless design," students are encouraged to think critically, question traditional and historical characteristics of furniture design, develop research skills and stimulate their creativity. The outcomes of this studio over the last three academic years have demonstrated the success of this approach, resulting in innovative, original, and functional furniture pieces with subversive yet aesthetically appealing features.

In design thinking, graphic recording skills are crucial in the ideation phase. However, using prompting in a classroom context can enable a significant transformation in the ideation process of students, thereby improving efficiency and productivity. The students can achieve this transformation when accelerating the generation of more ideas within a shorter period. Our study used the Midjourney AI programme as a prompting and imaging tool. The study found that the platform could understand the meaning of the text input and generate reliable results in the field of furniture design. However, mastering this tool requires understanding how to accurately describe what one wants to create and use the generator's grammar, such as defining styles, image quality, resolution, and URL seeds accordingly. Therefore, it encourages students to identify, clarify and express in clear language the fundamental elements of their concepts before starting an ideation step, which sometimes they do not take time to research properly. Furthermore, with the rise of artificial intelligence, specifically generative adversarial networks (GANs), computational visual representation has experienced a significant evolution. The prompting technique, which involves generating images from textual descriptors, has played a vital role in this development. This method has shown increased reliability and accuracy in maintaining semantic coherence between textual descriptions and the generated images.

The research delved into design concepts, specifically in furniture design, to help students understand the challenges involved in creating functional solutions with unusual aesthetic and formal characteristics. While the study focussed on generating visual outcomes for the ideation phase of design thinking, text-to-image prompting can also have relevant applicability in the other stages of design thinking. For instance, in the empathy phase, text-to-image prompting can be used to create images representing the user's demands and desires. This helps students empathise with the users and better understand their needs. In the problem definition phase, using text-to-image can help identify the problem to be solved by specifying prompts to create images representing solutions for each problem situation. In the prototyping phase, text-to-image can generate high-definition photo-realistic pictures of the solution found, producing hyper-realistic visuals of the final product or service, which can be very useful for client presentation purposes. Finally, during the testing phase, text-to-image can aid in gathering user feedback by creating images representing different variations of the final solution and testing with users.

In summary, incorporating prompt-based text-to-image tools in the design thinking process, particularly in furniture design studios, has the potential to support students and teachers in generating creative solutions. It can also enable them to generate more visual representation in less time, and get feedback from peers and users, and prospective clients. Further comparative studies will allow us to establish a complete protocol and collect a complete set of data to validate our preliminary findings.

References

1. Shelley G, Zielezinski MB (2022) Design thinking for every classroom: a practical guide for educators, 1st edn. Routledge
2. Lee D (2018) Design thinking in the classroom: easy-to-use teaching tools to foster creativity, encourage innovation, and unleash potential in every student. Ulysses Press
3. Gallagher A (2018) Design thinking for school leaders: five roles and mindsets that ignite positive change. Association for supervision & curriculum development
4. Spencer J, Juliani AJ (2016) Launch: using dsign thinking to boost creativity and bring out the maker in every student. Publishing Dave Burgess Consulting, Incorporated
5. Johansson-Sköldberg U, Woodilla J, Çetinkaya M (2013) Design thinking: past, present and possible futures. Creat Innov Manag 22(2):121–146. https://doi.org/10.1111/caim.12023
6. Liedtka J (2018) Why design thinking works. Harv Bus Rev 96(5):72–79
7. Arnold EJ, Clancey WJ (2017) Creative engineering: promotion innovation by thinking differently, 2nd edn. University of Texas Press
8. Early Design Research at the RCA, The Royal College of Art in the 1960s and 1970s. Accessed 28 Jan 2023 https://www.drs2016.org/exhibition/#ddr1
9. Simon HA (1996) The science of the artificial, 3rd edn. The MIT Press
10. Elsbach K, Stigliani I (2018) Design thinking and organisational culture: a review and framework for future research. J Manag 44(6):2274–2306
11. Panke S (2019) Design thinking in education: perspectives. Oppor Challenges Open Educ Stud 2019(1):281–306
12. Johansson-Sköldberg U, Woodilla J, Çetinkaya M (2013) Design thinking: past, present and possible futures. Creat Innov Manag 22(2):121–146
13. Qiao T, Zhang J, Xu D, Tao D (2019) Learn, imagine and create: text-to-image generation from prior knowledge. Adv Neural Inf Process Syst 32 (NeurIPS 2019)
14. Reed S, Akata Z, Yan X, Logeswaran L, Schiele B, Lee H (2016) Generative adversarial text to image synthesis. In: International conference on machine learning

15. Goodfellow, Pouget-Abadie J, Mirza M, Xu B, Warde-Farley D, Ozair S, Courville A, Bengio Y (2014) Generative adversarial nets. In: Advances in neural information processing systems
16. Xu T, Zhang P, Huang Q, Zhang H, Gan Z, Huang X, He X (2018) Attngan: fine-grained text to image generation with attentional generative adversarial networks. In: Proceedings of the IEEE conference on computer vision and pattern recognition
17. Zhang H, Xu T, Li H, Zhang S, Huang X, Wang X, Metaxas D (2017) Stackgan: text to photo-realistic image synthesis with stacked generative adversarial networks. In: Proceedings of the IEEE international conference on computer vision
18. Borji A (2022) Generated faces in the wild: quantitative comparison of stable diffusion, mid journey and DALL-E 2. In: Computer science, computer vision and pattern recognition, San Francisco. https://doi.org/10.48550/arXiv.2210.00586
19. Nielsen L (2023) Na advanced guide to writing prompts for Midjourney (text-to-image). Accessed 03 Mar 2023. https://medium.com/mlearning-ai/an-advanced-guide-to-writing-prompts-for-midjourney-text-to-image-aa12a1e33b6
20. Salama AM, Wilkinson N (eds) (2007) Design studio pedagogy: horizons for the future. The Urban International Press, Gateshead UK
21. Miller J (2005) Furniture. World styles from classical to contemporary. Dorling Kindersley Limited, London
22. Morley J (1999) The history of furniture, twenty-five centuries of style and design in the western tradition. Bulfinch Press. Boston
23. Orron J (2018) Chair anatomy, design and construction, 1st edn. Thames & Hudson
24. Kries M, Eisenbrand J (eds) (2019) Atlas of furniture design. Vitra Design Museum
25. Gorvett Z (2023) The useless design features in modern products, https://www.bbc.com/future/article/20171023-the-useless-design-features-in-modern-products. Accessed 26 Feb 2023
26. Caires CS (2001) O hiper no modernismo e na ficção. In: Xaquím Núñez Sabaris (ed) "Diálogos Ibéricos sobre a Modernidade". HUMUS Edition, 1st edn, pp 25–40. University of Minho, Braga
27. Lipovetsky G, Sebastien C (2004) Les temps Hypermodernes. Edition Grasset, Paris
28. Smith RC, Iversen OS, Hjorth M (2015) Design thinking for digital fabrication in education. Int J Child-Comput Interact 5:20–28. https://doi.org/10.1016/j.ijcci.2015.10.002
29. Noweski C, Scheer A, Büttner N, Von Thienen J, Erdmann J, Meinel C (2012) Towards a paradigm shift in education practice: developing twenty-first century skills with design thinking. In: Plattner H, Meinel C, Leifer L (eds), Design thinking research. Springer, Berlin, pp 71–94. https://doi.org/10.1007/978-3-642-31991-4_5
30. Orthel BD, Day JK (2016) Processing beyond drawing: a case study exploring ideation for teaching design. SAGE Open 6(3). https://doi.org/10.1177/2158244016663285
31. Ko H-K, Park G, Jeon H, Jo J, Kim J, Seo J (2023) Large-scale text-to-image generation models for visual artists' creative works. In: Proceedings of the 28th international conference on intelligent user interfaces, pp 919–933. https://doi.org/10.1145/3581641.3584078
32. Liu V, Vermeulen J, Fitzmaurice G, Matejka J (2022) 3DALL-E: integrating text-to-image AI in 3D design workflows. arXiv:2210.11603

Designing Physical and Virtual Speculative Walkshops to Explore Public Space Internet of Things

Nuri Kwon, Naomi Jacobs, Louise Mullagh, and Joe Bourne

Abstract This chapter explores the impact of digital technology on public spaces and the challenges policymakers face in keeping up with technological advancements. The development of connected places, such as Smart Cities, has been facilitated by digital technology breakthroughs since the 1950s, where real-time data is collected and analysed for decision-making. The chapter discusses a study that combined speculative design with a walking method and digital space to investigate the potential benefits, risks, and challenges of deploying connected technology in public spaces. The research methodology involved physical and virtual walkshops presenting fictional and real IoT sensor deployments. The findings highlight the opportunities and challenges of digitally connected systems, as discussed by the participants. The chapter concludes with insights, limitations, and recommendations for further research, emphasising the importance of diverse stakeholder involvement in technology implementation in public spaces and the role of speculative design in policymaking. Overall, this chapter contributes to the discussion on the role of speculative design in policymaking and the need for diverse stakeholder involvement in technology implementation in public spaces.

Keywords Speculative walkshop · Connected places · Internet of Things

N. Kwon (✉) · N. Jacobs · L. Mullagh · J. Bourne
Imagination Lancaster, Lancaster University, Bailrigg, Lancaster LA1 4YW, UK
e-mail: n.kwon1@lancaster.ac.uk

N. Jacobs
e-mail: naomi.jacobs@lancaster.ac.uk

L. Mullagh
e-mail: l.mullagh@lancaster.ac.uk

J. Bourne
e-mail: j.bourne@lancaster.ac.uk

N. Martins et al. (eds.), *Perspectives on Design and Digital Communication IV*, Springer Series in Design and Innovation 33, https://doi.org/10.1007/978-3-031-41770-2_8

1 Introduction

Since digital technology breakthroughs in the 1950s, society has become increasingly connected through smartphones, sensors, the Internet of Things (IoT), edge computing, and AI [37]. This digital connectivity has led to a "smart city" concept where sensors gather and analyse real-time data for decision-making [26]. Policy plays a significant role in this transition in presenting the development direction [33]. Meanwhile, policymakers encounter difficulties in keeping up with the swift pace of technological advancements, especially concerning cybersecurity [29]. Therefore, further research into methods that examine these changes and support policymaking needs to be advanced. Furthermore, while there is research exploring how invisible and pervasive digital technology will impact physical places, there is a need for further exploration of the methods used to investigate and support policymaking in this context.

This chapter presents diverse qualitative methods approach to investigate the impact of technology on society, culture, and the environment before the changes initiated by digital technology are established. First, we explore the potential of speculative design to examine the opportunities and challenges of physical and digital placemaking in public spaces. The research combines speculative design fiction with physical and virtual walkshops where participants can experience and question the implications of digitally connected systems. The chapter reviews existing literature on IoT sensors, connected places, speculative design methods, and digital placemaking practices. It then describes the design process and materials used to facilitate walking workshops ('walkshops'), presents the findings of the workshops, and concludes with insights, limitations, and recommendations for further research.

2 Background

The digital transformation of physical and built environments has created "connected places" (or smart cities) [26]. This transformation is not immediately visible but is becoming intertwined with our daily lives as people became more digitally connected [9]. The concept of connected places aims to use data collected from sensors to improve the distribution of civic resources and increase operational efficiency [26]. This data can include information such as traffic flows, weather conditions, parking space occupancy, and human activities in public spaces [15]. By incorporating IoT sensors to gather data, the automation process can help developers and place managers better understand how public spaces are utilised and monitored [19].

Whilst the connected technology industry, including IoT sensors, experiences rapid growth, there is growing criticism of connected places. The vast amount of real-time data collection in public spaces can cause security issues. Frustaci et al. [8] analyse diverse scenarios of physical and cyber-attacks for the IoT, such as physical damage to sensor nodes, data transit attacks, and data leakages. Hedestig, Skog and

Söderström [14] identify that IoT and sensor deployment in public spaces is still in its early stages, so there is confusion, misunderstandings, and a lack of awareness, about the function and purpose of deployments. Even though physical exposure to sensors can influence people's behaviour and actions, they are generally hidden in plain view through a lack of signage or labelling. A lack of public awareness of IoT systems in public spaces might lead to mistrust and misunderstanding among the public [14].

The implementation of IoT in public places has vital policy implications. However, the ways of implementing these technologies have been criticised for being too top-down and techno-centric and not considering the needs and perspectives of residents [34, 36]. Angelidou [2] analyses the Smart City initiatives of 15 major cities worldwide, revealing a significant lack of stakeholder involvement and a failure to consider local conditions. In order to address these challenges, it is essential to involve those impacted by the technology in the decision-making process to build and manage trust with the public [8] and to examine possible implementations before they are widely deployed. One possible approach to this is the use of speculative design.

2.1 *Speculative Design and Design Fiction*

Speculative design is a design practice that involves exploring and imagining future scenarios and possibilities [7]. In particular, the method of design fiction is becoming increasingly popular in evaluating the potential of technologies and encouraging critical thinking about their potential impacts [22]. It is seen as a branch of speculative design that merges science backgrounds and objects with design principles and practices, creating near-future scenarios that mix fact and fiction [3, 31]. Design fiction creates tangible objects to represent speculative ideas and involves people in considering the future of technology and its implementation [20]. Design fiction also explores technology's cultural, social, and ethical impacts [7]. It can engage people's imagination to consider what is and should be possible by integrating storytelling, technology, and design [4]. In current practices, speculative design fiction is used to provoke discussions about technology and involve the public in imagining future policy implications. Introducing design fiction in policymaking can provide insights into policy design and help the public understand technology better [35].

2.2 *Speculative Design and Walking Methods*

This study examines the use of speculative objects in the urban realm to garner insights into people's perceptions of those places. A place consists of physical attributes such as geographic location and built environment, as well as intangible aspects such as human activities and cultures [5]. These cognitive elements allow people to establish a lasting relationship with a place, such as a place meaning, identity, and attachment [1]. The practice of walking to interrogate technological phenomena located in a place was first written about by Greenfield and Kim [10]. The walkshop is described as a "learning experience that's equal parts urban walking tour, group discussion, and spontaneous exploration". In order to comprehend and reinforce these intangible aspects, walking has become a popular method that utilizes various technological components within a place [24].

Combining physical walking with speculative design offers a participatory approach which can enable local communities to participate in building imaginary futures in the context of a place. Stals, Smyth and Mival [30] employed the 'walking and talking' method to investigate a place's present status and residents' emotions about the place. They explored specific physical spaces with the residents and envisioned a hybrid city through speculative design fiction. Chopra et al. [6] combined the use of speculative design and walking methods to situate speculation at specific locations for the possible futures of food production in a neighbourhood. The situated speculation enabled the participants to question possibilities critically, more easily anticipating the consequences for the local environment and those within it, redefining everyday lives rather than creating diverse alternatives.

2.3 Digital Placemaking

In online contexts, digital platforms have given rise to the concept of digital placemaking, where digital media enhances people's sense of place. This digital interaction offers a playful way to learn about both the history and the future of the place [23, 28]. The COVID-19 pandemic caused people to re-evaluate the meaning of public and domestic "places" due to restrictions on physical mobility, which led to an increased reliance on digital conferencing platforms such as Zoom and Microsoft Teams [13]. Furthermore, emerging 'metaverse' platforms allow people to be digitally connected, through the utilisation of technologies which enable multi-sensory engagements with digital objects, virtual environments, and individuals [26]. However, incorporating digital interventions in place contexts remains challenging, due to the limited spatial experiences of digital platforms and the general unfamiliarity of people with digital platforms in place settings. There is a risk that these interventions may inaccurately represent the dynamics of the urban space and increase digital divides due to cultural and generational gaps, which pose an additional challenge for digital placemaking [32]. In addition, the integration of digital and physical environments has raised questions about the definition of "place" and how people can experience physical places with digital platforms [12].

3 The Workshop Approach

3.1 Physical Walkshop

The research activities, which included a physical walkshop and a virtual walkshop, were designed to discuss IoT's potential benefits, risks, and challenges in public spaces. The initial walkshop was conducted in person with participants who were Lancaster City Council employees. The participants were from various departments in the Council, ranging from tourism and marketing, and public realms, to planning and information governance.

The first step of designing the walkshop was determining the type and location of existing IoT and sensors in Lancaster City Centre. Some of the sensors were identified through discussions with council officers. At this point, the main challenge we encountered was the absence of a comprehensive record of deployment, resulting from the ownership of sensors by different government organisations and others. The choices of sensors that were included on the walk were made based on their proximity to the city centre. For instance, river sensors which measure water level in canals were initially considered but excluded as they were too far from the proposed walk area.

Several fictional scenarios were layered onto the existing deployments, inspired by connected place implementations in other cities. The process of setting the route through public spaces in the city involved observing ordinary objects and routines, and then merging them with the fiction. The fictional deployments were represented by design fiction objects, which were based in existing and often mundane features in the city centre. For example, ordinary streetlighting and bins were transformed into smart lighting and smart bins. Additionally, provocative speculative elements were considered to enable the research team to explore interesting and challenging questions, such as the implications of using AI monitoring and systems on a busy high street. Eight stops were selected in the city centre (Table 1).

3.2 Materials

The materials designed for the walkshop were created to help the participants speculate about future IoT deployments in the public space and demonstrate the possible scenarios. The materials also supported encouraging interaction and engagement among the participants.

Field Guide

The research team looked to various sources for inspiration on capturing data during the walkshop, and the research team developed a field guide. The inspiration included prior work by members of the team [24], similar field guides and a series of 'i-SPY'

Table 1 Fictional and real stops with or without signage

	Stop	Type of situated artefact or space	Fiction (F) or real (R)	Signage	Icons
1	Smart bin	Litter	F	○	
2	Air quality monitoring	Air pollution passive monitoring equipment	R		
3	Parking space monitoring	Parking space	F	○	○
4	Traffic monitoring	Bullard	F		
5	Covid bus station	Bus station	F	○	○
6	Restaurant contact tracing	Contact tracing QR code	R	○	
7	AI Security monitoring	Public signages	F	○	○
8	Smart lighting	Lamppost	F	○	

books for children which identify items in places. The guide included a list of each stop and six prompt questions to be answered by the participants:

1. What is it, and how does it work?
2. What data does it collect and why?
3. What are the potential benefits?
4. What are the potential risks?
5. Are there any security challenges?
6. And are there any ethical challenges?

The participants could use the guide to answer the questions and take notes of their thoughts about the design fictions. The field guides used by the participants were collected after the walkshop for data analysis (Fig. 1).

Stickers and a Map

The materials included stickers that provided additional information about the objects but were revealed only after the first two questions (above) were considered. These allowed participants to speculate about the attributes of the objects and guess whether they were real or fictional before this information was revealed. The answer to the first question in the field guide; '*What is it, and how does it work*', depended on the participant's speculation about the object without any intervention from the research team. After sharing their thoughts, the sticker was provided, which indicated whether an object was real or fictional and gave details about its functionality. The stickers were designed to enable participants to distinguish between real and fictional objects and to offer more informed consideration. A revised tourist map of Lancaster Centre was also provided which featured the stop locations and suggested a route to ensure participant safety. The final stop on the walk was not an individual design fiction object, but a highlighted area that allowed the participants to speculate on what other uses of IoT may be likely in this space.

Fig. 1 Field guide tool for the physical walkshop

Design Fiction Signs and Iconography

Signage was made for some of the fictional objects, and some were left without signage in order to replicate the potential for IoT sensors to be placed in public places without key information. For those objects with signs, the designs conveyed the function of the object. The level of transparency provided by these signs varied. The signs used a variety of visual typography and graphics, which represent diverse imagery and logos found throughout the city that are designed and placed by different stakeholders, such as traffic signs, public advertisements, and warnings.

Some of the signs included speculative iconography based on that created by Lindley et al. [21]. This iconography was designed to resemble clothes washing symbols and to help people speculate on the legibility of AI systems. Figure 2 shows the visual design of these signs, which covers concepts such as AI processing location, learning scope, data provenance, training data type, and intrinsic work for AI operators.

This work had approval from the Lancaster University Faculty of Arts and Social Sciences (FASS) ethics board. In order to address ethical concerns around potential deception, additional signs were placed under the design fiction signs to alert any passers-by to the fictional nature of the objects.

3.3 Digital Walkshop

The second walkshop was designed to be online to connect with IoT experts including participants from a governmental organisation dealing with cyber security. It was

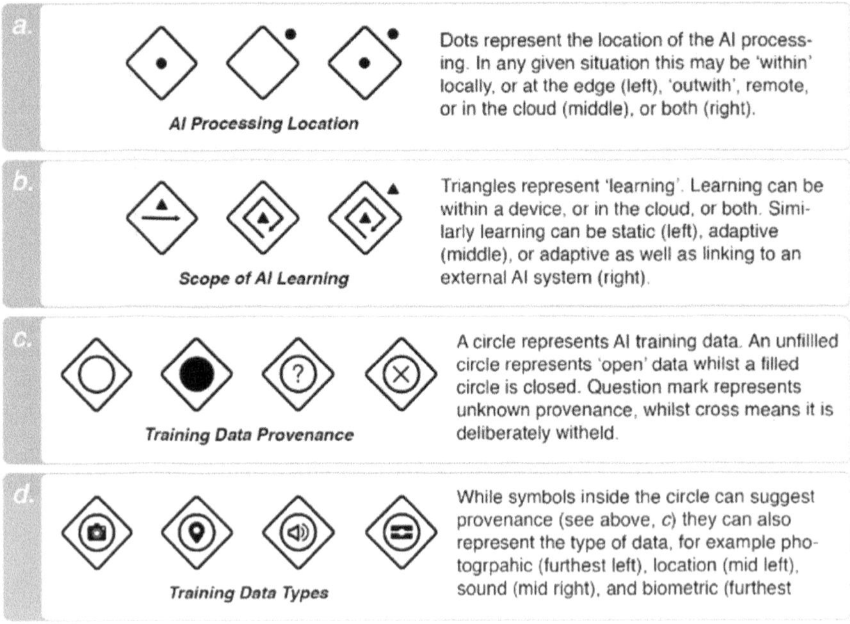

Fig. 2 Abstract icon style [21]

held during the COVID-19 pandemic when physical travel was limited. We used the Gather Town platform to create a 2D virtual space of Lancaster City Centre, the design of which was based on the tourist map used during the walkshop. Gather Town (https://www.gather.town/) is a virtual videoconferencing platform combining video chat function with gaming technology. In Gather, a user can create a customizable avatar which represents the user and move it around a virtual space using direction keys to join or leave conversations [17]. Gather also enables users to build their own 2D space, ranging from a traditional office to a cartoon. The platform became popular during COVID-19 as a remote working and socialising space. The research team aimed to replicate the physical walkshop experience on a digital platform, which was made possible with Gather's features (Fig. 3).

Photographs of specific public spaces and representations of other parts of the route were added to the Gather environment to create a virtual city experience. However, since making an utterly realistic portrayal of the entire city was impossible, the Gather representation was not an accurate depiction of all the actual sites on the route. This resulted in a somewhat selective experience of the places on the 2D map. To compensate for the lack of spatial experience, 360-degree videos of the nine stops were recorded to show the same real or imagined objects and associated signage as the physical walk. The videos were embedded in the Gather space so that participants could go to a stop and play the video, to get a visual idea of the location and the context in which the objects were located. The same materials used on the physical walk (printed field guide, map, and stickers) were sent to the participants by post.

Fig. 3 Screenshot of virtual Lancaster City Centre on Gather Town

The stickers were individually packaged and revealed only when prompted by the facilitator. The participants were asked to send back the field guides via post after the workshop, and a digital version of the materials was provided in case the mail did not arrive on time for the virtual walking workshop.

4 Findings

4.1 Participant Responses to the Objects

This section presents the participants' responses from each of the stops, for both the physical and virtual walks. Each stop is detailed, with key points taken from the field guides (physical and digital) and from field notes made by the research team (Fig. 4).

Stop 1: Smart Bin

The first stop, the Smart bin was implemented as a design fiction by adding signage. The fictional bin's functionality includes measuring the fill level of the bins, detecting smoke, and sending an alert to the council to arrange emptying and prevent vandalism. At the start, before any information was provided, the participants in both walkshops speculated that the bin could detect the materials deposited within it and recycle them as appropriate or calculate weights and fill levels which would trigger communication with the city council to schedule collection.

After handing out the stickers, participants discussed the benefits, risks, security, and ethical issues of the smart bin. In terms of benefits, the physical walkshop

Fig. 4 The stops and fictional and real objects on the map of Lancaster City Centre

participants highlighted operational efficiency. By understanding the use of bins, they decided this could be helpful in informing the future planning of new bins and scheduling collection routines.

The participants then went on to discuss the potential risks associated with the smart bins. One of the scenarios the participants drew out was about the materials placed into the bins, for instance, cardboard might weigh little, but take up space in the bin. They also speculated that different materials might cause the sensors to malfunction. Another participant was concerned that the weather might damage the sensors, especially on rainy days. Other risks identified related to people's behaviours towards the bins, such as vandalism and tampering with the data, or covering a sensor and smoking next to it. Such situations would threaten the efficacy of the sensors and potentially outweigh the identified benefits.

We found that the city council officers did not consider cyber security risks on the physical walk. In contrast, the IoT experts in the virtual walkshop pointed out that the data collected by the bin sensors might be at risk, depending on the means of communication by the sensors. They described the possibility of a side channel or a man-in-the-middle attack on the system [8], meaning data collected could be misused and cause issues for the council regarding information security. In this case, safeguarding the infrastructure (i.e., accessing the council's system) should be required rather than protecting from compromising the sensor itself. Manual waste collection scheduling could still be a fallback option even if the sensors malfunction.

One of the interesting findings from the physical walkshop was revealed by one of the participants who work in the public realm. The bins the research team had identified as being non-smart objects, in fact, contained sensors that monitor the amount of waste they contain, information that had not been shared by key contacts in the council with the research team during the design phase. This insight highlights the lack of knowledge in the wider city council regarding the existence of sensors in the public realm (Fig. 5).

Fig. 5 The bin with a sensor in the public realm

Stop 2: Air Quality Monitoring

The second stop, an air quality monitoring station, is an existing deployment, located in a small green space on the main road one-way system in the city centre. The information on the air quality monitoring for the Field Guide was taken from the City Council website. According to the website (Lancaster City Council, no date), the device is designed to gauge the amount of nitrogen dioxide in the air. Air is pumped continuously into each analyser, and the device records the level of pollutants in the air. These pollution readings are then stored in a data logger and can be accessed remotely using a computer and modem (Fig. 6).

In the first phase of the discussion, most participants in the physical walkshop recognised that it was measuring air quality. In contrast, many IoT experts in the virtual walkshop could not identify the device, speculating that the object might be a noise detector, a Smart bee counter or even 'a public shower for pigeons.'

The council officers described the benefits of the device, notably that collecting and monitoring data can be helpful for decision-making. They mostly agreed that no personal data is collected, so there are no ethical or cyber risks. However, the quality and veracity of the data is vital in enabling decisions to be made. Participants on the virtual walk discussed concerns regarding the deployment of the device. They thought the data could be easily tampered with, as it was in a cage and within reach of an adult's hand. Interestingly, one of the attendees pointed out that the device does not match the aesthetics of the public space, which is surrounded by historical buildings.

Both groups agreed that there were no ethical issues with the device. However, one of the experts criticised that there is no sign with information about what is measuring for the public.

Fig. 6 Air quality monitoring device

Stop 3: Parking Space Monitoring

This stop was based on an existing object, a parking meter, where the public pay using a card or coins. The added functionality given to this object in the design fiction featured car number plate recognition and the use of infrared sensors to detect occupied parking spaces. The imagined device also incorporates a predictive AI system that estimates when spaces may become vacant. This information is then made available through an app, which provides users with real-time information on available parking spaces. A sign was placed on a parking meter featuring the AI icons (discussed above) (Fig. 7).

The participants of the physical walk discussed the potential functionalities of the object, and that the data it collected consisted of car details. They described almost all the functions of the device as communicated on the fictional signage. However, the AI icons on the signage did not appear to be legible, with participants confused by their meaning. In the virtual walk, the participants also struggled to interpret the icons, and some participants speculated further that functionality could also be linked with CCTV and even users' phones around the area.

Physical walk participants identified operational and efficiency benefits, arguing that it would be cost-efficient to know which parking spaces are available so that they could plan better space management. The potential benefits for users were also

Fig. 7 Design fiction signage with AI iconography on the actual parking meter

discussed by virtual participants, such as the ability to find an available space more directly without wasting time. In addition, reducing unnecessary driving might help to reduce traffic emissions.

Both sets of participants agreed that the data collected would include personal details that might identify an individual through the correlation of location and time data. In the physical walkshop, different scenarios were mentioned, such as where data might be stored, how the data would be processed and the potential for information being used by a third party. In the virtual walkshop, some participants pointed out the possibility of false positives, which could occur when an algorithm indicates the presence of something when it is not there. The false positive can have consequences, such as incorrect fines by incorrectly indicating a user's presence at the location. Another possible scenario highlighted was price discrimination based on a car model, which might occur through the AI monitoring system. The participants imagined that the focus on current climate initiatives relating to vehicles, might enable discounts in parking for electric vehicles. The price differences could potentially be inequitable, with those on higher incomes and more likely to own electric vehicles received cheaper rates than those driving older vehicles with higher emissions [24, 25].

Stop 4: Traffic Monitoring

The fourth stop consisted of a bollard located on a pedestrian street adjacent to the main road through the city centre. Whilst the object itself is real, it currently has no IoT functionality. We imagined new functionality as an IoT object. The inclusion of this bollard builds upon an actual deployment in the Scottish city of Aberdeen. Like the real deployment, the bollard fiction had no signage [16]. The imagined fictional bollard contained sensors which count passing cyclists, and an infrared beam sensor to detect pedestrian numbers. The bollard transmits hourly counts of foot and cycle traffic to a central server.

Without knowing how the bollard worked, the participants in the physical walkshop guessed that it would be related to traffic but did not specify the functions. An officer with a visitor and tourism remit suggested that the traffic monitoring bollard might help control crowd dynamics in some events, such as festivals. Participants on the physical walk identified potential risks relating to data collected data from individual phones.

In the virtual walkshop, a participant thought the bollard might be a public Wi-Fi point that counts the frequency and number of connections made. In contrast, another participant recognised it as a traffic monitoring device because they have been installed in his home country. However, he pointed out that the example cited has clear information about the bollard and the ways in which it collects data. Some experts imagined the potential risk of the sensors or data being tampered with as an act of vandalism. Data that has been tampered with could potentially disrupt the decision-making of the city council.

Stop 5: COVID-19 Bus Station Sensors

This stop featured fictional signage for sensors within the bus station. It included various sensors to monitor and assess factors, including occupancy, ventilation, and adherence to COVID-19 safety measures. The imagined deployment consisted of infrared sensors to monitor occupancy and data-enabled AI algorithms to accurately count the number of people present, identification of mobile devices in the vicinity, measurement of CO2 levels to gauge air quality, and integration with the NHS COVID-19 app (See Stop 6). Given that the walkshop took place in 2021 adjacent to periods of national lockdown, COVID-19-related data collection was timely and emerging as an essential feature of life in a public space, with particular attention paid to deploying technologies to both monitor cases and warn against potential infection. This element allowed the participants to reflect on everyday life and future technology.

The signage featured AI icons and was placed next to the entrance door of the bus station with the message, "*This bus station is collecting data on occupancy and ventilation to protect health.*" (Fig. 8). Additionally, the research team created two new icons to represent the connection with the NHS COVID-19 app and the function of measuring CO_2 levels.

The discussion at this stop on the physical walk focussed on the AI iconography. The research team asked the participants what they thought the icons might represent, and most participants again found them challenging to understand. In the virtual walkshop, the participants thought the icons needed to be clarified. For example, some guessed that the NHS icon means the data collected would be transferred to NHS and recognised that the 'C' icon represents data that will be processed externally. Both groups agreed that the national standardisation of AI icons would be essential.

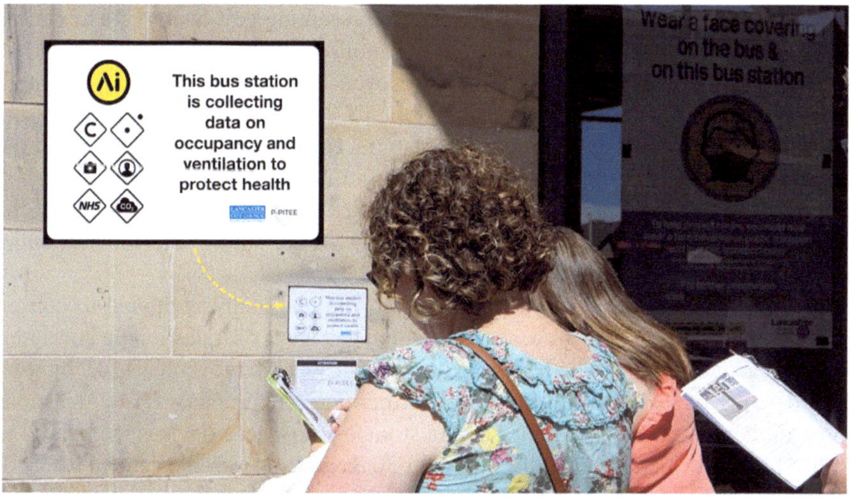

Fig. 8 Design fiction signage with AI iconography placed in front of Lancaster bus station

Stop 6: Restaurant Contact Tracing

This stop used existing signage on a coffee shop window, featuring the QR code that was linked to the UK NHS mobile contract tracing application during COVID-19 [11]. During this time, restaurant customers were asked to scan the QR code to register their presence to the NHS app on their mobile devices. Then, in the event of a notified case of COVID-19, the app sent a notification to the users who might have had close contact with the infected person.

In the physical walkshop, this case of the NHS app raised some questions from the participants, such as what are the criteria for the alerts? The nature of trust in the application was also discussed. Participants discussed the need for clarification about exactly what data were being collected, such as their location, and around data storage and use. In the virtual walkshop, participants discussed different behaviours exhibited by people scanning the track and trace QR code. For instance, many people hold their phones and pretend to scan them. Digital exclusion was also discussed relating to those who do not have a mobile phone and would be excluded from both the collection of data and the ability to be warned of contact with those infected with the virus.

Stop 7: AI Security Monitoring

This stop featured existing signage relating to CCTV in the pedestrianised centre of the city. We added fictional functionality to the system, which aimed to minimise criminal activity and anti-social behaviour. The functionalities added to the system were gait and body language analysis, using CCTV and AI processing. Videos were stored only if the predictive algorithm identified any behaviour considered to be potentially criminal or anti-social. Signage represented these functions through the

AI icons. The research team added an icon representing gait analysis, assessing a person's walking, running or other types of motion. Additionally, specific data is gathered to help determine the general noise level. The sign was located on the high street alongside other public signs, such as the CCTV and 'No Smoking' signs.

On the physical walk, some participants noticed the public signs placed on the street wall for the first time. However, it was still challenging for both walkshop groups to understand the iconography as it looked abstract. One participant of the virtual walkshop suggested the icons were vague, representing "stuff is happening here".

In particular, the council officers discussed the gait analysis related to current CCTV footage usage. They were sceptical about gait analysis, questioning how intelligent AI actually is and whether it can distinguish specific movements like fighting from dancing or people with disability. Furthermore, participants raised questions about the length and location of the data storage from these devices, compared to CCTV footage which is generally stored for 30 days. However, the general safety of the public and historic buildings in the vicinity using these systems was considered to be advantageous. Some participants stated that AI monitoring systems might be the next level of CCTV, which would require the implementation of national regulations. For instance, regulations are necessary on the length of data storage, to whom it will be shared and for what reasons.

The virtual walkshop participants discussed how AI might determine 'Anti-social behaviour', as this is an ongoing issue, identifying how we should view body language in AI and machine learning research. It was linked to the problem when an AI system became trapped in a contaminated feedback loop of instructions. Personal biases and prejudices of AI programmers can contaminate the feedback loop, which may lead to AI making biased and unequal decisions [27]. A participant pointed out that it could be dangerous when it cannot differentiate between an attacker manipulating "social behaviour" and an individual the adversary is aiming to victimise. In such a scenario, innocent people may be unjustly prosecuted or, at the very least, falsely accused.

Stop 8: Smart Lighting

This stop featured signage placed on non-smart lighting in the pedestrianised zone of the city centre. The additional fictional functionalities of the lights were smart light sensors that can detect both light intensity and individuals' proximity. The light will only be activated if both conditions are met, i.e., people are nearby, and the surrounding light levels fall below a specific threshold. The design fiction sign was placed on the lamppost in the middle of the busy area (Fig. 9).

The physical walk participants believed that the lighting would be cost-efficient and save energy and would reduce light pollution in the area. They also thought there would be a minimum risk with the lighting. On the other hand, the virtual walk participants were attentive to the surroundings of the Smart light, where a bank is visible near the lamppost. In addition, several participants considered various cyberattack scenarios that could impact the bank via the smart lighting system.

Fig. 9 Design fiction signage on the street lamppost

4.2 Findings from the Walks

Overall, we found that both the in-person and virtual walkshops created valuable insights into challenges and benefits of IoT deployments in public places. Key benefits included:

- The potential for efficiencies of finances and time. From the perspective of the physical walk participants who are engaged in public services, these deployments offered benefits such as understanding footfall within the city centre, which might be useful for event planning, and public health crises such as the COVID-19 pandemic.
- Data gathered from IoT devices in the public realm could also enable more focussed and targeted decision making, for example requirements of waste collection and recycling, use of car parking facilities, air quality interventions and public disorder prevention. This demonstrates the potential of connected technologies to help the public sector manage places more efficiently. This is particularly important for local governments working in resource constrained contexts.

Potential risks identified by participants of both walks related both to physical damage to IoT deployments, and the risks of cyber-attacks and data tampering.

- Physical damage to sensors might occur through extreme weather or vandalism.
- The council officers argued that the risks could be increased by processing and storing individual data and the approaches to collecting data.
- On the virtual walk, IoT experts discussed the issues from different, more technically focused angles.

- False positives or negatives might lead to false alerts, eventually wasting time and resources.
- Some locations might be excluded from monitoring.
- There is a risk of partial data and biased AI as a decision-maker.
- Digital exclusion might happen to citizens without the Internet in a digitalised public system and services.

After the walk, the research team discussed with the city council participants how they felt about the speculative walkshop. Overall, the walkshop participants felt the activity had opened a space of meaningful discussions and awareness of the connected technology in public space. The walkshop gave participants a holistic view of AI and IoT systems for public spaces. One said that "it will help further thinking and spotting interconnectivity and responsibility." The quote reflects the significance of the whole landscape of IoT deployments on the operational side. Also, by presenting the design fiction, participants had the opportunity to consider what is possible with IoTs, what commercial assets can be and who the externals are.

The use of AI agents as a control centre and issues relating to technological literacy were raised regarding the speculative AI symbols used on the signs. The icons are connected to transparency, public awareness, and acceptance of IoT technology. The signs caught the virtual walkshop participants' attention, but many found them challenging to understand, despite their familiarity with IoT. As a result, both groups suggested that a national standard for signs related to AI and other connected technologies should be established for better communication with the public. In addition, the physical walkshop participants, who are involved in different areas of policy making stated that once sufficient education opportunities for the signs and icons are provided to the public, it will boost public awareness and understanding of the technology, including what they are and what they do.

A further key issue that emerged was acceptance of sensors by the public. Through discussions about the signs, the council workers reflected on how the public might accept the technologies and what communication efforts the policymakers should consider. For example, one participant said, "People are busy, and they might not see the signs on the street." However, he speculated that the public would like to know and understand if they are affected by these technologies. At this point, interacting with the public is important in understanding and managing diverse acceptance levels. Throughout the reflections from the participants, it seems evident that the speculative walkshop method has the potential to enlighten both policymakers and citizens to think about possible IoT deployments and ways of interaction with the public.

5 Discussion

5.1 The Challenges for IoT Deployments in Public Spaces

The IoT experts on the virtual walk highlighted, on the one hand, that current challenges with implementing IoT devices and systems in public spaces relate to the infrastructure and issues such as bias being embedded within AI systems. On the other hand, the participants in the physical walk focused on communication and engagement with the public. This highlighted the importance of consulting and communicating with the public when considering the implementation of digital technologies in public spaces. A further key finding related to the need for careful development of standardised AI and IoT signage at a national level. Furthermore, literacy around these issues is crucial so that the public can understand the functions of any sensors in public places.

We also found no centralised map or log of sensors or IoT devices in the city centre. This issue was discussed on the physical walk when it was identified that the bins (Stop 1) which we thought were non-smart did, in fact, contain sensors. The complex nature of governance causes this issue within the urban realm. For example, some traffic sensors will be owned and monitored by the county council, whilst others, such as river monitors, will be owned and managed by the Environment Agency. This is further complicated when private organisations use sensors in the public realm. When this issue was discussed on the physical walk, the participants agreed that a centralised map of sensors, including what data was collected, who held it and what it was used for, would be useful. This leads to questions about data sharing and the ownership of such data gathered in public places.

5.2 Speculative Design in Policymaking as a Community Engagement Strategy

Speculative design approaches and techniques can potentially address societal and technological concerns. They have been used to inform the public sector and policymakers to provide in-depth insight and understand technology [35]. In this study, we combined speculative design fiction with a physical walk to provide a place-specific experience of a future interconnected environment. Individuals can use their imagination by placing fictional objects in real locations, and blending real and fictional objects can create immersive realities. We found that participants of both walkshops mentioned the potential of speculation activities in the emerging area of policymaking. From the physical workshop, the participants from the city council described that it was helpful for further thinking and spotting the system's interconnectivity, responsibility within the city council departments and relationship with external partners.

We also found that a design strategy should be considered to bring different stakeholders as participants to articulate issues and themes of emerging technology. In the walkshops, two groups of participants focused on specific areas according to their interests. The two groups generally agreed that sensors, IoT, and edge computing would benefit the local government in terms of cost and efficiency. However, the focus of each group was slightly different as the participants had different backgrounds and expertise. For instance, the council staff interpreted the possible impacts of IoT sensors and systems through their interest lenses, such as public realm management, planning policy, tourism, and information governance. They described how the technology would be helpful for operational efficiency and safety management in public spaces. The second group covered broad topics based on their expertise and data security knowledge. The experts discussed data collection and storage, and the system and infrastructure. They were concerned about accessibility for those who might not have the Internet or digital devices. This difference highlights a need for collective intelligence towards issues around technology and policymaking. Policymakers can potentially consider speculative design as an approach for community engagement in the future.

6 Conclusion

This chapter demonstrates a study that tested various qualitative methods combining speculative design with a walking method and then utilising a digital space, Gather Town, for wider audiences. We observed that these methods and approaches enabled the participants to discuss the potential benefits, risks, and challenges of deployments of connected technology in public spaces such as IoT sensors. In particular, the 'walkshop' method could offer realistic encounters physically and virtually with a connected place through familiar but imaginary objects. These contextual but situated objects allowed participants to connect themselves with design fictions and envision possible deployments, the place, surroundings, and people.

At the same time, we have demonstrated that speculative design approaches have value in policymaking by provoking discussions and raising understanding and interest in technology implementations. We also found that participants had differences in experience and understanding. This observation leads to consideration for local government to involve diverse stakeholders, including the public, as a new community engagement strategy. However, the short pilot study was time-limited, so our following research stage, the *Taking IoT for a Walk* project, aims to conduct physical and virtual walkshops working with four cities in the UK as public engagement activities. The activities will allow us to reveal each place's specific needs and challenges towards connected place initiatives, learn how design practices are used to engage with the public and gather insights from various stakeholders and communities.

Acknowledgements We would like to thank the participants who took part in our physical and virtual walkshops. This research was funded by PETRAS (National Centre of Excellence in IoT and Cybersecurity) and Research England, Expanding Excellence in England (E3).

References

1. Ahmed M, Zeile P (2020) 'Places' representation on social media'
2. Angelidou M (2017) The role of smart city characteristics in the plans of fifteen cities. J Urban Technol 24(4):3–28
3. Bleecker J (2009) Design fiction: a short essay on design, science, fact and fiction. In: Near future laboratory [Preprint]
4. Blythe M (2014) Research through design fiction: narrative in real and imaginary abstracts. In: Conference on human factors in computing systems—proceedings. https://doi.org/10.1145/2556288.2557098
5. Canter D (1977) The psychology of place. St Martin's Press
6. Chopra S et al (2022) Negotiating sustainable futures in communities through participatory speculative design and experiments in living. In: Conference on human factors in computing systems - proceedings. Association for computing machinery. https://doi.org/10.1145/3491102.3501929
7. Dunne A, Raby F (2013) Speculative everything: design, fiction, and social dreaming, speculative everything: design, fiction, and social dreaming. https://doi.org/10.1093/jdh/epv001
8. Frustaci M et al (2018) Evaluating critical security issues of the IoT world: present and future challenges. IEEE Internet Things J 5(4):2483–2495. https://doi.org/10.1109/JIOT.2017.2767291
9. Greenfield A (2017) Radical technologies: the design of everyday life. Verso Books
10. Greenfield A, Kim N (2011) Systems/layers: how to run a walkshop on networked urbanism. http://diffusion.org.uk/?p=2364
11. Guinchard A (2021) Our digital footprint under Covid-19: should we fear the UK digital contact tracing app? Int Rev Law Comput Technol 35(1):84–97
12. Halegoua G, Polson E (2021) Exploring "digital placemaking." Convergence 27(3):573–578
13. Hardley J, Richardson I (2020) Digital placemaking and networked corporeality: embodied mobile media practices in domestic space during Covid-19. In: Convergence, p 1354856520979963
14. Hedestig U, Skog D, Söderström M (2020) Challenges in using IoT in public spaces
15. Hviid Trier S, Jenkins T (2020) Participatory sensing in the speculative smart city. In: 32nd Australian conference on human-computer interaction, pp 462–470
16. Jacobs N, Markovic M, Cottrill CD, Edwards P, Corsar D, Salt K (2020) Made-up rubbish: design fiction as a tool for participatory internet of things research. Des J 23(3):419–440
17. Jacobs NJ, Lindley J (2021) Room for improvement in the video conferencing "Space". In: AoIR selected papers of internet research [Preprint]
18. Lancaster City Council (no date) Air quality: How we monitor, assess and coordinate control over local air pollution. https://www.lancaster.gov.uk/environmental-health/environmental-protection/air-quality. Accessed 6 Mar 2023
19. Lau BPL et al (2018) Sensor fusion for public space utilization monitoring in a Smart City. IEEE Internet Things J 5(2):473–481. https://doi.org/10.1109/JIOT.2017.2748987
20. Lindley J, Coulton P (2016) Pushing the limits of design fiction: the case for fictional research papers. In: Conference on human factors in computing systems—proceedings. Association for computing machinery, pp 4032–4043. https://doi.org/10.1145/2858036.2858446

21. Lindley JG et al (2020) Signs of the time: making AI legible. https://doi.org/10.21606/drs.2020.237
22. Malpass (2017) Critical design in context: history, theory, and practices. Bloomsbury Academic, an imprint of Bloomsbury Publishing Plc.
23. McClure CD, Williams PN (2021) Gather. town: an opportunity for self-paced learning in a synchronous, distance-learning environment. Compass: J Learn Teach 14(2):1–19
24. Mullagh L (2022) Design principles for more-than-data walks. [Doctoral Thesis, Lancaster University]. Lancaster University. https://doi.org/10.17635/lancaster/thesis/1770
25. Mullagh L (2022) The little book of design policy for equitable net zero. Lancaster University https://eprints.lancs.ac.uk/id/eprint/170188/1/NERC_the_little_book_of_DESIGN_POLICY_for_equitalbe_net_zero.pdf
26. Mystakidis S (2022) Metaverse. Encyclopedia 2(1):486–497
27. NCSC (2021) Connected places: cyber security principles
28. van der Niet AG, Bleakley A (2021) Where medical education meets artificial intelligence: "Does technology care?" Med Educ 55(1):30–36. https://doi.org/10.1111/medu.14131
29. Pang C et al (2020) The role of a location-based city exploration game in digital placemaking. Behav Inf Technol 39(6):624–647
30. Singer PW, Friedman A (2014) Cybersecurity: what everyone needs to know. oup usa
31. Stals S, Smyth M, Mival O (2019) UrbanixD: from ethnography to speculative design fictions for the hybrid city. In: ACM international conference proceeding series. https://doi.org/10.1145/3363384.3363486
32. Sterling B (2005) Shaping things, p 152. October, 39. https://doi.org/papers2://publication/uuid/9ECCBCC1-ADE4-4E3C-B376-0E8DE64E4044
33. Stokes B et al (2021) Urban furniture in digital placemaking: adapting a storytelling payphone across Los Angeles. Convergence, p 1354856521999181
34. Stoneman P, Diederen P (1994) Technology diffusion and public policy. Econ J. https://www.jstor.org/stable/2234987?seq=1&cid=pdf
35. Tomitsch M (2016) Communities, spectacles and infrastructures: three approaches to digital placemaking. In: What urban media art can do: why when where and how, pp 339–347
36. Tsekleves E et al (2017) Co-designing design fictions: a new approach for debating and priming future healthcare technologies and services. Arch Des Res 30(2):5–21
37. de Waal M, Dignum M (2017) The citizen in the smart city. How the smart city could transform citizenship. It-Inf Technol 59(6):263–273
38. WEF (2016) The fourth industrial revolution: what it means and how to respond | World Economic Forum. https://www.weforum.org/agenda/2016/01/the-fourth-industrial-revolution-what-it-means-and-how-to-respond. Accessed: 27 April 2020

Design and Democracy: A Case Study of a Project for the European Parliament

Caio Marcelo Miolo, Natália Debeluck Plentz, Melissa Pozatti, and Rita Assoreira Almendra

Abstract Democracy refers to the condition where citizens can exercise self-determination, thereby allowing for the realization of their aspirations beyond the mere formal right to vote. Design for democracy primarily involves improving the mechanisms of participation in politics, such as through the interpretation of the needs of social groups or the creation of artifacts. We present a case study of EU4ALL, an initiative developed by a Portuguese social NGO, co-founded by the European Parliament (EP). EU4ALL's principal aim was to develop communication strategies that make the ten democratic values highlighted by the EP more attainable to communities in situations of information vulnerability, particularly inmates and those living in social housing. To achieve this, the Double Diamond design model was used, which involved a bibliographic search on democratic values, synthesizing of research, validation sessions with Portuguese Euro Deputies (MEPs), development of communication materials, testing with users, and finally, the delivery of materials and awareness sessions with vulnerable groups. Based on the conducted process, it is evident that design holds immense value in its potential to bolster democracy by creating inclusive and accessible communication materials and promoting transparency and accountability through the visual representation of data and information.

Keywords Democratic values · Information vulnerability · Strategic communication · Human-centered design

C. M. Miolo (✉) · N. D. Plentz · R. A. Almendra
CIAUD, Research Centre for Architecture, Urbanism and Design, Lisbon School of Architecture, Universidade de Lisboa, Pólo Universitário do Alto da Ajuda, Rua Sá Nogueira, 1349-063 Lisboa, Portugal
e-mail: caiomarcelo.mo@gmail.com

M. Pozatti
Universidad Técnica Federico Santa María, Valparaíso, Chile

N. Martins et al. (eds.), *Perspectives on Design and Digital Communication IV*, Springer Series in Design and Innovation 33, https://doi.org/10.1007/978-3-031-41770-2_9

1 Introduction

Although democracy is not flawless, it remains the most effective system for safeguarding human rights, promoting freedom, and ensuring peace and stability around the world [1]. Data shows that democracies tend to be richer than non-democracies, as well as better at fighting corruption and being less likely to go to war [2]. Furthermore, democracy is worth defending because it provides individuals with a voice in their governance and the ability to hold their leaders responsible for their decisions [1]. It can also guarantee that every citizen is treated equally under the law, regardless of social status or background.

For a considerable stretch of time, we inhabited a planet that, notwithstanding its predicaments, was still dedicated to democratic ideals such as upholding human rights, and basic liberties, and fostering individual growth; however, today this landscape has undergone a profound transformation [3]. There are clear signs that democracy is under attack in several countries [3, 4] and even established democracies are showing flaws in the system that are becoming increasingly visible, creating a widespread sense of disillusionment with politics [2]. Threats to democracy in European countries have become a concern for its institutions, such as the European Parliament. The rise of populism and authoritarianism in some countries, as well as widespread polarization, which becomes worse through the dissemination of fake news, are major concerns.

Democracy is a core principle of the European Union and one that the Union is obligated by law to promote on a global scale. The European Parliament (EP) plays an active role in supporting the Union's efforts to advance democracy and other fundamental values, such as the protection of human dignity, freedom, equality, the rule of law, and human rights. This commitment extends not only to the EU, where the European Parliament strives to safeguard the basic rights of all inhabitants but also to areas beyond the Union's borders, where the European Parliament has become a dedicated and proactive champion for democracy [5].

In this context, efficiently and engagingly communicating information about politics and European democratic values to society represents a significant challenge faced by the EP in the contemporary era. Transparent communication promotes social cohesion by encouraging dialogue, cooperation, and compromise among diverse groups, leading to a more inclusive and harmonious society. It can also make institutions more transparent, allowing citizens to access information on the workings of the government, which enhances accountability and reduces corruption. Despite utilizing a range of digital communication vehicles to disclose the decision-making processes undertaken during plenary assemblies (and thus demonstrate Parliament's progress towards a more cohesive and developed Europe), the information disseminated remains largely inaccessible to those groups experiencing what is called 'informational vulnerability'.

This vulnerability is primarily experienced by individuals with lower levels of education and those enduring social exclusion [6]. Furthermore, two additional factors exacerbate the issue of informational vulnerability. First, the limited access to the internet among these vulnerable groups, and second, the prevalence of misinformation, which not only influences public opinion but also has deleterious consequences for society as a whole. Misinformation seeks to manipulate and shape public opinion to serve private interests and such poses a significant threat to the dissemination of accurate information.

Design is an activity that can not be neutral to social and political issues since it influences and is influenced by the diverse interests of social groups involved in the process [7]. Therefore, diverse ideas regarding what constitutes democracy, its themes of inquiry and courses of action should have an impact on design practice and design studies [8]. The professional designer's formative role as a creator of both physical and non-physical realities [9] enables them to contribute significantly to reducing information vulnerability and promoting democratic values, ultimately leading to stronger and healthier democracies through the facilitation of information. To ensure effective communication for individuals experiencing different types of information vulnerability, it is essential to carefully plan the dissemination of information in a way that is accessible and comprehensible to these groups. In this sense, the Human-Centered Design (HCD) approach becomes quite appropriate, as it draws on anthropology and participatory design tools and methods to better understand its recipients and their diverse contexts and is particularly concerned with the way they interpret, perceive, and interact with the developed communication channels [10–12].

Against this backdrop, this study presents a case study of an initiative co-funded by the European Parliament and designed by the NGO "Aproximar, Cooperativa de Solidariedade Social": EU4ALL is an abbreviation for "European Union for All," and this initiative was created and implemented between January 2021 and July 2022. The primary objective of EU4ALL was to design a series of communication strategies utilizing the HCD (Human-Centered Design) approach that effectively communicates the ten democratic values emphasized by the European Parliament. The initiative's particular focus was on individuals experiencing information vulnerability, including those in detention centers and those residing in social housing. These individuals are often from ethnic and minority groups, nomads, and refugees, among others. The ten democratic values identified in the project were: Social Justice; Solidarity; Rule of Law; Research and Innovation; Health and Environment; Freedom; Education; Economic Development; Human Rights and Equality; and Valorization of European Culture. It is worth noting that this initiative was proposed and submitted by the Portuguese NGO and approved by the European Parliament.

While acknowledging the existence of numerous democratic values, the decision to limit the number to ten was made to enable greater accessibility and assimilation of the information by the target audience. The EU4ALL project was developed using the "Double Diamond" design model [13], which consists of four design phases: (1) Discover, in which a bibliographic review of democratic values was conducted based on the "Plenary insights" reports (as per indicated by the European Parliament itself); (2) Define, in which an analysis and synthesis of the values found and validation sessions with Portuguese MEPs were carried out; (3) Develop, in which the broad communication strategy and communication materials were designed; and

(4) Deliver, in which tests were conducted with the target audience, and improvements were made to the materials, resulting in the final communication strategy, which involved the production/delivery of materials and awareness sessions with the vulnerable groups mentioned above.

Given the information presented above, this paper aims to provide a literature review that investigates the relationship between democracy and design, the context of information vulnerability, the application of human-centered design (HCD) methodologies, and the creation of meaning and relevance through design.

2 Design and Democracy

Democracy was born in Athens, around the fifth century B.C.E., and lay dormant until the Enlightenment, becoming more widespread in the nineteenth century; since then, democracy is based on nation-states and parliaments where people elect representatives who govern for a determined period of time [2]. But even with a long history, the concept of democracy is neither given nor fixed, changing through time and having differences between countries, with its ideals and practices varying widely even within a single nation-state system [8].

While more people than ever before live in countries holding free and fair elections, the momentum of progress in a democracy that characterized the late twentieth century has come to a halt in the twenty-first century–and may have even reversed since even nations that are traditionally democratic are presenting a trend of slipping towards autocratic forms of government. Numerous nations give the impression of being democratic by holding elections, yet they lack the crucial rights and institutions that are indispensable for a truly functional democratic system [2]. This implies that democracy is facing an assault in numerous countries, including places where its fundamental principles appeared unshakable [3].

Over the past decade, there has been an increasing interest in the idea of 'design for democracy', and this idea not only carries significant implications for design methodologies but also provides new paths for research in the field of design [8]. One common interpretation of democracy in the prevalent discourse on 'design for democracy' has not been yet defined [8], since democracy is constantly changing. However, we do not need to share identical beliefs about democracy, in order to regard it as a fundamental principle [3].

The typical assumption is that democracy entails obtaining consensus through structured deliberation, and that 'design for democracy' primarily entails improving mechanisms for political participation. While these endeavors are significant, they do not encompass the complete spectrum of democratic actions and criticisms that design practitioners or scholars can explore [8]. According to Manzini and Margolin [3], there are at least four critical domains in which design and democracy overlap and where design can contribute: (i) designing democracy, through the enhancement of democratic processes and institutions; (ii) designing for democracy, by enabling broader participation in the democratic process, particularly through technology; (iii) designing in democracy, by promoting access, transparency, and equity in institutions; and (iv) designing as democracy, through participatory design that enables diverse groups to shape our present and future in a fair and inclusive way.

Democracy is usually safeguarded by public institutions more than private organizations. This happens because a public sector institution is obligated to act in the best interest of all its constituents, while private organizations may prioritize the interests of their shareholders [14]. It is important to consider the perspectives of stakeholders and communities, but equally important is to take a holistic approach to understand the challenges faced by social systems and their underlying causes, rather than focusing solely on individual perspectives [15].

Democracy can only be built and maintained through open dialogue between all members of society, however, vulnerable populations often suffer from vulnerability also in information, being excluded from the democratic process.

3 Misinformation and Information Vulnerability

The concept of information vulnerability is defined as a state of propensity to harm caused by either excessive exposure or a lack of access to information [16]. The constant production of communication materials, exacerbated by the digital age, has resulted in an overwhelming amount of information received through various channels, such as social media, television, and pay-per-view channels. The rapid production of such content has led to a lack of purpose and cultural, political, and social values in the messages conveyed, resulting in a trivialization of interfaces and leading society towards indifference [17]. Consequently, society is consumed by mass culture, where all forms of expression are subjugated to commercial slogans and frivolous entertainment, creating a self-feeding system of intense consumption of trivialized visual signs and messages [18]. Therefore, in compliance with these authors, the messages produced are not only numerous but also often inaccessible, lacking clear objectives, and being unable to be assimilated by the intended receivers.

Furthermore, the lack of information is associated with the inability to reach individuals due to a lack of access to communication channels, such as the internet, audio-based, or printed media. Marginalized communities are disproportionately impacted by this problem, such as those who are subordinate within hierarchies, older adults, those living with dementia or in assisted living facilities, individuals receiving government assistance, unemployed persons, ethnic and racial minorities, individuals experiencing homelessness, nomadic groups, refugees, and incarcerated individuals [6].

A strategic communication plan was proposed to address the issue of disinformation, also known as "fake news," which the European Parliament has tackled in several publications on their social media platforms, including Instagram, Facebook, Twitter, and their official page. Disinformation relates to the spread of misleading information through various channels, with digital networks being the primary source of information today. To penalize this type of news, the EP has established a sanctions regime [19].

Considering these challenges related to information vulnerability, a communication strategy was developed to promote the democratic values of the EP by applying Design methodologies. This can help safeguard democracy, when we use the definition of Bonsiepe [20], considering it a form of participation where citizens who were previously dominated become active subjects, creating space for self-determination and allowing for the pursuit of personal projects.

4 Development of the EU4ALL Project

In the next subsections, each of the stages will be presented along with what was accomplished in each phase. The EU4ALL initiative, developed between January 2021 and July 2022, is co-founded by the European Parliament and designed by "Aproximar, Cooperativa de Solidariedade Social", the Portuguese NGO.

To guide the project, the "Double Diamond" design model was employed. The Double Diamond design diagram was officially introduced by the British Design Council in 2005 and has since become a popular system for designers to follow in their creative process. Essentially, the Double Diamond model is a framework that guides designers through a series of stages to generate innovative solutions [13] (Fig. 1).

It is important to note that the Double Diamond model is not a linear process. Instead, designers are encouraged to move back and forth between stages to gain a deeper understanding of the problem and find the most effective solution or improve

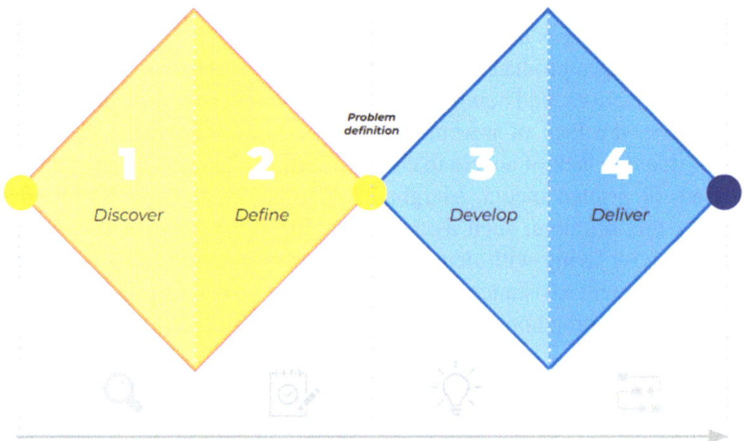

Fig. 1 Double Diamond model [13]

an existing one. This model employs divergent thinking to explore an issue broadly or deeply, followed by convergent thinking to take focused action [13].

The Double Diamond has four stages:

- Discover, the first stage of the design process, is focused on understanding the problem by involving stakeholders who are impacted by the issues or have experience with the situation being examined. This phase entails gathering information and insights on the different factors that affect the problem and its possible solutions [13].
- Define, what involves shifting through the information gathered during the Discovery phase and refining it. This process may include identifying areas of inefficiency or wasted resources, uncovering hidden opportunities, or creating a list of things the design team should avoid. The Definition stage aims to provide a comprehensive understanding of the initiative's internal and external context while also considering the team initiative's capabilities [13].
- Develop, which is the creative phase where answers are devised for the defined question. There is a significant amount of cross-disciplinary collaboration between designers and stakeholders. One of the most significant advantages of this stage is that it expedites problem-solving by bringing together experts from various fields [13].
- Deliver, involves conducting the last round of product testing, obtaining official sign-off for production, and launching the product/service. The final testing stage is a crucial step in ensuring that the product/service is free of any defects or issues. This typically involves subjecting the product/service to regulatory and legal standards, damage testing, and compatibility testing [13].

In the following topics, each of the stages will be presented along with the results achieved in each of them.

4.1 Discover and Define

The EU4ALL initiative, which began in January 2021, was created through a partnership between the European Parliament and a Portuguese NGO and employs the "Double Diamond" design methodology. This model involves exploring an issue thoroughly through divergent thinking, followed by convergent thinking to identify specific actions. The design process consists of four stages: Discover, Define, Develop, and Deliver.

The Discovery stage began with a review of the monthly reports available on the European Parliament's website, EP Insights. The objective of this research was to identify the topics related to democratic values that have been discussed the most in recent years. Additionally, to develop empathy towards the target audience, comprising social housing people and inmates, research was planned. However, given the NGO's experience of over ten years working with these groups in various European projects, this research was conducted elsewhere. The team planned to involve users in subsequent phases of the initiative.

In the Define stage, a thematic network technique was employed to analyze and synthesize the themes identified in the literature review. This technique helps in understanding and organizing words into central themes and subtopics, creating a network of words and convergence. Following this, individual online sessions were conducted with Portuguese Euro Deputies (MEPs) to present the initiative, and the identified keywords, and gather their insights on democratic values. The objective was to identify the ten most relevant values. Although all 21 MEPs were invited, only five of them and one parliamentary adviser could participate in the 30-min sessions due to their busy schedules.

During the sessions, infographics were generated with the primary values recognized by each MEP, while they shared their insights. An example of one such infographic is presented in Fig. 2, where the thematic network is displayed on the left and an infographic created during one of the sessions is presented on the right. The team used pre-drawn illustrations connected to the researched topics and incorporated the topics mentioned by MEPs along with their explanations and established relationships, where applicable.

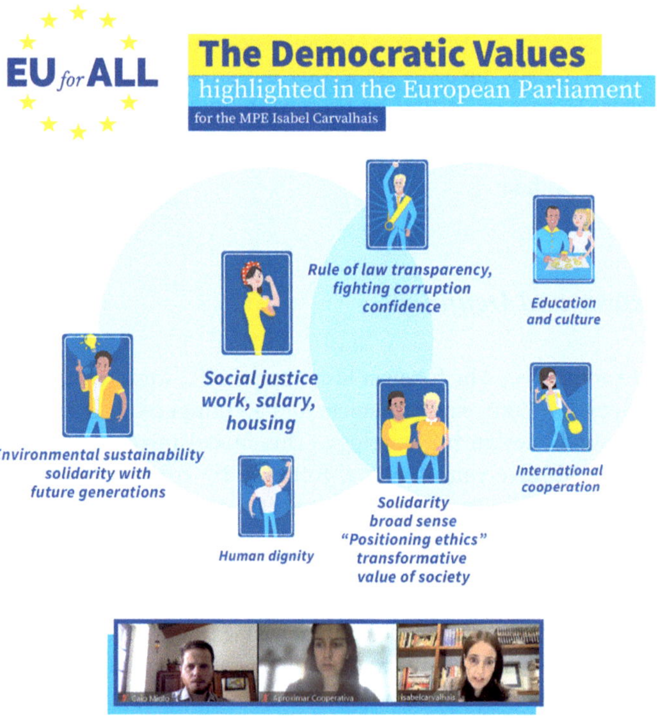

Fig. 2 Session with Portuguese MEPs

Following the six sessions, the discussed topics were analyzed, and the most frequently cited ones were highlighted. An effort was made to consolidate closely related topics to arrive at only 10 values. For instance, "health and environment" were grouped together, as they were frequently cited as intrinsically related by MEPs.

Two values that were notably pointed out by MEPs, namely "Social Justice" and "Appreciation of European Culture," were not included in the initial thematic network. Therefore, the following values were randomly enumerated as the 10 most important ones: the Rule of Law; the Social Justice; the Solidarity; the Health & Environment; the Research & Innovation; the Freedom; the Education; the Human Rights & Equality; the Economic Development; and the Valorization of the European Culture.

The MEPs who took part in the initiative demonstrated a strong interest and recognized the importance of both the EU4ALL initiative and the design approach employed by the NGO.

4.2 Develop

Succeeding the selection of the ten most pertinent democratic values, a strategy was formulated, which was partially predetermined in the co-financing application, along with the design of a visual identity and the creation of printed and digital materials.

Two communication approaches were delineated: a comprehensive one, involving the delivery of communication materials to ten prison establishments and ten social housing units in Portugal through a door-to-door approach (or distribution in key locations such as community centers). The plan was to distribute around 10,000 printed materials as part of this initiative. The other approach was more strategic and involved interactive and engaging awareness-raising sessions for groups of 15–30 people. Ten sessions were planned in each of the ten prisons and ten social housing units, respectively, with the materials created being distributed during these sessions.

An A4 folder containing infographics on the ten values was designed for distribution. The idea was to present the values in an appealing manner that could be reused, unlike other materials like flyers, which are used for storing papers and not as practical for the receiver. Although the objective was to minimize printing materials, it was necessary to reach vulnerable groups that may lack internet access.

A letter search game was also included in the folder, to make the content more interesting and memorable. Pens were to be distributed in the 20 sessions, and T-shirts were to be provided to the NGO employees who acted as session facilitators.

Each value in the EU4ALL initiative was depicted with an illustration featuring characters representing diverse gender, age, cultural, and ethnic backgrounds. Alongside each character, the respective value was written along with an explanation of its practical applications. These topics were based on the conducted research as well as MEP inputs.

The duration of the sensitization sessions was approximately one hour and 30 min, which was divided into multiple segments. A 20-min segment was dedicated to presenting the NGO, the EU4ALL initiative, and historical data on the European Union and the EP. A 10-min segment was reserved for the letter search activity, encouraging participants to search for the ten values and share their findings. A 15-min segment was allocated to the NGO team for presenting the ten highlighted values, followed by a 5-min presentation of the videos that animatedly explained each value. Also included in the sensitization sessions was a 15-min recap activity, during which the ten images representing the democratic values were displayed, and participants were instructed to place a post-it note on each image corresponding to the represented value. Afterward, each participant was given five minutes to answer the question of whether they would choose these values or suggest others (Fig. 3).

Fig. 3 Composition of images of the folder prototypes

4.3 Deliver

Following the planning of communication actions and material design, a validation and testing session was conducted with Portuguese citizens in vulnerable situations, in order to identify inaccessible information and potential areas for improvement. The recruitment of participants was facilitated through another NGO project designed to assist individuals facing similar vulnerabilities. The insights gleaned from these sessions were valuable, with several participants identifying the need to display the numerical order of values prominently on the material, starting on the cover and continuing in sequence on the inside and back. Additionally, some participants expressed dissatisfaction with certain visual elements, which they found confusing.

Following user testing (Fig. 4), modifications were implemented to the folder and video content intended for the awareness sessions, and digital publications about each of the values were produced. While the initiative focuses on in-person communication, leveraging the potential of digital media was deemed advantageous for greater outreach.

The resources were developed in Portuguese, which is the main used language in the targeted locations, as English may not be accessible due to educational levels. However, English editions were also prepared in case foreign individuals have a better grasp of the language. Figure 4 depicts an amalgamation of images showcasing the folder's interior and exterior, as well as the "letter search" insert for each material (Fig. 5).

Following the development of the materials, the last stage of the initiative entailed the printing of graphical materials and the execution of 20 awareness-raising sessions, that occurred until June 2022. A noteworthy point is that the sessions are still being conducted to date. Figure 6 displays documentation from one of the initial

Fig. 4 Usability user tests of the resources

Fig. 5 Composition of images of the folder and the "letter search" contained within. [21]

Fig. 6 Awareness session in social housing

sessions held at a social housing complex, which received positive participation and engagement from the attendees, demonstrating a strong interest in the presented topics.

5 Conclusion

Based on the findings of the applied research carried out, the effectiveness of the approaches used for designing a communication strategy has been confirmed. In particular, the Human-Centered Design (HCD) approach was found to be highly valuable, emphasizing the importance of engaging people with lived experience throughout the entire process, from the conceptualization of communication content to the prototype testing phases. The "Double Diamond" design model proved to be particularly useful for the EU4ALL project, as it facilitated a creative process with constructive synthesis and generative processes that encouraged stakeholder involvement throughout the various stages. Both approaches have the potential to enhance information dissemination, a key factor in preserving democratic systems.

Furthermore, we believe that the European Parliament initiative, as described, exemplifies two of the concepts involving design and democracy mentioned by Manzini and Margolin: (iii) designing in democracy, since it helps to translate and share the democratic values promoted by the organization to populations that usually have less access to them; and (iv) designing as democracy, by providing access to such information, these vulnerable groups can participate in the democratic process and help shape its ideas. According to Manzini [22], the absence of dialogue processes presents a challenge to discussions, the creation of shared visions for the future, and the establishment of foundations. This obstacle impacts both the broader debates on political and economic movements at the societal level, as noted by Bittencourt and Freire [23] and Manzini [22]. Faced with this, transdisciplinary training and the ability to interpret and integrate languages from various fields of knowledge makes professional designers able to deal with these kinds of challenges, since they can develop creative syntheses that facilitate the dialogue itself [24].

The conducted research and initiative highlights the significant role of design in strengthening democracy. By creating communication materials that are inclusive and accessible, designing user-friendly interfaces, and representing data and information transparently, design can promote transparency, accountability, and civic engagement. It is, therefore, essential to prioritize design as a tool for fostering democracy and ensuring that it is used effectively to promote equitable and participatory societies.

In conclusion, this study provides a valuable example of the manner in which design processes can be used to communicate relevant yet complex information to vulnerable groups. As such, this article may serve as a foundation for other social domains, such as psychology, social work, anthropology, sociology or public policy, which often need to convey complex, socially relevant content.

Acknowledgements This research was supported by national funding from the FCT—Fundação para a Ciência e a Tecnologia, I.P., under the Strategic Project with references UIDB/04008/2020 and UIDP/04008/2020. The authors also express their gratitude to the European Parliament for co-founding the EU4ALL initiative and to Aproximar—Cooperativa de Solidariedade Social for their contributions to the development of the initiative.

References

1. Altman D, Pérez-Liñán A (2002) Assessing the quality of democracy: Freedom, competitiveness and participation in eighteen Latin American countries. Democratization 9(2):85–100
2. Economist (2014) What's gone wrong with democracy? The Economist. https://www.economist.com/essay/2014/02/27/whats-gone-wrong-with-democracy

3. Manzini E, Margolin V (2017) Open letter to the design community: Stand up for democracy. Diseña 11:11–17
4. Levitsky S, Ziblatt D (2018) How democracies die. Crown Publishing Group, New York
5. European Parliament. Global democracy support. 10 years of EP's democracy support. https://www.europarl.europa.eu/globaldemocracysupport/en/home/home. Last accessed 24 Mar 2023
6. Macklin R (2004) Bioética, vulnerabilidade e proteção. In: Garrafa V, Pessini L (Org.). Bioética: poder e injustiça. Loyola, São Paulo, pp 59–70
7. Izidio LL, Moraes D (2019) Design as democracy: The democratic potential of design. designing sustainability for all. In: Ambrosio M, Vezzoli C (eds) Proceedings of the 3rd lens world distributed conference, vol 3, pp 727–731
8. DiSalvo C (2010) Design, Democracy and Agonistic Pluralism. In: Durling D, Bousbaci R, Chen L, Gauthier P, Poldma T, Roworth-Stokes S, Stolterman E (eds) Design and complexity—DRS international conference 2010. Design Research Society, Montreal, pp 7–9
9. Manzini E (2008) Design para a inovação social e sustentabilidade: comunidades criativas, organizações colaborativas e novas redes projetuais. E-papers, Rio de Janeiro
10. Steen M (2008) The fragility of Human-centered design. Delft University, Amsterdam
11. Sbordone M (2008) Human-centered design: Sustainable ideas and scenarios for the development of projects and products based on knowledge and human abilities. Revista Internacional de Sostenibilidad, Tecnología y Humanismo, 127–157
12. Krippendorff K (2011) The diagonal reality of meaning. Am J Semiot. EE.UU
13. Council D (2007) Eleven lessons: Managing design in eleven global companies-desk research report. Design Council
14. Staszowski E, Brown S, Winter B (2013) Reflections on designing for social innovation in the public sector: a case study in New York City. In: Manzini E, Staszowski E (eds) Public and collaborative: exploring the intersection of design, social innovation and public policy. DESIS Network, pp 27–34
15. Janzer CL, Weinstein LS (2014) Social design and neocolonialism. Des Cult 6(3):327–343
16. Vitorino V (2018) A competência em informação e a vulnerabilidade: construindo sentidos à temática da vulnerabilidade em informação. Ciência da Informação, 47(2)
17. Baudrillard J (1990) A transparência do mal: ensaios sobre fenômenos externos. Papirus, Rio de Janeiro
18. Eco U (1993) Apocalípticos e integrados. Editora Perspectiva, São Paulo
19. European Parliament. Atualidade. Como combater a desinformação. https://www.europarl.europa.eu/news/pt/headlines/priorities/desinformacao. Last accessed 1 May 2022
20. Bonsiepe G (2006) Design and democracy. Design issues, 22(2):27-34
21. Miolo de Oliveira CM, Almendra RA, Leitão T, Lourenço AR (2023) EU4ALL: For a strategic communication on the democratic values highlighted by the European parliament to groups vulnerable to information. In: Martins N, Brandão D (eds) Advances in design and digital communication III. DIGICOM 2022. Springer Series in design and innovation, vol 27. Springer, Cham, pp 345–356
22. Manzini E (2003) Scenarios of sustainable wellbeing. Design Philosophy Papers 1(1):5–21
23. Bittencourt GB, Freire KM (2021) The usage of scenarios in the search for innovation regarding the model of rare disease associations: an empirical experience and its findings. Strat Des Res J 14(3):508–515
24. Santos A, Chaves LI, Santos AS, Mazzieiro AT, Cavalcante ALB, Pazmino AVPM, Vilela APX, Cavalcanti ALS, Gomés CRP, Lima FLN, Prado G, Murphy GCR, Perez IU, Ribeiro JP, Castillo L, Schmitz M, Alves MC, Mourão NM, Lepre PR, Engler RC, Cavalcanti T (2019) Design para a sustentabilidade: dimensão social. Insight, Curitiba

Words of Woundedness. Designing for Behavior Change Through the Power of Vulnerability

Renata Maia Arezes, Joana Quental, Anabela Pereira, and Raquel Guimarães

Abstract Vulnerability has become a key concept in modern society, relating to all dimensions of human existence, and the main determinants of health. Vulnerability is subject-dependent and so a subjective condition deeply rooted in selfhood. Addressing vulnerability means talking about resistance, resilience, and strength, essential characteristics of adaptation, the main requisite for survival. Affecting both physical and emotional bodies, chronic disease is a main cause of vulnerability. Any comprehensive effort to understand and treat illness should take a holistic view of health, thinking in wholeness and interconnectedness terms, and focusing on the interactions between mind, body, and behavior. Design impacts people's lives and influences human behavior. New approaches to promote and facilitate behavior change can be created through design for behavior change. This work seeks to understand the relevance of vulnerability as an analytical frame of reference and a multidisciplinary concept and find its contributive potential to designing for behavioral change in chronic health. As a symbolic interaction mechanism, stories can play an important role in strengthening and building resilience, regaining trust, and favoring forgiveness. Through design for behavior change, stories with a higher level of resonance can be created, thus more success-prone for person engagement and behavior change.

R. M. Arezes (✉)
DeCA, University of Aveiro, Campus Universitário de Santiago, 3810-193 Aveiro, Portugal
e-mail: arezesrenata@gmail.com

Media and Culture (ID+), DeCA, Research Institute for Design, University of Aveiro, Campus Universitário de Santiago, 3810-193 Aveiro, Portugal

J. Quental
DeCA, Department of Communication and Art, University of Aveiro, Campus Universitário de Santiago, 3810-193 Aveiro, Portugal

A. Pereira
DEP, Department of Education and Psychology, University of Aveiro, Campus Universitário de Santiago, 3810-193 Aveiro, Portugal

R. Guimarães
São João Hospital Breast Centre, Alameda Prof. Hernâni Monteiro, 4200-319 Porto, Portugal

N. Martins et al. (eds.), *Perspectives on Design and Digital Communication IV*, Springer Series in Design and Innovation 33, https://doi.org/10.1007/978-3-031-41770-2_10

Keywords Design for behavior change · Vulnerability · Storytelling · Chronic disease · Cancer

1 Introduction

Is one to die voluntarily or to hope in spite of everything? [1][1]

Vulnerability is a modern society's Zeitgeist [2]. In the days we are living, vulnerability is becoming a mainstream term, widely used across all sectors of society and human life dimensions, both as a transitory or permanent state or condition. Now seen as a relevant object of study, researchers from the most diverse fields of knowledge (sociology, psychology, psychotherapy, biology, genetics, economy, political and economic sciences, and justice, among others), are digging deeper into the concept of vulnerability. These are times of uncertainty and ambiguity (Bauman [3]), where everything seems possible and yet non-permanent—even selfhood, a self-reflexive and multiple constructs. However, this fluidity and emphasis on change may bring along new opportunities for adaptation and survival. This work seeks to understand the relevance of vulnerability as an analytical frame of reference and a multidisciplinary concept, along with its function as a human condition, and find its contributive potential to design for behavioral change in chronic health.

2 Vulnerability

The word vulnerability comes from the Latin *vulnus* [4, p. 3736], which means 'wound'. As a state or condition, vulnerability indicates the possibility of being physically or emotionally wounded, attacked, defeated, harmed, or injured. When referring to the individual, vulnerability pertains to the idea of fragility, weakness, helplessness, defenselessness, and susceptibility to being mistreated, taken advantage of, or physically or emotionally exploited.

The concept of vulnerability is extensive, being widely used nowadays across all sectors of modern society and applied to every human dimension. This does not come as a surprise since its complexity and multidimensionality make it a transversal multidisciplinary concept. Largely used in social, political, and economical discourses, this is a well-embedded notion in people's lives and is widely disseminated by mass media and social media. More than a descriptive word, vulnerability facilitates critical evaluation and understanding and can expand comprehension of changing relationships between opposite parts, enhancing knowledge and understanding of the individual's

[1] Camus (1995) The Myth of Sisyphus and Other Essays. Knopf Doubleday Publishing Group, New York, p. 26.

experience by favoring reflection around equality, justice, and resilience, either individually or collectively [2]. This suggests vulnerability might have an important role in quality-of-life improvement through its inner and outer relevance and significance, both within the individual, his experience, or his context.

Studies on vulnerability propose different forms of categorization, based on each author's main research field and work dimensions.

Based on the most recent definition of health [5], this study searches for a holistic approach to chronic health. On this, Misztal's sociological model [2] offers a relevant perspective. According to this author, vulnerability may be divided into three forms, each defined through different interdependent relationships made by the person over three dimensions of human life: social (the subject and the others: people/ family, community, society); biological (the subject and his capacity to respond, survive and evolve); and personal (the subject and his actions in time, responsibility, understanding, and forgiveness).

The first form is *dependence on others*, entrenched in the core dependency of human beings, and stated in terms of value and trust [6, p. 235]. If those are affected by other person choices or actions, to whom we have limited or inexistent control, vulnerability will emerge [7, p. 143] accordingly with the individual degree of dependence and/or independence on another. Dependency is a universal experience, as human beings are social creatures and by that physically and emotionally mutually dependent on one another "to fulfill their humanity' (Phillips and Taylor, 2009, as cited in Misztal [2]).[2] In modern society the level of interdependence is continuously increasing, not only between people and social-political institutions but also between individuals, as this new era of social networks and connections feeds from vulnerability but is able, at the same time, of leveraging ethical decisions and responsibility taking [8], from one towards the other or the community. This form of vulnerability is particularly relevant to this study as selfhood and personhood are constructed and defined upon dependency on others [9], through comparison and reflection, and social connection enhances individual access to material resources, information, and emotional support [10].

The predicament of unpredictability is the second form of vulnerability proposed by Misztal through what Arendt calls an "infinite improbability" of action (Arendt, 1958, as cited in Misztal [2]),[3] rooted in the human being's uncertainty about the future. This is a core form of vulnerability since it pertains to the physical dimension of existence of the self and, ultimately, the whole species, dependent on one's or the group's capacity to act, respond, evolve, and survive, accordingly to the biological imprint and the social construct. As so, it deeply relates to value, either measured by learning experiences (comprising failure and/or success) or attained success (focus on achievement) through fixed or progressive perspectives of evaluation [11]. Deeply

[2] Phillips, Taylor (2009) On Kindness. London: Hamish Hamilton. In Misztal [2] The Challenges of Vulnerability: In Search of Strategies for a Less Vulnerable Social Life. Palgrave Macmillan, Basingstoke.

[3] Arendt (1958) The Human Condition. Chicago: University of Chicago Press p. 245. In Misztal. B. (2011) The Challenges of Vulnerability: In Search of Strategies for a Less Vulnerable Social Life. Palgrave Macmillan, Basingstoke p. 75.

connected with the instinct for survival, a fundamental driver for change that is many times accompanied by fear, a primal emotion, or with enthusiasm, an action-driving emotion, this form of vulnerability is particularly relevant for this work for linking vulnerability to change.

The third form of vulnerability, *the predicament of irreversibility*, is based on past traumas, sufferings, and wrongdoings imprisonment, from which a person cannot release himself (Arendt, 1958, as cited in Misztal [2]).[4] It's the irreversibility of past actions and experiences, which weighs down the individual, tied to fear, repentance, rage, worthiness, bitterness, shame, or other emotions or feelings that limit his capability for self-protection, connection, and cooperation. This happens within the subject, even though it can assume a collective dimension, spreading to the community or even the entire humanity, and evolves around concepts like responsibility, understanding, redemption, and forgiveness, the key to freedom and human hood retrieve.

Misztal proposes different remedies for each form of vulnerability based on trust-related mechanisms since vulnerability is a precondition to trust [2, p. 116]—responsibility, promise, and forgiveness. Responsibility implies strong, affective, or bonding trust, in the present time. Promise allows for the building and establishment of trust, working towards the future. Forgiveness enables reframing the past and is future-oriented as it sees the past not as a model, but as a warning for the present. Forgiveness allows for renewal, restoration, justice, and the re-establishment of balance.

The three forms of vulnerability may exist independently or co-related to one or both the other ones and all have a different relation with the dimension of time—present, future, and past, respectively, as each remedy—, a fundamental vector of structure and construction in human life and all its dimensions.

Concerning health, Mackenzie et al. [12] recognize three different aspects of vulnerability, that transversally relate to Misztal's sociological model.

The first aspect proposes that being human implies inescapable vulnerability since the biological and social nature of human beings makes them dependent at the core of their constitution as "embodied individuals with needs and emotional dependencies" [13, p. 2]. The second refers to the circumstances and the contextual affairs of life that may be responsible for situational vulnerability, which may come as physical, emotional, or both. Finally, the third one addresses the political and institutional aspect of vulnerability, supported when a policy impedes or excludes already vulnerable people from a specific activity.

Mackenzie's work defends re-conceptualizing vulnerability as an ontological condition of our embodied humanity linked to autonomy and dissociates the concept from negative connotations of victimhood, helplessness, neediness, and pathology. To conceive of vulnerability and autonomy as oppositional fosters authoritarian and coercive forms of intervention to protect those identified as vulnerable [14].

---

[4] Arendt (1958) The Human Condition. Chicago: University of Chicago Press p 238. In Misztal [2] The Challenges of Vulnerability: In Search of Strategies for a Less Vulnerable Social Life. Palgrave Macmillan, Basingstoke p.75.

These two models identify vulnerability as a core condition in any person-centric study. Allowing an understanding of the challenge that vulnerability represents in human beings' life, at the same time Misztal and Mackenzie's work show us the potential it can offer to behavior change through responsibility, promise and forgiveness, 'corrections' that will favor the state of autonomy through newly gained or restored empowerment and control.

In the context of health, talking about illness and vulnerability means addressing stress and potential stressors, as illness represents a challenge and a menace to quality of life. Health-related stressors for individuals can be genetic or biochemical, disease-specific, or physical, as well as psychosocial [15, 16]. Since vulnerability is a condition that differs between each person, accordingly to their own physical, mental, and emotional health, along with their social, cultural, financial, and political context, the impact on each individual and his response when confronted with challenges will be both unpredictable and diverse, according to personal *generalized resistance resources* [17]. "Perhaps this is related to the amount of control that they feel over their situation. The less control one has over a situation, the more vulnerable one feels" [18]. The variables that influence vulnerability to stress-related pathology and stress resilience depend on interindividual differences in stress responses that result from multiple factors like age, sex, personality type, genetic influences, and/or previous exposure to stressful experiences. These, among others, contribute to the psychological, emotional, cognitive, and biological condition of the individual that will determine or influence his response.

When addressing illness or dealing with stressors, talking about vulnerability implies naming resilience also, as a positive response to challenge through some type of change. Vulnerability "typically refers to the susceptibility of a person, group, or even a whole society to disturbances that develop in response to particular environmental or social challenges". In contrast, resilience is "the ability or disposition to recover readily from illness, but in the context we're dealing with, it can also be taken to mean that an individual has the capacity to limit or preclude the detrimental effects of a stressor" [19, p. 13].

Relating to chronic health, this study aims to find the possibility of tackling any form of vulnerability through the intervention of design, before, during, or after stressors' exposure, so that resilience may be built as a response either by limiting the damaging effects of a stressor or by applying responsibility, promise and/or forgiveness to reframe the given experience.

3 Chronic Health and Vulnerability

The World Health Organization (WHO) defines health as "a resource for everyday life, not the objective of living" [5]. How well a person lives depends on their health, relating to physical capacities and social and personal resources, on which participation in society relies. Health is "an ecological understanding of the interaction between individuals and their social and physical environment" [20, p. 39]. Living is interacting. Through interaction, human beings experience action and mutual influence. Acknowledging the physical, mental, social, and spiritual dimensions through which human beings experience health and disease [6–8], the patient is assigned an active and participative part in the center of health care, encouraging empowerment and patients' autonomy and capability for action [21, 22].

In this century, developed countries are beginning to feel the burden represented by the care and management of chronic conditions known as chronic health. WHO defines chronic or long-term conditions (non-communicable diseases) as health problems not mainly caused by an acute infection (from person to person), that are of long duration (over a period of years or decades) and slow progression, requiring ongoing management (treatment and care) [23]. Cancer is identified by WHO as one of the four main types of chronic diseases [24]. With the progressive increase of the older population in developed countries comes to a consequent raise in the number of people with at least one long-term disease. Facing this "wicked problem" [25], a "socially embedded, difficult or impossible to solve" problem, "where the information is hill-formulated and confusing" (Churchman, 1967, as cited in Prendiville [26],[5] governments are focusing their actions on public health around chronic health, working around prevention, diagnosis, care, management, and support of people with chronic disease.

Any comprehensive effort to understand and treat illness should take a holistic view of health, thinking in terms of wholeness and interconnectedness, and focusing on the interactions between mind, body, and behavior [27 p. 174]. Through this holistic approach, and when addressing chronic health, individual and social vulnerability should be integrated and carefully considered, since long-term conditions make a person more vulnerable or vulnerable-prone. According to Miksch, the continuous between health and disease could be explained from a salutogenic perspective by regarding vulnerability as a coefficient between both [28].

Dahlgreen and Whitehead [29] present a complex model of health determinants that is still in use to this day, indicating that discrepancies, imbalances, and shortages in individual aspects (e. g. genetical or biological), individual lifestyle and behavior factors (e. g. physical activity, health habits), social and community networks (e. g. family, friends, neighborhood), and living and working conditions (socio-economic, cultural, environmental conditions) contribute to health-related vulnerability.

Vulnerability is subject-dependent, and so a subjective condition. Bengel et al. propose that even the concepts of 'disease' and 'health' are both subjective notions, developed in every individual during socialization within a particular social context and climate [30]. Illness and disease are generally related to pain, weakness, limitations, impairment, dependence, frailty, and fear. On the opposite, health may be defined as well-being, happiness, coping mechanisms, energy, vitality, pleasure, peace, and autonomy. The level of perceived 'illness' or 'health' is also influenced by social or individual judgments or appraisals. The perception of health is also

[5] Churchman (1967) Wicked problems. Management Science, 14(4):B141—2 In Prendiville, A. (2017) Communication design in chronic health. In Tsekleves, E., & Cooper, R. (Eds.). (2017). Design for Health (1st ed.). Routledge. [p. 204].

addressed by Antonovsky's work, through what he names *Sense of Coherence*, determined as "a global orientation that expresses the extent to which one has a pervasive, enduring though dynamic feeling of confidence that (1) the stimuli from one's internal and external environments in the course of living are structured, predictable, and explicable; (2) the resources are available to one to meet the demands posed by these stimuli; and (3) these demands are challenges, worthy of investment and engagement" [31]. Formed by three components, comprehensibility (the ability to perceive information as structured, ordered, and consistent even during extreme life events), manageability (a profound belief in coping mechanisms, either from internal or external resources), and meaningfulness (belief in the meaning of challenges, worthy of engagement and emotional and cognitive energy investment), the Sense of Coherence and health status are deeply related. Someone that possesses a solid Sense of Coherence is capable of activating the necessary resources to manage a given challenge and might even benefit from positive experiences within his own life [32].

Health and well-being reflect the physical and psychological dimensions of the human being. Holistic approaches that integrate both dimensions can positively work towards resource balance and consequent resilience achievement, mindfully handling health-related vulnerability.

"Health can be activated through the empowerment of individuals" [27, p. 204]. These theories are particularly relevant when looking for opportunities to address the impact of threatening factors on individual health, opening the possibility of intervention on preconceived models and concepts towards change and resilience-building mechanisms.

4 Design for Behavior Change

According to Buchanan, design may be applied to any human experience and its scope is potentially universal [33]. Design is fast adapting to deal with the accelerated rise of complex and systemic challenges presented today, through rethinking, reconceiving, and reframing its own nature. The progression of design's role in chronic health clearly shows how this 'rethinking' is taking place over time, illustrating the gradual evolution stated in Buchanan's *Four Orders of Design* [34], from graphic and product design to interaction design[6] and immaterial systems design.[7] This question about design in an increasingly complex and intangible world is also approached similarly by Krippendorf's *trajectory of artificiality*. Krippendorf suggests a transition in the practice of design from products to immaterial and abstract manifestations of design so that users can find meaning in material or social artifacts, larger communities

[6] The third order of design focuses on interaction design and how "humans relate to each other through the mediating influence of products, emphasizing that the products are more than physical objects". In Buchanan [33] Design Research and the New Learning. Design Issues 17:3–23 [p. 11].

[7] The fourth order of design focuses on immaterial systems, "human systems, the integrations of information, physical artifacts, and interactions in the environments of living, working and playing". Buchanan [33] Design Research and the New Learning. Design Issues 17:3–23 [p. 12].

may be aided and society can be supported in the process of self-reconstruction in unseen ways [34]. In a resembling way, Jones [35] proposes four levels of design's involvement in healthcare, from design 1.0 (traditional design), 2.0 (product/service design), 3.0 (organizational transformative design), to 4.0 (social transformative design), illustrating design's evolution to different levels of complexity that reflects design's engagement in chronic health. All three models show design's progressive involvement, undertaking greater challenges with each level and increasing contribution to wider social-cultural transformative processes, acting from the individual towards the collective.

Design is all around people and it is all about people. Design is closely related to both behavior and change. By transforming existing situations into preferred ones [36, p. 111], design has an impact on people's lives and can influence human behaviour—any material or social artifact created has a function (either physical, symbolic, or social) that implies an action by the user [37], and thus capable of creating change by restraining or enabling behaviors [38]. Behavior design, one of the four traditional design disciplines, incorporates theories of behavioral change from psychology, sociology, and neuroscience, among others, to focus on how human behavior can be shaped or influenced by design. More recently, design for behavior change has emerged as a new field of research [39–46], seeking to understand the role that design can have as transformative change dominant's figure [47] in promoting and facilitating behavior change, through attempting to *understand* and *influence* user behavior.

Design for behavior change involves a multidisciplinary approach and borrows models from psychology and behavioral science and economics, like the Theory of Planned Behavior Model [48], the Health Belief Model [49, 50], or the Stages of Change/Trans-Theoretical Model [51, 52], among others. Even though there is no 'right' model, all the different cross-domain and cross-disciplinary concepts bring useful insights to design for behavior change research and practice.

Although there is no accepted unified model of human behavior and it is never possible to accurately predict people's behavior, as they may act contrary to the intended or expected by designers, it is useful to have users' mental models that will allow for the construction of actions and to explain the results of a particular action [53]. Daae and Boks explore the missing link between understanding the user and knowing how design may affect behavior, and propose nine dimensions in which designers can operate for deeper insight and understanding—Control, Obstructiveness, Direction, Empathy, Meaning, Encouragement, Importance, Timing, and Exposure [54]. Understanding user behavior is made through research methods used in other design disciplines that focus on the person and the person's behavior, such as focus groups, prototyping, and surveys (contextual inquiries, interviews), and methodologies such as co-design, participatory design, and design ethnography.

The shift from patient-centered to person-centric healthcare is gradually happening [55–58]. Design for health should address the challenge and opportunity of designing for a person, rather than a patient, aiming to improve quality of life, besides health. In behavior design, this means focusing both on the person's internal and external behavior, considering their needs, concerns, interests, and motivations, and seeking to reduce or remove the hurdles and barriers that prevent health-promoting behaviors. Person-centric design requires a holistic approach to designing for holistic healthcare, addressing all the challenges of well-being and mental health, along with social interaction.

Human behavior is based on decision-making, either conscious or unconscious. Addressing individual behavior for propagating change is then understanding and trying to influence the decision-making process. Design for behavior change mainly applies three strategies—*enabling* (facilitating the user's intended behavior), *motivating* (trying to convince the user of performing/not performing a certain behavior), and *constraining* (hampering the user's not desired behavior) [59]. Within these, different cross-disciplinary strategies and tools may be used accordingly to the nuances of each project and the intended target, addressing either cognition (the person), the context (the environment), or both. A holistic view should be applied in domains such as health, where decisions, attitudes, and practices influence and are influenced by both the individual and the environment. Thus, design for behavior change should equally focus on designing for responsible decision-making of the user (motivation), promoting accountable reflection. Niedderer [60–62] suggests mindful design as a promising method through which design can work along emotions on different levels of content, choice, and complexity. By raising the user's attention and awareness through disrupting their consciousness (disruption) and re-directing attention (thematization) to conscious-making decisions and problem-solving, the mindful design method may lead to change. The importance given to emotions in this method is particularly relevant in cases of looming cognitive style and looming vulnerability [63].

This means design has the potential and the possibility to wider its action into designing for raising awareness and challenging stigma, prejudice, and preconceived attitudes towards healthcare, including the concepts of health and chronic disease, rather than focusing solely on the immediate and tangible needs of a person living with a long-term condition.

5 Storytelling

Our world is made of multidimensional stories' multiplicity. Narratives are a universal source of knowledge, insight, and awareness, where we look for example, inspiration, guidance, and reassurance. Our brain is a "story processor" (Haidt, 2012, as cited in Storr [64]),[8] constantly working to deconstruct, decompose and analyze the appearing random, chaotic, and unexpected events of daily life, trying to find order

[8] Haidt (2012) The Righteous Mind. Allen Lane p. 281. In Storr [64] The Science of Storytelling. Abrams, New York [p.14].

and meaning amid change and challenging, or disruptive experiences. "Seeking an understanding or explanation for worrisome physical or emotional symptoms is a fundamental human need" [65, p.173]. Stories bring definition, meaning, purpose, and a sense of belonging. In healthcare, this may be of enormous value, as stated in Frankl's words: "In some ways, suffering ceases to be suffering at the moment it finds a meaning, such as the meaning of a sacrifice" [66].

Since we live immersed and embedded in stories, they have the power to shape our individual development and our proximal and distal context—also a factor of "impact [on] the individual's experience of health, disease, and the illness experience" [20, p. 107].

From ancient times storytelling has been a valuable resource for individual and collective survival, evolving through different eras and adapting to technological advances and challenges as a byproduct of change [64, 67, 68]. Through stories we survive. As a common or individual legacy, stories connect us to past, present, and future times, defying finitude and oblivion, and projecting us towards innovation—we survive time. As a source of meaning and knowledge, stories inspire action and behavior—we survive challenges. As a way of emotional connection, stories tell us about others as much as ourselves—we learn. As boundless organisms, stories allow us to navigate more easily between the physical, psychological, spiritual, and moral worlds, inviting for deeper awareness—we live the multidimensionality of our existence.

The structural importance stories have in human beings' life suggests storytelling may be a useful tool in promoting coping and empowerment in chronic disease patients. Storytelling is included in health management techniques in Patient-Centered Care (PCC)—psychotherapy uses narrative therapy, for optimizing coping [69–71]. In storytelling, the main role belongs to the person. Stories may help patients to find purpose, reconstruct a solid sense of will, reclaim control, ownership, and courage, and reconnect with others. By recognizing patients' distress, suffering, and difficult situation, dignity is restored, and hope reborns [72, 73]. "Story's gift is the hope that we might not be quite so alone, in that dark bone vault, after all" [64, p. 196].

Requiring patient activation, PCC implies self-care as the person manages disease through their own life's management. Patients become (co)responsible for their health maintenance, meaning patients must adopt behaviors, not acting. Requiring capability for action or patient agency, this relies primarily on personal and biological autonomy,[9] and empowerment,[10] which combined create competence—the fundamental freedom and skills to successfully take action—, and self-efficacy [22].[11] When confronted with difficult situations, people with self-efficacy persist in problem

[9] Autonomy: "patients' capacity and freedom to make health care decisions." In Ibid. [p. 192].

[10] Empowerment: "activating patients on their own behalf so they can better meet their own goals." In Ibid. [p. 192].

[11] Self-efficacy: "The belief that one is capable of performing the behaviors required to produce a desired outcome." In Sullivan [22] The Patient as Agent of Health and Health Care: Autonomy in Patient-Centered Care for Chronic Conditions. Oxford University Press p. 194.

resolution by being able to use available resources and coping strategies [74]. Bandura suggests patients "are agents of experience rather than simply undergoers of experience." (Brandura, 1977, as cited in Sullivan [22])[12] This change in perspective introduces patients as self-reflexive beings capable of adjusting plans and actions according to their sense of self, through built-in morals and desired efficacy. "Everything can be taken from a man but one thing: the last of the human freedoms—to choose one's attitude in any given set of circumstances, to choose one's own way" [66].

To Sullivan "health is experienced by socially connected and functioning biological individuals" [22, p. 382] that take upon concrete opportunities created by social resources to exercise agency. Stories play a fundamental role as a resourceful tool to inspire patients to take action and shape their lives in a personally meaningful way. Through resonance, stories can be a resourceful tool in encouraging, developing, and sustaining health behaviors and people/patient action [9].

The continuous increase of chronic illness in developed countries and the consequent strain caused on healthcare delivery [75] favors the search for PCC approaches that promote patient behavior with both health and illness' positive effects.

In cases of cognitive vulnerabilities, defined as "faulty beliefs, long-previously developed cognitive patterns, or structures that are hypothesized to set the stage for later psychological problems" which need addressing for maintaining long-term therapeutic improvements [76, p. 8], storytelling can be an effective tool on promoting behavior change. In the same way, for people with a looming vulnerability style (the degree to which people tend to interpret, represent, and imagine ambiguous threats as rapidly and dynamically growing, and by so being more vulnerable to anxiety), that theoretically complements anxiety familiar and other belief-based constructs [63, p. 97], the usage of storytelling towards behavior change may be a successful tool for addressing and strengthening their Sense of Coherence. This can be made by using tools from cognitive behaviors psychotherapies through cognition-emotion interaction [77, p. 16], among others. This is of particular importance when designing for chronic health since anxiety about cancer (and other long-term diseases) may be influenced by people's inflated perceptions of the rapid spread and progression of disease [78, p. 16], and exaggerated perceptions of rapid gains in diseases that, along with hot cognition (cognition related to emotion) [77, p. 15], can potentially affect the course of disease and impact health outcomes and the quality of patient's lives.

This represents a wide variety of opportunities for design for behavior change. Designing artifacts that could plausibly increase patients' adherence to medical recommendations in cancer treatments, but also working towards reducing their perceptions of the dynamic looming threat and their fear, or in encouraging the implementation of therapeutical strategies like forgiveness [2] to increase personal strength and psychological and physiological well-being [2, p. 201], could be some of them, as a way to implement behavior change towards patients' health outcomes and quality of life improvement.

---

[12] Bandura (1977) Social Learning Theory. In Ibid. [pp. 194–195].

Storytelling, as a *performative narrative* [79], can be a powerful and effective tool when addressing health and behavior change, as shown by Weg [80], Brewster [71], Ferro [81], Robertson, Clegg and Huntley [82], Herman [70], Hardy and Summer [67], Allen and Krebs [83] or Hammel [69]. Yet, in the traditional approach, storytelling doesn't achieve its full potential due to its clinical feel [64, p. 16], since they are built focusing on structure. We believe designers may have a fundamental role in targeting stories to their fullest potential. Using research methods such as focus groups, prototyping, and surveys (contextual inquiries, interviews) and through methodologies such as co-design, participatory design, and design ethnography, designers can observe with *empathic resonance* and have access to the patient's perspective on the experience of disease, illness, sickness, health, healing, and the wholeness that allows the creation of an *archive of perception* [9]. This will allow them to create true and rich characters, and to build stories that will focus on the character rather than the events [64, p. 16], and thus with a higher level of resonance.

6 Conclusion

Vulnerability is an inevitable state of the human condition due to our biological and social nature. However, vulnerability is also a result of social constructs through acquired values about physical safety, economic welfare, autonomy, and psychological well-being, along with political independence and territorial integrity, among others [84], that shape the individual sense of security, defined by a "low probability of damage to acquired values" [85]. This is of particular importance since vulnerability should be addressed in the context of reflexive modernity. As an intrinsic characteristic of any social unit or system and a result of modernization, vulnerability is gradually becoming a global phenomenon in modern society, which in turn contributes to its progress through unsafe conditions in social, political, and economic systems.

The Aristotelian model of strength of character, strength, and vulnerability are both sides of a good character. Understanding that human beings can be individually and collectively vulnerable and yet collectively strong through intrinsically caring about others may be the root of the psychology that fits them for survival [86], restores dignity and favors the absurd. "Living is keeping the absurd alive [1]."[13] This is particularly relevant in chronic health, since the difference between life and death may be dependent on the person's beliefs, choices, and willingness to change.

Stories and storytelling can play an important part in strengthening and building resilience, regaining trust, and favoring forgiveness, working as a symbolic interaction mechanism for attaining conscious identity through the symbolic representation

[13] Camus (1995) The Myth of Sisyphus and Other Essays. Knopf Doubleday Publishing Group, New York, p. 70.

of the other [87], either the individual or the collective. Human beings are mutually dependent and thus vulnerable to each other's actions [88], not only existing in relation to others but also being shaped by the history of such relationships [89].

This represents an opportunity for future multidisciplinary work between designers and health professionals in designing stories for behavior change. Listening with emphatic resonance, the designer has access to the patient's perspective on the experience of disease, illness, sickness, health, healing, and the wholeness that allows the creation of an *archive of perception* [9]. This allows on having precious material and symbolic elements to be used in creating character-focused stories addressing comprehensibility, manageability, and meaningfulness with a high level of resonance and thus more prone to succeed towards the person's engagement and transformation. In tales about vulnerability and strength, words of wounds can be a valuable resource in behavioral change toward health and contribute to developing patients' coping mechanisms and empowerment.

Acknowledgements This work is financed by national funds through the FCT – Fundação para a Ciência e a Tecnologia, I.P., under the scope of the project SFRH/BD/144177/2019.

References

1. Camus A (1995) The Myth of Sisyphus and Other Essays. Knopf Doubleday Publishing Group, New York
2. Misztal B (2011) The challenges of vulnerability: In search of strategies for a less vulnerable social life. Palgrave Macmillan UK
3. Bauman Z (2013) Liquid modernity. Polity Press
4. Instituto António Houaiss (2001) Dicionário Houaiss da Língua Portuguesa. Círculo de Leitores, Rio de Janeiro
5. Constitution of the World Health Organization. https://www.who.int/about/governance/constitution. Accessed 27 Jun 2022
6. Baier A (1986) Trust and antitrust. Ethics 96:231–260
7. Goodin RE (1985) Protecting the vulnerable: a reanalysis of our social responsibilities. Hastings Cent Rep 15:46
8. Bauman Z (2007) Liquid times: Living in an age of uncertainty. Wiley
9. Arezes R, Quental J, Pereira A, Guimarães R (2023) Embroidering behaviour. The power of storytelling in patient activation. In: Martins N, Brandão D (eds) Advances in design and digital communication III. Springer Nature Switzerland, Cham, pp 435–444
10. Cornwell EY, Waite LJ (2009) Social disconnectedness, perceived isolation, and health among older adults. J Health Soc Behav 50:31–48. https://doi.org/10.1177/002214650905000103
11. Dweck CS (2007) Mindset: The New Psychology of Success. Random House Publishing Group
12. Mackenzie C, Rogers W, Dodds S, et al (2013) What is vulnerability, and why does it matter for moral theory? In: Vulnerability: new essays in ethics and feminist philosophy. Oxford University Press, p 0
13. Matthews S, Tobin B (2016) Human vulnerability in medical contexts. Theor Med Bioeth 37:1–7. https://doi.org/10.1007/s11017-016-9357-9
14. Mackenzie C (2013) 33The importance of relational autonomy and capabilities for an ethics of vulnerability. In: Mackenzie C, Rogers W, Dodds S (eds) Vulnerability: New essays in ethics and feminist philosophy. Oxford University Press, p 0

15. Schneiderman N, Ironson G, Siegel SD (2005) Stress and health: psychological, behavioral, and biological determinants. Annu Rev Clin Psychol 1:607–628. https://doi.org/10.1146/annurev.clinpsy.1.102803.144141
16. Besser A, Flett GL, Hewitt PL (2010) Silencing the self and personality vulnerabilities associated with depression. In: Jack DC, Ali A (eds) Silencing the self across cultures: depression and gender in the social world. Oxford University Press, p 0
17. Antonovsky A (1979) Health, stress, and coping. Jossey-Bass Publishers, San Francisco
18. Rogers AC (1997) Vulnerability, health and health care. J Adv Nurs 26:65–72. https://doi.org/10.1046/j.1365-2648.1997.1997026065.x
19. Anisman H (2015) Stress and your health: From vulnerability to resilience. Wiley & Sons, Ltd
20. Belle Brown J, Stewart M, Weston WW et al (2012) Exploring Health, Disease and the Illness Experience. In: Belle Brown J, Stewart T, Stewart M (eds) Challenges and solutions: narratives of patient-centered care. Radcliffe Publishing, United Kingdom, pp 35–55
21. Mullaney T, Pettersson H, Nyholm T, Stolterman E (2012) Thinking beyond the cure: A case for human-centered design in cancer care. Int J Des 6(3):2012
22. Sullivan M (2017) The Patient as agent of health and health care: autonomy in patient-centered care for chronic conditions. Oxford University Press
23. World Health Organization. Noncommunicable Diseases and Mental Health Cluster (2002) Innovative care for chronic conditions: building blocks for actions: global report. Des soins novateurs pour les affections chroniques: éléments constitutifs: rapport mondial
24. World Health Organization (2014) Global status report on noncommunicable diseases 2014. World Health Organization, Geneva
25. Rittel HWJ, Webber MM (1973) Dilemmas in a general theory of planning. Policy Sci 4:155–169
26. Prendiville A (2017) Communication design in chronic health. In: Tsekleves E, Cooper R (Eds) (2017) Design for Health (1st ed). Routledge, p 204
27. Miksch A (2009) Exploring Vulnerability. In: Kabat-Zinn J, Hanh TN (eds) Full catastrophe living: using the wisdom of your body and mind to face stress, pain, and illness. Random House Publishing Group, pp 207–214
28. Wisner B, Adams J (2002) Environmental health in emergencies and disasters: a practical guide
29. Dahlgren G, Whitehead M (1991) Policies and strategies to promote social equity in health. Background document to WHO—Strategy paper for Europe. Institute for Futures Studies, Arbetsrapport 14
30. Bengel J, Strittmatter R, Willmann H (1999) What keeps people healthy? the current state of discussion and the relevance of antonovsky's salutogenic model of health. Federal Centre for Health Education
31. Antonovsky A (1987) Unraveling the mystery of health: how people manage stress and stay well. Wiley
32. Eriksson M, Lindström B (2006) Antonovsky's sense of coherence scale and the relation with health: A systematic review. J Epidemiol Community Health 60:376–381. https://doi.org/10.1136/jech.2005.041616
33. Buchanan R (2001) Design research and the new learning. Des Issues 17:3–23. https://doi.org/10.1162/07479360152681056
34. (2007) The semantic turn: A new foundation for design. Artifact 1:56–59. https://doi.org/10.1080/17493460600844157
35. Jones PH (2013) Design for care: Innovating healthcare experience. Rosenfeld Media
36. Simon HA (1996) The sciences of the artificial, 3rd edn. MIT Press
37. Ligo LLR (1984) The concept of function in twentieth-century architectural criticism. UMI Research Press
38. Norman DA (1988) The psychology of everyday things. Basic Books
39. Daae J, Chamberlin L, Boks C (2018) Dimensions of behaviour change in the context of designing for a circular economy. Des J 21:521–541. https://doi.org/10.1080/14606925.2018.1468003

40. Strömberg H, Selvefors A, Renström S (2015) Mapping out the design opportunities: pathways of sustainable behaviour. Int J Sustain Eng 8:163–172. https://doi.org/10.1080/19397038.2014.1001469
41. Niedderer K, Mackrill J, Clune S, et al (2014) Joining Forces: Investigating the influence of design for behavior change on sustainable innovation
42. Wendel S (2020) Designing for behavior change: applying psychology and behavioral economics. O'Reilly Media
43. Daae JZ, Boks C (2014) Dimensions of behaviour change. J of Design Research 12:145
44. Wever R, van Kuijk J, Boks C (2008) User-centred design for sustainable behaviour. Int J Sustain Eng 1:9–20. https://doi.org/10.1080/19397030802166205
45. Bhamra T, Lilley D, Tang T (2011) Design for Sustainable behaviour: using products to change consumer behaviour. Des J 14:427–445. https://doi.org/10.2752/175630611X13091688930453
46. Lilley D (2009) Design for sustainable behaviour: strategies and perceptions. Des Stud 30:704–720. https://doi.org/10.1016/j.destud.2009.05.001
47. Suchman L (2011) Anthropological relocations and the limits of design. Annu Rev Anthropol 40:1–18. https://doi.org/10.1146/annurev.anthro.041608.105640
48. Ajzen I (1991) The theory of planned behavior. Organ Behav Hum Decis Process 50:179–211. https://doi.org/10.1016/0749-5978(91)90020-T
49. Rosenstock IM, Strecher VJ, Becker MH (1988) Social learning theory and the health belief model. Health Educ Q 15:175–183. https://doi.org/10.1177/109019818801500203
50. (2016) Theoretical foundations of health education and health promotion. Jones & Bartlett Learning
51. Prochaska JO, DiClemente CC (1983) Stages and processes of self-change of smoking: Toward an integrative model of change. J Consult Clin Psychol 51:390–395. https://doi.org/10.1037/0022-006X.51.3.390
52. Prochaska JO, DiClemente CC, Norcross JC (1993) In search of how people change: applications to addictive behaviors. J Addict Nurs 5
53. Carroll JM, Olson JR (1988) Chapter 2—Mental models in human-computer interaction 11 This chapter appeared in its entirety and is reprinted from mental models in human computer interaction: Research Issues about What the User of Software Knows, J.M. Carroll and J.R. Olson, Editors,-The report of the workshop on software human factors: Users mental models, Nancy Anderson, chair, sponsored by the Committee on Human Factors, Commission on Behavioral and Social Sciences and Education, National Research Council, published by the National Academy Press, 1987. In: HELANDER M (ed) Handbook of Human-Computer Interaction. North-Holland, Amsterdam, pp 45–65
54. Daae J, Boks C (2017) Tweaking interaction through understanding the user. In: Niedderer K, Ckune S (eds) Design for behaviour change: theories and practices of designing for change. Routledge, London, pp xix, 277
55. Hummers-Pradier E, Beyer M, Chevallier P, et al (2010) The research agenda for general practice/family medicine and primary health care in Europe. Part 1. Background and methodology1. Eur J Gen Pract 15
56. Milson-Hawke S, McMillan M, Peek C et al (2007) Towards innovation: the development of a person-centred model of care for older people in acute care. Contemp Nurse: J Aust Nurs Prof 26:164
57. Eaton S (2016) Delivering person-centred care in long-term conditions. Future Hosp J 3:128. https://doi.org/10.7861/futurehosp.3-2-128
58. (2014) Person-centred care made simple What everyone should know about person-centred care. Health Foundation, London
59. Lockton D (2017) Design, behaviour change and the Design with Intent toolkit: Theories and practices of designing for change. pp 58–73
60. Niedderer K (2007) Mapping the meaning of knowledge in design research. Des Res Q
61. Niedderer K (2013) Mindful design as a driver for social behaviour change

62. Niedderer K, Clune S, Ludden G (2017) Design for behaviour change: theories and practices of designing for change
63. Riskind JH, Rector NA (2019) Looming vulnerability: theory, research and practice in anxiety. Springer New York
64. Storr W (2021) The science of storytelling. Abrams, New York
65. Oben P (2020) Understanding the patient experience: a conceptual framework. J Patient Exp 7:906–910. https://doi.org/10.1177/2374373520951672
66. Frankl VE (Viktor E 1905–1997 author (1962) Man's search for meaning : an introduction to logotherapy. Beacon Press
67. Hardy P, Sumner T (2018) Cultivating compassion: how digital storytelling is transforming healthcare
68. Fullford R (2011) The triumph of narrative: Storytelling in the age of mass culture. House of Anansi Press Inc., Toronto
69. Hammel S (2018) Handbook of therapeutic storytelling: stories and metaphors in psychotherapy, child and family therapy, medical treatment, coaching and supervision
70. Herman D (2013) Storytelling and the sciences of mind. MIT Press
71. Brewster A, Zimmerman R, Fernandopulle R, Young R (2022) The healing power of storytelling: using personal narrative to navigate illness, trauma, and loss. North Atlantic Books
72. Lloyd-Williams M, Reeve J, Kissane D (2008) Distress in palliative care patients: Developing patient-centred approaches to clinical management. Eur J Cancer 44:1133–1138. https://doi.org/10.1016/j.ejca.2008.02.032
73. Spaulding AE (2011) The art of storytelling: telling truths through telling stories. Scarecrow Press, Incorporated
74. Pereira MG (2016) Caleidoscópio de Olhares em Psicologia Oncológica: Teoria e Intervenção
75. Wilson EV (2008) Patient-centered E-health. Med Inf Sci Ref
76. Mathews AM, MacLeod C (2005) Cognitive vulnerability to emotional disorders. Annu Rev Clin Psychol 1:167–195
77. Tiba A (2020) Embodied hot cognitive vulnerability to emotional disorders. From Theory Treat. https://doi.org/10.1007/978-3-030-53989-4
78. Levin TT, Riskind JH, Li Y (2007) Looming threat-processing style in a cancer cohort. Gen Hosp Psychiatry 29:32–38. https://doi.org/10.1016/j.genhosppsych.2006.10.005
79. Langellier KM, Peterson EE (2004) Storytelling in daily life. Temple University Press
80. Weg AH (2011) OCD Treatment through storytelling: a strategy for successful therapy. Oxford University Press, USA
81. Ferro A (2006) Psychoanalysis as therapy and storytelling. Routledge
82. Robertson C, Clegg G, Huntley J (2023) Storytelling in medicine: how narrative can improve practice. CRC Press
83. Allen R, Krebs N (2006) Dramatic psychological storytelling, using the expressive arts and psychotheatrics
84. Baldwin DA (1997) The concept of security. Rev Int Stud 23:5–26. https://doi.org/10.1017/S0260210597000053
85. Forbes-Mewett H (2018) The new security : individual, community and cultural experiences. Palgrave Macmillan, London
86. Badhwar N, Harris G (2001) Dignity and Vulnerability: Strength and Quality of Character. Philos Phenomenol Res 63. https://doi.org/10.2307/3071100
87. Mead GH (1929) The nature of the past. In: Coss J (ed) Essays in honor of John Dewey. Henry Holt, New York, pp 235–242
88. Molm L, Schaefer D, Collett JL (2009) Fragile and Resilient Trust: Risk and Uncertainty in Negotiated and Reciprocal Exchange*. Sociological Theory 27. https://doi.org/10.1111/j.1467-9558.2009.00336.x
89. Elias N (2001) Society of individuals. Bloomsbury Publishing

Branding Design

TechnoFashion2Sustainability. Digital Tools, Modular Strategies and Participatory Design Approach to Brand Engagement and Customized Solutions

Catarina Marquês, Ana Margarida Ferreira, and Fernando Oliveira

Abstract Participatory design practices allow users to understand and experience products more deeply. How, what, and why emerge as meaningful questions to answer collaboratively, strengthening the link between creators, companies, and consumers. As one of its results, companies have been adapting to the consumer market by allowing users to participate in the design of their own products. In this chapter, as an ongoing research project, two case studies are analyzed, helping to understand how companies promote the consumer's enrolment in the design, and their engagement with the creative part of future products and services through technology. Using a qualitative methodology, we seek to understand how modular design allows consumers to create more innovative, creative, fun-allocated pieces and how it could work as a design strategy for more sustainable marketable solutions and user behaviors. The main conclusion shows us a new creative sphere where the consumer has a critical role in developing future things emotionally bonded.

Keywords Design · Modularity · Technology · Innovation

C. Marquês (✉) · A. M. Ferreira · F. Oliveira
UNIDCOM/IADE, Unidade de Investigação em Design e Comunicação, Av. D. Carlos I, 4, 1200-649 Lisbon, Portugal
e-mail: catarinateixeiramarques@gmail.com

F. Oliveira
Faculdade de Design, Tecnologia e Comunicação, Universidade Europeia, IADE, Av. D. Carlos I, 4, 1200-649 Lisbon, Portugal

A. M. Ferreira
Faculdade de Artes e Letras, Universidade da Beira Interior, Rua Marquês d'Ávila e Bolama, 6201-001 Covilhã, Portugal

Labcom - Comunicação e Artes, Universidade da Beira Interior, Rua Marquês d'Ávila e Bolama, 6201-001 Covilhã, Portugal

N. Martins et al. (eds.), *Perspectives on Design and Digital Communication IV*, Springer Series in Design and Innovation 33, https://doi.org/10.1007/978-3-031-41770-2_11

1 Methodology

This chapter is based, as the paper that it follows (Modular Design and Technology for Diversity and a More Sustainable Fashion. The RTFKT X Nike and CLO Case Studies), in literary review and in qualitative research (Nonmetric Data) [24, 28] and on cases studies [8, 14], as a way to comprehend the connection between design by modules and technology, and how they can be engaged to the brands and the selected case studies, based in our own perception [8], after the study of the information of the cases, we will analyze how those brands proceeded to be able to offer solutions to the consumer, and consequently, to the market in general.

To realize the approach of the market, we picked the union between RTFKT and Nike, an elaborate new approach taken by a world-renowned fashion brand, and by a technological program, more properly, RTFKT, that joined the Metaverse technology to generate a unique and fantastic avant-garde fashion product, and the CLO software, a means to create techno-fashion products in 2D and 3D, without the waste of brand new resources, that permits innovation of fashion products/clothes. These cases reach a particular level, because they bring design and technology together, touching the design by modules, and allowing to have an alternative approach, with the capacity to create more design possibilities in a much more enjoyable way, as already said. For us to be capable to do their study, in order to have a full and better understanding of their performance in these actual society's business model, and also their future approaches. As so, a multicase [36] research and analysis was conducted, and a qualitative approach [33], as said before, creating the means in which those companies are acting, and how are they creating some impressions on our community.

2 Modular Design

Design by modules is a design systems form that is based on an easy way of decoding systems: modularity means that when we subdivides them, it is considered well-constructed when any change in a certain component doesn't affect another one in any way [10]. Thus, it is assumed that through an anticipation of the future, allowed through modularity [46], it is possible to significantly improve the quality of products in terms of a number of aspects, among which: a faster update in their appearance/design, a reduction in costs, greater innovation—factors that contribute to a very favorable environmental impact [43]—and also, and above all, a greater diversity and/or variety of products [35].

In order to a product be perfectly designed in a marriage between modular design and sustainability, it must be based, according to [43], on 4 phases. These are the pre-production phase, where the prototype(s) is or are designed, and where the best way to produce it is studied. The goal is to have the greatest number of variations with the fewest possible resources expended from the initial (prototyping) phase. One of the basic concepts is the increase in the interchange of components and sub-components of products, which by being modular, and preferably, if not ideally, of the same material, substantially increase the life cycle of the product in question and reduce both the number of materials and components used and the disposal of the product and its materials; the production phase, that is the phase that the author indicates that the reduction of times, from setup to unnecessary transportation, are important focuses of the impact of modulation. An extremely important aspect to achieve the goal of modular design system is the way of communicating to the user how to use the object in order to get the most out of it, and also the possibility of exploring other senses besides the visual (e.g. touch), due to modular creation, such as the LEGO brand [40, 43, 45]. The third phase is the use phase, where solutions are implemented that aim, as previously mentioned, to solve aspects related both to the product itself and to the user's daily life, who, as he goes through changes caused by the most varied factors, finds in a modular product an alternative that by its adaptability—functional, aesthetic, symbolic—becomes capable of solving a more varied "range" of problems. Finally, the phase of (ideally) reuse, or disposal. Here, the "modular" factor has much of its protagonism, because it is in this stage that one can (re)use some component (life cycle extension) or discard only certain components, instead of the whole piece. It is also possible to separate these components, if they are made of different materials, as happens in some articles, which helps at the recycling level.

There are countless projects that are globally known, but that we sometimes fail to see the concept, the core of them. For example, LEGO, already mentioned above, and Tangram.

LEGO, as you can see in Fig. 1, is an example of how modular design system works, allowing the user to design anything he wants, starting with just a single module. These modules are used by kids and grown-ups alike, serving for many different forms of artistic expression—fashion, sculpture, interior design, architecture, among others [17]. The famous LEGO's resulted from a series of events that, despite appearing to be a certain tragedy, would result in one of the most iconic modular brands on the globe. The creator of the brand lost his business and turned it into a brand that has been passing from generation to generation since the mid-1950s until our days [2, 26, 54].

Another example is the tangram (Fig. 2), which is a puzzle that is based in geometric shapes with seven pieces that are called "tans". The goal is to use all the pieces without overlapping them, in order to form the most varied figures (thousands of possibilities). There are some legends that revolve around this puzzle, however, and the message that is retained is that creativity and the amount of possibilities go beyond the limits that most think they have. Therefore, it is an element used in the study of geometry, as a facilitator of the understanding of geometric shapes, a creative and logical thinking fosterer [34, 55]. Tangram's puzzle has already been used by AnRealAge (Fig. 3) as inspiration to create a collection called "Block", in 2020 [19, 48, 49].

Modular design, in fashion, is already used, as said before. It is a system, and the expected is, that it could be even more revolutionary in creativity, economy, sustainability an so on. There are some designs already used in fashion that are modular besides their "camouflage", for example, pants that can be converted in

Fig. 1 LEGO piece. *Source* 27 Gen (2016)

shorts, jackets that have detachable sleeves and without them, they are vests, puffers that can be shortened and the part left become a tote bags (Fig. 4) [53], etc. They are easily constructed and deconstructed through several types of connecting mechanisms such as Velcro, zippers, buttons, magnets, among many others. The principal purpose is to manage to change the structure of the object according to its purpose, or the to the look craved by its user, allowing the pieces to have the definition of "multifunctional", "convertible" or "adaptable" [52].

3 Technology

Recently, companies around the globe, and in all areas are improving all sectors due to technology. The fashion area is not an exception, and have also presented lots of technological solutions that, at the level of the creative process and to be a support for the sustainable fashion's field [9]. Those solutions can show in the environmental issue by the reduction of textile waste, by making some changes in clothes characteristics, such as its fabrics, colors, ornaments, among others features [42].

Fig. 2 Tangram and its figures. *Source* Escola Kids [34]

Digital transformation has caused the entire production machine to speed up—doing more with less impetus—and has enabled startups to achieve goals not previously possible. It was such an impactful change that large companies began to move towards finding strategic solutions through technological advances (Fig. 5).

Since design is a systematic way of seeing the world, of bridging the gap between a need and a practical way to meet it, because it is an area where form serves function, whether by aesthetics, ergonomics, or process simplicity, when included in a business strategy, it can be a powerful source of insights for new paths to success in the future.

Design Thinking can help by allying the two concepts, by solving problems and creating more practical products. Thus, the union of design with technology may be the way to make the company a reference in a "new market", focused on digital and volatile transformation. "Design thinking" is not about a methodology, but a way to approach issues in the work routine that would normally be ignored because they are too "out of the box". The 4 main steps for its ideal realization are immersion, which arises when the professional tries to understand a problem—either in the public demand or in internal processes; the analysis and synthesis, when this problem is interpreted and synthesized in a central issue that needs to be attacked; the ideation,

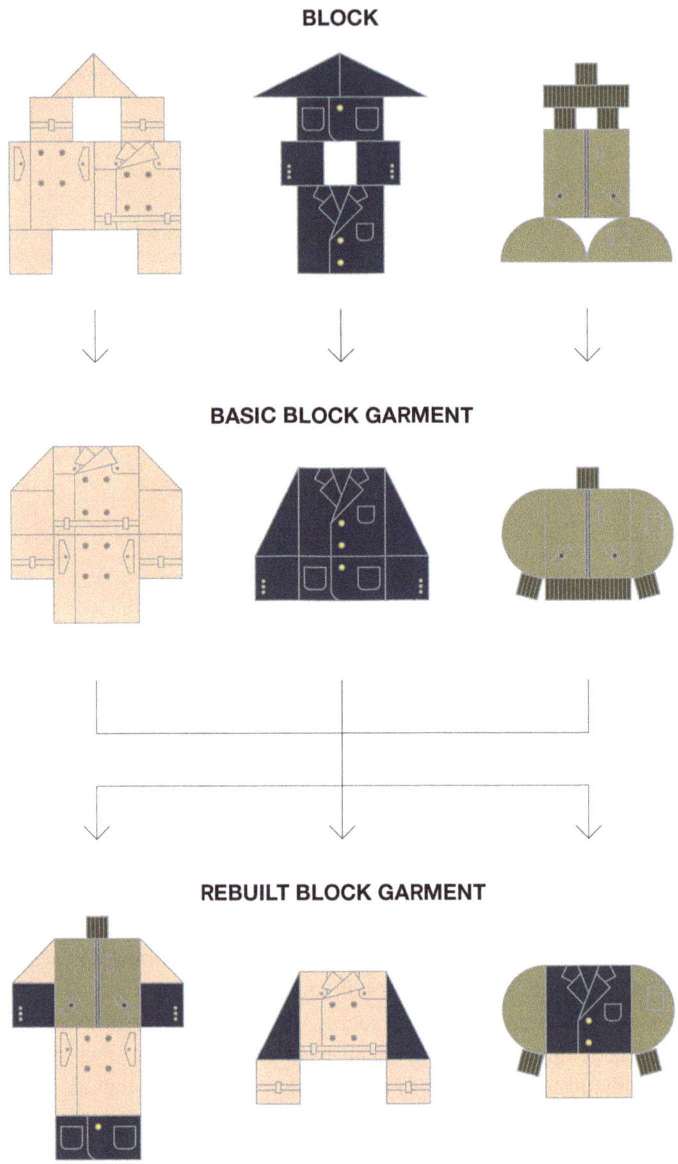

Fig. 3 "Block" by AnRealAge. *Source* Humble [19]

when the professional or team in charge raises all the available technological tools and proposes ways of interaction between them that create new disruptive outputs for the problem; and the prototyping, when the idea goes to the testing, validation and iteration stage until it reaches the expected result [31, 38, 44].

Fig. 4 Hey Nineteen Twenty's Jacket. *Source* NineteenTwenty [32]

Fig. 5 Virtual closet. *Source* 3DLOOK [1]

In short, the union of technology and design speeds up development processes, problem solving, and several other issues in various areas. We can describe some of them as, for example: new solutions for old problems, that is, as what moves technology in the corporate world is to find faster and more efficient ways to perform tasks and deliver more quality to the client, and what moves design is to get out of the common sense and discover more optimized ways that lead to the solution of a problem, by joining both concepts, the company gains new tools, such as software, processes, and services that can give the necessary technological advantage to beat the competition. This approach to problems can even result in the emergence of totally disruptive business models; an increase in data security, because, there are difficulties in dealing with data security, in an environment threatened every day by new forms of attacks, invasions and leaks, what IT departments need, to be one step ahead of schemes that corrupt the internal security of companies and businesses, is exactly to create innovative solutions that they cannot yet foresee, that is, to think about technology in new ways that bring new solutions and functionalities that do not exist yet; new ways of working, because, despite not being a methodology, Design Thinking is the inspiration for several productive models that contribute to the agility and quality of the work inside the companies, so that they can constantly use the structure of immersion, analysis, ideation and prototyping to constantly iterate on their products and services, and obtain innovative solutions for their clients cost reduction, which, by obtaining more agile processes, technologies shaped to fit the function, and creative solutions to persistent problems, leads to greater savings of time and money in the operation and management of the business; and finally, a new digital culture, because this should be a motivator for a business that hopes to see itself consolidated in the future, in order to have alignment between design and technology, since this will be the center of change in thinking in all sectors of various companies. In short, Design Thinking has a lot (or everything) to do with the search for innovation. This is the central goal when it comes to combining technology and design [31, 38, 44].

Technology can help to clarify certain problems, whether through Artificial Intelligence or 3D printing, along with others. Consequently, technology events allied themselves to fashion and clothing (and will for sure continue to be allied), where topics such as those mentioned above are addressed. As well as Fashion [16] said, it will be taken into consideration, from creation to digital and ecological transition, the digital process relative to the consumer, not forgetting the strategic adoption relative to social networks, the integration between e-commerce and consequent (traditional) distribution, and the potential relative to blockchain and Metaverse.

Once again, we can bring a LEGO's case back to show you something. Not a so avant-gard show, as we are already used to see nowadays,—as coperni's show, for example, when the brand started a performance, and joined fashion with technology, to create a spray-painted dress (Fig. 6), "made of Fabrican—a liquid fiber bound together with polymers, biopolymers, and greener solvents that evaporates when the spray makes contact with a surface" [20], that has been directly applied in the models body, in the runway [22]—happened in 2008, when Jean Charles de Castelbajac in partnership with LEGO and Agence Four H created a runway show, as seen in Fig. 7,—SPRING/SUMMER 3001by JC de Castelbajac—only with LEGO's as sets, audience and models. This show was broadcasted during Paris Fashion Week on French channels [27, 37]. Another example has happen this year, more properly, during Paris Fashion Week, when AnRealAge presented us a white collection. And the most amazing thing was that all the clothes, "made of *faux* fur, velvet, lace, knit, jacquard, and satin, suddenly become colored with bright pinks, blues and greens, in a multitude of patterns thanks to "experimental, photochromic materials" that the

Fig. 6 Coperni's spray dress. *Source* The New York Times [47]

brand made itself" [20]. Those clothes where scanned by an UV light that permitted the color to appear, thanks to the technology that the brand has been incorporating into his designs since its inception [20].

4 Case Studies

4.1 RTFKT X Nike

RTFKT ("artifact"), in 2021, was bought by the Nike. It (RTFKT) is a virtual fashion platform, known due to its capacity to create fashion virtual products—digital avatars and **metaverse** collectibles—[21, 41], in this case, sneakers [6, 18, 39]. RTFKT was created because their fathers (Benoit Pagotto, Chris Le and Steven Vasilev [30]) realized that in games that they used to play, such as Counter-Strike, League of Legends, among many others [30], the gamers buy skins to be used in games as a representation of what they want for themselves. Most of these players are fans of the sneaker's culture, in real life. Then, Pagotto, Le and Vasilev saw an opportunity, have put their ideas together, and developed RTFKT [3]. RTFKT was inspired by different cultures, but all deeply involved in areas like gaming (for sure), fashion, luxury and streetwear. The RTFKT creators understood that there have been an

Fig. 7 LEGO + Jean-Charles de Castelbajac's toy runway. *Source* TrendHunter [5]

evolution, because, years before, all was related with music culture and sports, but with the advancement of time and technology, people started to be more involved with gaming and all around that [30], so they turned their ideas for that and started to work.

Even before the "partnership" with Nike, RTFKT developed an avatar collection composed of 20000 "clone" avatars. This collection was launched in 2021 and it is the biggest and the most remarkable part of the brand. RTFKT started their NFT projects before its acquisition by Nike in December 2021, but in middle 2022, RTFKT was set up as an independent brand, like Converse or Jordan. RTFKT maintains full control of its own brand, including the creative direction and NFT's roadmap, but they continue to exchange knowledge and resources with Nike, and both brands works on its own projects. And, besides that, "RTFKT released the commercial rights for its clones, meaning that NFT holders will be able to take control of the clones, create brands and earn revenue from their creations" [55].

In 2019, Nike, that had some partnership with Roblox (a games platform) to create some virtual characters, and worked with the Fortnite company to put one of their most epical model's—Jordan—accessible to be a skin for players [6].

Some fresh news from the partnership between those brands, Nike and RTFKT, was the launch of a futuristic line of NFT sneakers, more properly, a collection named "RTFKT X Nike CryptoKicks NFT: a RTFKT X NIKE DUNK GENESIS CRYPTOKICKS Sneaker Powered by DRM OS and Skin Vial Tech". It is made by a range of 200000 pairs of NFT's sneakers for the metaverse, which can be used

Fig. 8 RTFKT X Nike Cryptokicks iRL. *Source* Sneaker News [23]

on a platform called "NikeLand", and also as a skin/filter in cellphone apps, such as Snapchap [15]. The Cryptokicks iRL (Fig. 8), the futuristic line, it's a match between Nike's sneakers innovation and RTFKT's vision to merge both digital and physical worlds [21].

This aggregation of sneakers has the peculiarity that the models do not have the aesthetic part, more exactly, the fabrics/non-wovens, the colors or/and the patterns, basically, all parts that customize the models. Just a limited number of sneakers are able to be sold with this singularity, the others will be personalized after, by their owners, with compatible products (skins) that are also available in the metaverse [25]. The change makers "tools" are called "Skin Vials" and they consist of a bunch of NFT's that are placed on the sneakers and that will give them the appearance contained in that Vial, as we can see in Fig. 9. These Vials are sold in terms of rarity and the rarer they are, the more expensive they will be [51]. With the union with Nike, RTFKT Studio is able to bring these virtual designs (NFT's) into the real world by releasing the first collection of these physical Cryptokicks—The RTFKT x Nike Cryptokicks iRL (Fig. 8). Besides all the modular features presented before, these real-life Cryptokicks present a number of impressive other features too, and "each pair have a wireless Power Deck that can connect to the Cryptokicks iRL app via Bluetooth, for example. The sneakers are equipped with RTFKT's latest WM (World Merging) NFC (Near Field Communication) chip that enables users to authenticate each physical copy with its digital counterpart to verify legitimacy" [21], for example, among many other awesome features.

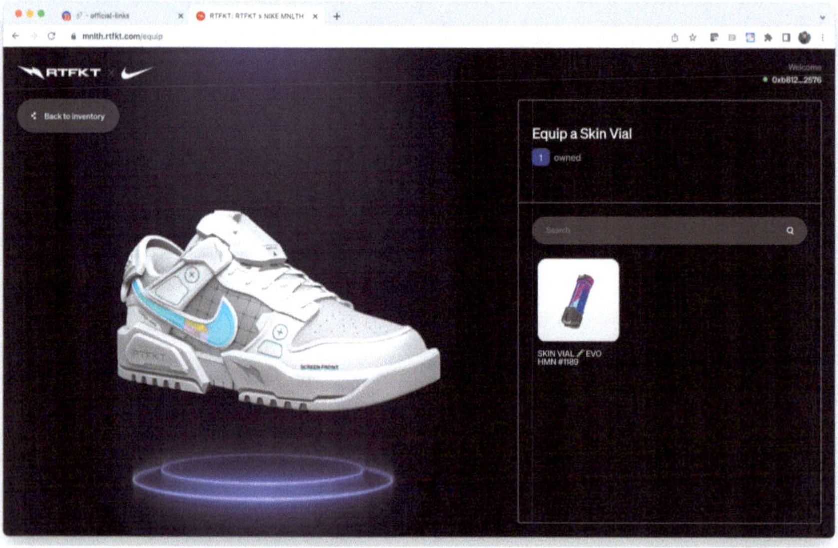

Fig. 9 Nike Dunk genesis. *Source* It's nice that [15]

The RTFKT creators, in "RTFKT is the Creator-led Studio bringing Art, Fashion & Nike into the Metaverse" [30] said an amazing thing, that can, for sure, kind of represents the engagement of the user in the design process of a product: "Speaking to the founders brings with it a rush of energy, optimism, and the expertise to create not only a new type of brand, but to imagine the new spaces where that brand will exist. "We actually forbid the word 'consumer.' And sometimes we need to tell Nike, 'No, no, we don't use that word,'" says Pagotto. 'Because when you understand that it's just people consuming what you decide to sell them, and not active collectors, artists, and co-creators, you completely kill the way you think of your brand.'"

These example shows us how technology, design systems, modular design and fashion can have an impressive marriage, and how they can create new ways to evolve and to involve the users. Basically, this is an example of how, in the future, industries and ecosystems will work, and how that work will create "new templates for brand-building in the digital age" [30]. These sneakers, and that kind of technology, allows to join fashion and modular design, even when we can't perceive it. It's playful, fun and a way to stay creative.

4.2 CLO

This program, CLO, is a 3D apparel design software that is used by fashion designers (and others), by small companies and/or businesses, and by other investors in the fashion industry too, in order to achieve an attractive digital business flow [12] and, according to Masterkey [29], "CLO is the world's most preferred 3D Fashion Design Software CAD Program creating virtual, true-to life garment visualization with cutting-edge simulation technologies for the Fashion, Clothing, Garment Business and Textile industries".

CLO, was created as a result of a doctoral thesis in the area of Computer Engineering. His creator initially developed a prototype of a program that permitted his colleagues to create 3D clothes, and, because it was so fun to use, he make it public. After that, one of its first users, a cosplay artist, understood that the program had the capacity to be revolutionary. CLO quickly became very popular, and then, another program have been created: "Marvelous Designer". This one is specific for virtual cosplay artists, and, as it usually happens, made a path for cracked versions to appear, which generated some concern about its viability. Weta Digital, that was one of Clo's first customers, became aware of Marvelous Designer because of Computer Graphics (CG) community, as many other companies, who became Clo's partners subsequently. After all that, and as a change brought about by users of the program, CLO began to have a more user-focused mindset [7, 12].

After, the program growth in the GC's community, another specific program was developed, but now, to create designs and solutions for the fashion industry, with all the solutions that professionals in this area needs, to facilitate the process from creation to production, allowing to simplify all the development of the garments from the very beginning [7]. Both CLO and Marvelous Designer have grown over the past 10 years and they have been developed everyday a little bit more due to the engagement with the user, and as a result, CLO-SET [11], another program related with both two, was developed. This one allows a 3D simulation of clothing created by himself. The team responsible for the creation of these programs continues to work and develop new projects, always being focused on the user and on the industry [7, 12].

Clo3D has been adopted as a tool in classes, and has been proved to be essential, both for learning and for teaching. Kang, as assistant Professor, said that Clo3D it's an "innovative technology" they're using at Parsons that is a very helpful tool to students because they "can see their designs in real time and see the variations (...) can do virtual pattern making, virtual sewing, and virtual rendering" [50]. They say in the article [50] that CLO and it's 3D capabilities, along with other design features, make it a digital tool unlike any other used by students before. Basically, before, students could only use software's that can make some works in 2D formats, and with Clo3D they can experience a new dimension and, with that, be much more creative, as we can see in Fig. 10, for example.

Kang and Sammaritano, in "CLO 3D: A Partner on Designing for the Future" [11] said that they "believe Clo3D is the future of fashion: Along with its usability in person or online, it also provides a more sustainable way to design. "Clo3D saves on resources and time," said Sammaritano. "You can do a lot more with a lot less waste." Kang agrees, adding, "I don't need a dress form, a sewing machine, paper, or material to teach. All I need is the Internet, a computer, and the software, and all students need is their sketchbook and the concept of what they want to create"'.

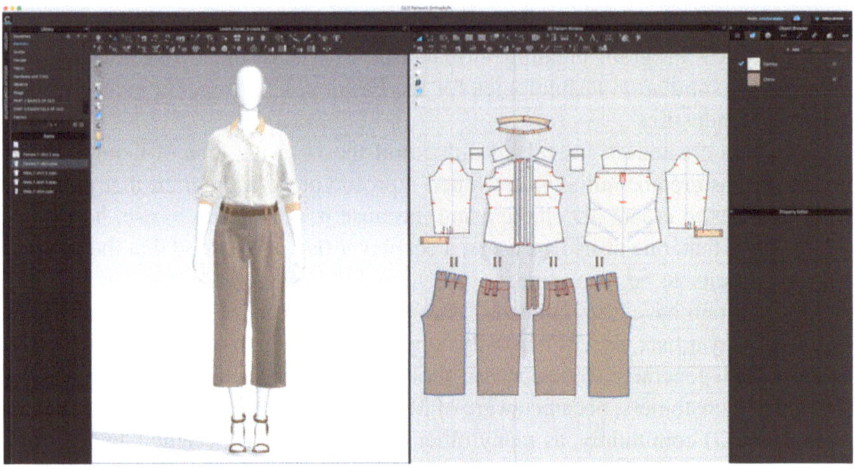

Fig. 10 Drafting 2D pattern and 3D prototype in CLO3D. *Source* TECHPACKER [4]

There are many things that makes CLO one of the most amazing software's of the fashion industry. Basically, according to Masterkey [29], CLO's software allows us to use 2D Pattern and 3D Simulation windows on one screen, it is compatible with all 2D Pattern CAD programs, it is possible to export existing patterns in DXF-ASTM/AAMA format to CLO, virtual male, female, child avatars and dummies are available default within the software and can be realistically scaled, and we can also use our own virtual dummy and avatar. It is possible to make Virtual Fitting, Garment Fit Map, Check Fit, Stress Map with real body measurements, it accurately simulates drape-sensitive fabrics, including Pantone Color Palette default, it is possible to work in unlimited colorways and variants (Fig. 11). Clo is compatible with Adobe Photoshop and Illustrator. Allows to create styles with countless layers and intricate details, to design a variety of garments, from a simple blouse to technical outerwear with complicated pattern pieces and construction, create anything constructed with materials, fabric, leather and fur vb. including hats, bags, wallets, lingerie, swimsuits, industrial workwear and more, Virtual Fashion Shows, Runways, Animations, Image-Video Turntable and High Quality Render features come with CLO by default, it has unlimited virtual accessory options (button, trim, zipper, hardware, etc.), unlimited sewing options too, 3D Teck Pack and BOM (Bill of Materials) feature, a Virtual fabric library consisting of the most used fabrics in the industry and unlimited creation of our own virtual digital fabric library with the CLO Fabric Kit. It is possible to visualize immediately how many variations as we want of the clothes, no matter how they look—simple or the more eccentric they may be. And besides all those amazing features that CLO allows us to use, the use of these software helps to reduce the "sample delivery time from 1 month to 27 h, and you can increase the sample adoption rate from 15 to 55% or more on average" [29].

Fig. 11 CLO3D. *Source* CLO [13]

5 Conclusion

To better understand what we do want to perceive, we first saw how modular design and technology can work by themselves and how they can work together, and together with fashion, in order to create products that can please us and to create solutions to a better tomorrow. We can never forget that besides the fact that we are growing old, we all have an inner child, and we need to feed her and to motivate her to stay with us, and that's why we need to see through things that brings us joy.

As so, through the analysis of what modularity and technology are, and how they can create a different perspective to various communities and in our lives, and based on the cases studies, more properly, on RTFKT X NIKE and CLO, it was possible to see how our world and the interest of people are constantly being changed and how they can adapt to newness. The use of design by modules, as already more than presented (to change objects and to create diversity with fun and joy), and technology, can be used to bridge the gap between information and the creation of a more combinations range. RTFKT and Nike, through their new collection, have made a connection between fashion design and modular design, using technology, making it feasible to alternate certain features of sneakers. CLO, conversely, possibilities to create fashion design pieces/clothes, using the creative part, giving the user total freedom to create and to change whatever and whenever he wants to. Thus, we conclude that by joining a modular approach in design and a technological approach can be a strong bond and a strong contribution for the future, allowing to create solutions focused in practicality, diversity and sustainability, while making the experience enjoyable and fun.

References

1. 3DLOOK (n.d.) Sustainable fashion technology for eco-friendly production in 2022. https://3dlook.me/content-hub/7-sustainable-fashion-technologies/. Accessed 28 Mar 2023
2. Adams B (2016) Consider the LEGO brick… |. https://27gen.com/2016/04/04/consider-the-lego-brick/
3. Alsever J (2021) RTFKT is selling NFT cybersneakers to the metaverse—Utah Business. https://www.utahbusiness.com/rtfkt-made-millions-off-of-nft-cybersneakers/
4. Baukh O (2022) What is 3D fashion design and how does it impact the future of fashion? https://techpacker.com/blog/design/what-is-3d-fashion-design/
5. Bayas LS (2008) LEGO fashion shows: Jean-Charles de Castelbajac's Toy runway. https://www.trendhunter.com/trends/lego-fashion-shows-jean-charles-de-castelbajacs-lego-runway
6. BoF (2022) Nike acquires virtual fashion start-up RTFKT. https://www.businessoffashion.com/articles/technology/nike-acquires-virtual-fashion-start-up-rtfkt/
7. BOF TEAM (2020) At CLO virtual fashion, digitising the design process to drive transformation | BoF. https://www.businessoffashion.com/articles/technology/at-clo-virtual-fashion-digitising-the-design-process-to-drive-transformation/
8. Brown J (2020) Literature review qualitative research methods and assumptions. https://www.linkedin.com/pulse/literature-review-qualitative-research-methods-jeff-brown-mba/
9. Camuri M (n.d.) Unlocking AI's potential to transform fashion
10. Catel F, Monateri J (2006) Strategic perspectives on modularity to cite this version: HAL Id : halshs-00097733
11. CLO-SET (n.d.) CLO-SET. https://style.clo-set.com/aboutus. Accessed 25 Mar 2022
12. CLO (n.d.-a) CLO | 3D fashion design software. https://www.clo3d.com/en/company/about-clo. Accessed 10 May 2022
13. CLO. (n.d.-b). CLO | 3D fashion design software. https://www.clo3d.com/en/clo/features. Accessed 29 Mar 2023
14. Davies M (2011) Case studies: research methods. In: Study skills for international postgraduate students. https://doi.org/10.1007/978-0-230-34553-9
15. Davis G (2022) Nike Dunk genesis cryptokicks now available! Here's how to purchase these NFT sneakers | Tech Times. https://www.techtimes.com/articles/274797/20220427/nike-dunk-genesis-cryptokicks-now-available-heres-purchase-nft-sneakers.htm
16. Fashion Network and Muret D (2022) Da Kenzo à Gucci, o luxo rende-se aos encantos do Metaverso. https://pt.fashionnetwork.com/news/Da-kenzo-a-gucci-o-luxo-rende-se-aos-encantos-do-metaverso,1378071.html
17. Foong H (2015) LEGO HEAD. http://harngsays.com/2015/01/11/lego-head/
18. Hernandez J (2021) NIKE, Inc. Acquires RTFKT, Preps For Metaverse. https://sneakernews.com/2021/12/13/nike-rtfkt-acquisition/
19. Humble Magazine (2020) Anrealage A/W 2020–21 Collection "Block"—Humble magazine. https://www.humblemagazine.com/artfashion/anrealage-aw-2020-21
20. Issawi D (2023) Remember the spray-on dress? This season, another magic show. https://www.thecut.com/2023/03/remember-the-spray-on-dress-this-season-another-magic-show.html
21. Katsikopoulou M (2022) Nike brings RTFKT's sneaker NFTs to life. https://www.designboom.com/design/nike-rtfkt-sneaker-nft-real-life-cryptokicks-irl-12-06-2022/
22. Kessler A (2022) Bella Hadid Has Her Coperni Slip dress spray-painted on live on the runway. https://www.vogue.co.uk/fashion/article/coperni-ss23
23. Le M (2022) RTFKT Nike cryptokicks iRL release date | SneakerNews.com. https://sneakernews.com/2022/12/07/nike-cryptokicks-irl-release-date/
24. Leavy P (2017) Research design—Quantitative, qualitative, mixed methods, arts-based, and comunity-based participatory research approaches. The Guilfoord Press
25. Ledger Insights (2022) Nike, RTFKT launch CryptoKicks NFT customizable sneaker collection—Ledger insights—Enterprise blockchain. https://www.ledgerinsights.com/nike-rtfkt-launch-cryptokicks-nft-customizable-sneaker-collection/

26. *LEGO* (n.d.). http://www.lego.com/. Accessed 14 Mar 2023
27. Luxuo (2008) JC de Castelbajac × Lego Virtual Fashion Show. https://www.luxuo.com/style/fashion/castelbajac-lego-virtual-fashion-show.html
28. Marczyk G, DeMatteo D, Festinger D (2005) Essentials of research design and methodology. Wiley. https://doi.org/10.18488/journal.10/2017.6.3/10.3.48.57
29. Masterkey (n.d.) The most Robust 3D fashion design software CAD program in the world. https://masterkey.com.tr/en/masterkey/clo/. Accessed 21 Mar 2023
30. Maughan P (n.d.) RTFKT is the creator-led studio bringing nike into the metaverse. https://www.highsnobiety.com/p/rtfkt-interview/. Accessed 22 Mar 2023
31. Mootee I (2013) Design thinking for strategic innovation. Wiley, pp 1–204
32. NineteenTwenty (n.d.) NineteenTwenty | Bold and practical. https://www.heynineteentwenty.com/. Accessed 28 Mar 2023
33. Novikov AM, Novikov DA (2013) Communications in cybernetics, systems science and engineering—Research methodology from philosophy of science to research design. In: Kybernetes, vol 41, no 5/6. Taylor & Francis Group. https://doi.org/10.1108/k.2012.06741eaa.005
34. Nunes D (n.d.) Tangram | Ensino Fundamental I—Escola Kids. https://escolakids.uol.com.br/matematica/tangram.htm. Accessed 28 Mar 2023
35. Nunes F, Júnior J, Mauro R (2014) Modularização—Conceitos, Abordagens e Benefícios: Uma Revisão Teórica
36. Ozcan P, Han S, Graebner ME (2018) Single cases. In: The Routledge companion to qualitative research in organization studies, pp 92–112. https://doi.org/10.4324/9781315686103-7
37. Pathier F (2008) [Art video] Jean Charles de Castelbajac Lego runway. https://www.youtube.com/watch?v=-gsTMUk5QGE
38. Positivo Tecnologia (2018) Entenda os benefícios da relação da tecnologia e design. https://www.meupositivo.com.br/panoramapositivo/tecnologia-e-design/
39. pplware (2021) Nike aposta no metaverso com a compra da RTFKT. https://pplware.sapo.pt/internet/nike-aposta-no-metaverso-com-a-compra-da-rtfkt/
40. Ribeiro L, Miguel R, Duarte P (2009) Aplicação experimental do co-design e da modularidade no design de moda. Congresso E_design—Visões Para o Ensino Na Europa Nos Novos Contextos Ambientais e Económicos
41. RTFKT (n.d.) RTFKT. https://rtfkt.com/. Accessed 22 Mar 2023
42. Ruppert-Stroescu M (2009) Technology and creativity: fashion design in the 21st century, pp 1–169. https://mospace.umsystem.edu/xmlui/bitstream/handle/10355/10763/research.pdf?sequence=3&isAllowed=y
43. Santos A, dos Razera DL, Sampaio CP, Karam R (2008) Modularização como estratégia para ampliar o ciclo de vida de produtos
44. Santos V (2018) Quantificação da Criatividade no Processo de Design—Inovação e Ideação de Artefactos Através doos Processos Analógico e Digital. Universidade Europeia
45. Silva MT, da (2017) Moda e versatilidade: peças modulares como meio de otimização da vida útil do vestuário
46. Staib G, Dorrhofer A, Rosenthal M (2008) Components and systems: modular construction. In: Detail, pp 1–21
47. Testa J (2022) Bella Hadid's spray-on dress at fashion week—The New York Times. https://www.nytimes.com/2022/10/02/style/bella-hadid-coperni-paris-fashion-week.html
48. The Cutting Class (2020a) A modular Anrealage Silhouette through blocks—The cutting class. https://www.thecuttingclass.com/a-modular-anrealage-silhouette-through-blocks/
49. The New School Parsons (n.d.) CLO 3D: a partner on designing for the future. https://www.newschool.edu/parsons/story/rendering-a-new-concept-for-design/. Accessed 21 Mar 2023
50. Tran V (2022) Nike and RTFKT introduce cryptokicks sneakers with skin Vial Tech | Coin Culture. https://coinculture.com/au/business/nike-and-rtfkt-introduce-cryptokicks-sneakers-with-skin-vial-tech/
51. Vaid M (2021) Modular fashion—The next big thing? | Features | News. https://www.ecotextile.com/2021063028006/features/modular-fashion-the-next-big-thing-2.html

52. Van Soest S (2022) This puffer jacket zips into a tote, so lemme secure that bag. https://www.theedge.co.nz/home/whats-good/2022/11/this-puffer-jacket-zips-into-a-tote-so-lemme-secure-that-bag.html
53. Wikipédia (n.d.-a) História do Lego. https://pt.wikipedia.org/wiki/História_do_Lego. Accessed 14 Mar 2023
54. Wikipédia (n.d.-b) Tangram https://pt.wikipedia.org/wiki/Tangram. Accessed 14 Mar 2023
55. Zwieglinska Z (2022) RTFKT's Benoit Pagotto: quality and community are key in NFTs' evolution. https://www.glossy.co/fashion/rtfkts-benoit-pagotto-quality-and-community-are-key-in-nfts-evolution/

Brand Engagement and Creative Digital Advertising: A Case Study of Galo Portuguese Brand

Sara Santos, Pedro Espírito Santo, and Sónia Ferreira

Abstract Advertising has evolved over the decades. Knowing consumers are increasingly avoiding advertisements, creating campaigns that attract attention and stay in memory is essential. Advertisers' main goal is to get closer to consumers, generating visibility, attractiveness, and engagement with the brand. Furthermore, this topic has aroused the academy's interest, and studies show the importance of advertising and its connection with brand engagement. Thus, this study (n = 244) was carried out through a research model, tested through the PLS-SEM methodology, which showed that informativeness, narrative structure and narrative transportation have effects on brand engagement and advertising stimulation play a mediating role in this relationship. Therefore, it is suggested that brands create video ads with information and relevant structures to attract consumers that consider themselves part of these ads.

Keywords Informativeness · Narrative structure · Narrative transportation · Brand engagement · Advertising stimulation

---

S. Santos (✉)
Research Centre in Digital Services (CISeD)—Polytechnic of Viseu, Viseu, Portugal
e-mail: ssantos@esev.ipv.pt

P. E. Santo
Centre of Applied Research in Management and Economics (CARME), School of Technology and Management, Polytechnic of Leiria, Leiria, Portugal
e-mail: pedroes@ua.pt

S. Ferreira
Center for Studies in Education and Innovation (CIDEI), School of Education, Polytechnic of Viseu, Viseu, Portugal
e-mail: sonia.ferreira@esev.ipv.pt

N. Martins et al. (eds.), *Perspectives on Design and Digital Communication IV*, Springer Series in Design and Innovation 33, https://doi.org/10.1007/978-3-031-41770-2_12

1 Introduction

Following the global trend, the growth of digital advertising investment in Portugal is undeniable and inevitable, given the numbers of digital adoption and use. According to data from DataReportal [1], in its study dedicated to the digital ecosystem, in early 2023, the number of Internet users in Portugal was 8.73 million, with a total population of 10.26 million people. The data from the same report, for the same period, reveal that 7.43 million are YouTube users, making this platform one of the most popular in Portugal.

Knowing this reality, companies and entities increasingly bet on promoting products and services through online video. Creating a video and waiting for the users and the platform to act is not enough for this. It is necessary to consider characteristics such as the informativeness of the ad, the structure, and the transportation to the narrative. These characteristics play a relevant role for brands in generating brand engagement.

The main objective of this study is to analyse narrative transportation, narrative structure, and informativeness as antecedents of advertising stimulation, which, in turn, impacts brand engagement.

First, the theoretical background on the importance of video and ad informativeness, structure, narrative transportation, advertising stimulation, and brand engagement is analysed. Then, the methodology (data collection and sample) is presented. Subsequently, the results are demonstrated through the measurement model and hypothesis testing. Finally, the discussion and conclusions are offered.

2 Theoretical Background

2.1 *Video Ad: Informativeness, Structure, and Narrative Transportation*

An ad is informative when it offers essential information or facts about the product or service in a way that is understandable to consumers [2, 3] and helps them make purchase decisions [4, 5].

According to Dao, Le, Cheng and Chen [6], in their study on the value of social media advertising and its effect on online purchase intention, informative ads are considered more relevant and of more value when compared to less informative or creative ones. However, on social networking sites, the effects are weaker than on content community websites. A communication form that covers valuable information about the product or service determines the effectiveness of the message and enhances customer awareness [7], causing a positive response to the strategy. In addition, it also reduces the need for the potential consumer to research the product or service, making it more trustworthy [3].

However, the reverse is also true. Chan's [8] study regarding Chinese viewers' perception of a sample of over 60 ads with informational and emotional advertising revealed that the 160 respondents considered informational ads boring, uninteresting, and informative. In contrast, emotional ads were described as attractive, engaging, and original. Also, Petrescu and Korgaonkaer [9], in their study on viral advertising, realised that an informative ad is less likely to be shared, and creativity is an essential factor. Along with informativeness and entertainment, according to Taylor, Lewin and Stru-tton [10], creativity is also a builder of consumer attitudes and associations and, for [11], fundamental to the effectiveness of video advertising. The study revealed that a brand is better remembered and recognised when advertised creatively. Already in 2013, Rosengren et al. [12] studied whether advertising creativity could benefit more than the advertiser and concluded that advertising creativity has a positive influence on consumers' creativity and their perception of the value of the media vehicle.

When it is necessary to think about the transmission of information through ads, it is crucial to think carefully about the structure of the narrative.

According to Delgadiilo and Escalas [13], chronology and causality are fundamental elements in the narrative structure. The authors explain that, although there are many theories about the essential elements to build a story, they consistently agree on the importance of a chronological dimension and a relationship between story elements. The narrative organises events in time, configured as episodes, each with a beginning, middle and end. In addition, it structures relationships between story elements allowing causal inference to occur. The authors present, "To begin, an event, or series of events, initiates a psychological reaction and activates goals in a main character. […] The protagonist's psychological state and goals provide reasons for his/her subsequent actions. These actions lead to an outcome or result." (p. 187). Since these narrative elements are organised temporally, causal inferences can happen.

For Lee, Huang, and Marron [14], all stories are grounded in the background and have a core representing their essence. They mention that the structure integrates context, concept, climax, action, reversal, resolution, and learning. For Van Laer et al. [15], narratives are also based on a plot—a logical and natural succession of the essential storytelling elements—with characters to whom conflicts and events happen, with incidents and surprises, and with a climactic moment where the resolution of the story takes place. Regardless of the narrative elements, a successful story enhances the purpose of the communication strategy—consumer persuasion.

According to Brock and Green [16], based on Gerrig's principle, transportation can be conceptualised as a distinct mental process, a fusion of attention, imagination, and feelings. It is a convergent process in which all capacities and cognitive systems focus on the narrative. As a consequence, parts of the world become inaccessible, i.e. the receiver needs access to actual world facts at the expense of accepting the narrative world that has been created. Furthermore, he or she experiences strong emotions and motivations, even knowing that some elements may not be real and may return from transportation altered by the experiences. Brock and Green [16] argue that transportation can thus make the recipients less likely to disbelieve or counter-argue. For the author, when transported, they also tend to create stronger feelings toward the characters and the story. Dunn and Sinclair [17] corroborate this relationship and assume that transportation into the narrative can be a way for the audience to trust as they share values and trust with the characters in the story. Van Laer et al. [15] even states that narrative transportation can "cause affective and cognitive responses, beliefs and attitude and intentional changes" (p. 800).

2.2 Advertising Stimulation

Advertisements, in addition to seeking to increase the visibility of products/services, must be creative and generate consumer engagement. In addition to capturing the public's attention, advertisements must persuade to create a return on investment [18].

Persuasion is trying to alter, modify or convert the values, beliefs and actions of others, admitting that all social life is dominated by conscious or unconscious, energetic or superficial attempts at persuasion [19]. Persuasion is fundamental in advertising, since advertising texts can lead to reflection and decision-making, which is the central role of persuasion [20].

Advertising is also related to creativity, originality, innovation and problem-solving [21]. Ads that are novel and meaningful manage to connect with what people see and make them feel positive and remember the brand [22]. By provoking positive feelings in people, consumer engagement is more likely [23].

Through creativity and humour, brands manage to evoke different emotions in consumers. These have positive responses and remember the brands [24, 25].

As more and more consumers tend to avoid ads, it is critical to stimulate advertising and use features that attract consumers' attention.

Advertisers use different communication channels to get closer to consumers, generating visibility, attractiveness and engagement with the brand.

2.3 Brand Engagement

Consumer Brand Engagement is defined by Hollebeek et al. [29] as "a consumer's positively valenced brand-related cognitive, emotional and behavioural activity during or related to focal consumer/brand interactions". This definition implies three dimensions of consumer engagement; cognitive, emotional, and behavioural engagement. Cognitive brand engagement refers to brand-related thought processing in consumer-brand interaction [26]. Emotional or affective brand engagement means consumers are more attached to the brand and more time and effort they are willing to invest [27]. Lastly, behavioural or activation brand engagement refers to the time and effort consumers are willing to spend in a consumer–brand interaction [29].

On the other hand, Yang et al. [28] define brand engagement as "the consumers' behavioural manifestation toward a brand—beyond purchase—resulting from motivational drivers, which is captured through the interactive behaviours between consumers and brands".

With the advancement of the internet and social media, advertising engagement through social media also gained importance in the marketing literature [29, 30].

To generate engagement with consumers, brands' use of social media is a top priority [31, 32]. Thus, brands can use social media advertisements to influence the purchase intention of products/services [33].

Some studies suggest that consumer engagement is the antecedent of brand choice, and advertisers strategically use different media types to generate engagement [34]. When consumers engage with the media vehicle, they are more likely to respond to brand advertising [35, 36].

The engagement of consumers with brands has several consequences, such as the development of positive feelings towards the brand and reinforcement of brand strength [32], increased brand knowledge [36], emotional attachment [38] as well as loyalty intentions [29].

In order to evaluate brand performance and advertising strategy effectiveness is crucial to analyse consumers' social media brand engagement [39]. It is the principal method for raising the success of advertising messages [28].

Conversely, social media adverting impacts brand engagement and purchase intention [39].

Through analysis of the literature review, it was constructed the following research hypotheses:

H1: Informativeness has positive impacts on narrative transportation;
H2: Narrative structure has a positive influence on narrative transportation;
H3: Informativeness has positive impacts on advertising stimulation;
H4: Narrative structure has a positive influence on advertising stimulation;
H5: Narrative transportation has positive effects on advertising stimulation;
H6: Advertising stimulation has positive effects on brand engagement.

From the research hypotheses, it was built the following conceptual model (Fig. 1).

3 Methods

3.1 Data Collection

A video advertisement was presented for the information collected so the participants could watch it before answering the questionnaire. After watching the video, participants completed the questionnaire through their equipment (smartphones, tablets or laptops).

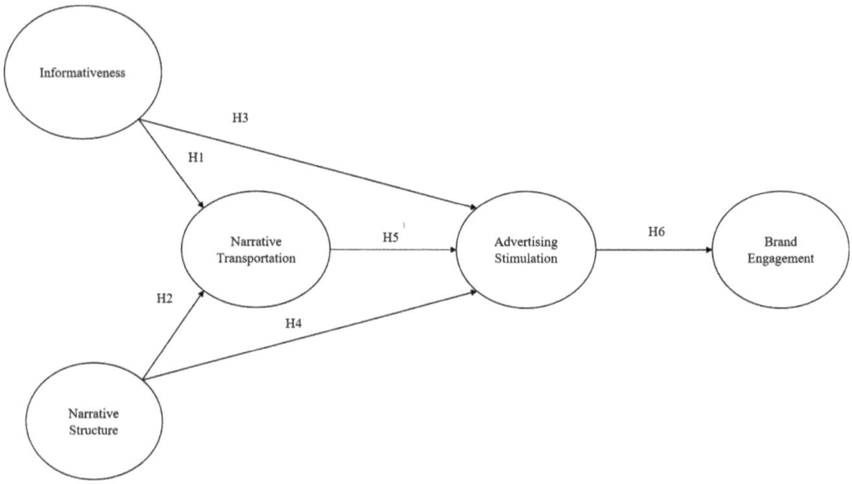

Fig. 1 Conceptual model

For the development of the questionnaire, a self-administered questionnaire was prepared to consist of 18 items that were structured for this research. The questionnaire was composed of two parts.

The first part included a video advertisement of the Gallo brand, which is available on YouTube at the following link: https://www.youtube.com/watch?v=Zd09954cJsM.

The papers used to evaluate informativeness were adjusted by Lee and Hong [40]. The evaluation of narrative transportation was measured through 4 items from Solja's research [41]. The items used for advertising stimulation were developed by four items of Solja [41]. The measurement items used for brand engagement were adjusted from Santos, Santo, and Ferreira's research [42, 43]. Also, it included three items to assess narrative structure, which was adjusted by Santos, Santo, and Augusto [42].

The second part of the questionnaire consisted of collecting information to characterise the sample.

3.2 Sample

The data collected is considered valid 244 observations were obtained from individuals between 18 and 75 years old. These individuals usually know the Gallo brand (97.9%). The respondents, in most cases, are male (69.3%), and most are under 30 years of age (Table 1).

Table 1 Sample

Variable	Dimension	N	%
Gender	Male	169	69.3
	Female	75	30.7
Age	<20	40	16.4
	20–29	119	48.8
	30–39	40	16.4
	40–49	33	13.5
	50–59	10	4.1
	≥60	2	0,8

4 Results

After analysing the sample, it was considered that this sample allowed the use of structural equations considering the 5:1 model. Thus, it was used to test the conceptual model and structural equations through Partial Least Squares. This methodology allows for studying causal relationships between concepts and is appropriate for exploratory research [45].

In this context, PLS-SEM is developed in two stages. In the first stage, the aim is to assess the measurement model's reliability and validity. In the second stage, this methodology seeks to analyse the causal relationships between the variables, as suggested in the literature [40]. The PLS-SEM algorithm was run on SMART PLS v3.3.4 software [41].

4.1 Measurement Model

The common method bias was analysed prior to the data analysis. This research collected data from individuals with similar characteristics, which may bias the results. Therefore, the questionnaire was tested previously to avoid vague or complex concepts. Furthermore, the respondents were informed that the questionnaire was anonymous and that there were no correct or wrong answers. Then Harman's factor was analysed, finding that the first factor explains 25.32% of the variance.

In this previous data analysis, it was also checked the VIF (Variance Inflation Factor) value of each item and found that the values were below the value recommended by the literature (VIF < 5) [47].

After common method bias analysis, it was assessed the measurement model.

To analyse the measurement model, its viability was evaluated using the indicators presented in Table 2 and Table 3. The data obtained in these tables concluded that the standardised coefficients shown in Table 2 are above the recommended value ($\lambda > 0.7$). In addition, Table 3 indicates that the composite reliability (C.R.) and the

Table 2 Measurement model

	λ	*t* values	*p* values
INF1	0.921	61.382	0.000
INF2	0.826	30.956	0.000
INF3	0.929	80.518	0.000
ADST1	0.879	55.252	0.000
ADST2	0.821	27.093	0.000
ADST3	0.760	21.167	0.000
ADST4	0.797	26.143	0.000
ENG1	0.889	43.811	0.000
ENG2	0.895	47.778	0.000
ENG3	0.861	41.688	0.000
ENG4	0.854	34.769	0.000
ENG5	0.916	71.069	0.000
ENG6	0.879	41.606	0.000
NATX1	0.754	17.932	0.000
NATX2	0.901	68.316	0.000
NATX3	0.919	71.245	0.000
NAST1	0.967	170.658	0.000
NAST2	0.955	79.090	0.000

Notes λ = Standardised Coefficients; α—Cronbach's alpha; AVE—Average Extracted Variance; C.R. —Composite reliability. ADST = Advertising Stimulation; ENG = Brand Engagement; INF = Informativeness; NATX = Narrative Transportation; NAST = Narrative Structure.

average variance extracted (AVE) are above the values suggested in the literature (AVE > 0.5; C.R. > 0.7) [42].

Table 3 Constructs validity and reliability

	Cronbach's Alpha	ρA	Composite reliability (C.R.)	Average variance extracted (AVE)
ADST	0.831	0.843	0.888	0.665
ENG	0.943	0.947	0.955	0.779
INF	0.871	0.872	0.922	0.798
NAST	0.918	0.933	0.960	0.924
NATX	0.825	0.863	0.895	0.741

Notes Diagonal entries are the square root of AVE values; all correlations are significant at level 1%; ADST = Advertising Stimulation; ENG = Brand Engagement; INF = Informativeness; NATX = Narrative Transportation; NAST = Narrative Structure.

5 Discriminant Validity

The discriminant variance was analysed in the second phase of the analysis of a measurement model. First, the Fornell and Larcker [48] criterion was analysed in Table 4.

Notes ADST = Advertising Stimulation; ENG = Brand Engagement; INF = Informativeness; NATX = Narrative Transportation; NAST = Narrative Structure.

To confirm the discriminant analysis, the HTMT ratio was evaluated, shown in Table 5, and it reveals that these HTMT values are below the recommended value of 0.9. [44].

In summary, this research confirms that the concepts under study have validity and can be assessed in a structural model.

5.1 Hypothesis Testing

The hypothesis test of the causal model between the constructs was analysed. The results reveal that six hypotheses were confirmed, and Table 6 presents the values of the corroboration of the hypotheses.

Table 4 Fornell and Larcker criteria

	ADST	ENG	INF	NAST	NATX
ADST	0.815	–	–	–	–
ENG	0.629	0.883	–	–	–
INF	0.561	0.402	0.893	–	–
NAST	0.595	0.256	0.531	0.961	–
NATX	0.723	0.514	0.521	0.593	0.861

Table 5 Heterotrait-Monotrait (HTMT) ratio

	ADST	ENG	INF	NAST	NATX
ADST	–	–	–	–	–
ENG	0.699	–	–	–	–
INF	0.651	0.439	–	–	–
NAST	0.677	0.269	0.592	–	–
NATX	0.855	0.581	0.607	0.654	–

Notes ADST = Advertising Stimulation; ENG = Brand Engagement; INF = Informativeness; NATX = Narrative Transportation; NAST = Narrative Structure.

Table 6 Hypothesis testing

	β	*t* values	*p* values
ADST – ENG	0.629	18.311	0.000
INF – ADST	0.195	3.531	0.000
INF – NATX	0.287	3.929	0.000
NAST – ADST	0.190	3.366	0.001
NAST – NATX	0.441	6.660	0.000
NATX – ADST	0.509	9.741	0.000

Notes β = Standardized path ADST = Advertising Stimulation; ENG = Brand Engagement; INF = Informativeness; NATX = Narrative Transportation; NAST = Narrative Structure.

6 Discussion and Conclusions

In the analysis of the results, this research found that it is informativeness influences the immersion in the narrative, and these results corroborate the theory that evidence that it is informativeness a factor that attracts people to get gripped to the video and consequently be transported to the scenes of that video [7]. Therefore, hypothesis H1 is corroborated by this study.

In the analysis of hypothesis H2, it was also considered that this study corroborated it because, as the authors, Delgadiilo and Escalas [13] state, a pleasing structure with appropriate characters and well-structured video scenes attracts individuals to the story. Consequently, individuals enter within the scenes of the presented advertisement. These results are consistent with the literature reviewed.

This research also evaluated informativeness as an antecedent of advertising stimulation. The H3 hypothesis was tested, showing that advertising stimulation can be accomplished through more information in the videos.

Furthermore, advertising stimulation can be done through video ads that have a good structure. Hypothesis H4 complements this analysis since videos consumers perceive as having good structure stimulate their senses. Therefore, this research considers that the hypothesis of this study corroborates H4.

This study also reveals that narrative transportation is an essential antecedent of ad stimulation because immersion in ads activates consumers' senses and influences stimulation in ads [25]. Hypothesis H5 is also corroborated in this study.

Finally, it was evaluated that stimulating video ads influences brand engagement. According to the literature, this research reveals that hypothesis H6 is corroborated by Wang [23].

In summary, this study analysed the role of narrative structure and emotion in the narrative in its relation to emotional marketing. This research also reveals that the feelings generated in advertising stimulate more significant engagement with brands.

As a practical revelation, we recommend that brands consider it fundamental to create video ads with information and relevant structures so that consumers are attracted to the ad and consider themselves part of these ads, stimulating their senses in this way.

Despite considering that these conclusions are significant for brands, some limitations and challenges can be analysed with some reflection. First, it is important to evidence that individuals who answered this survey are young and mostly male. New studies should be conducted with another sample type, and comparative analyses between groups should be possible. Furthermore, this research did not include some variables that may complement this type of study. Additionally, this research did not have brand equity or analyse the impact of brand image. In this sense, further studies may explore these concepts. Furthermore, conducting this study in another activity sector or with another brand in the food sector is also essential.

Acknowledgements This work is funded by National Funds through the FCT—Foundation for Science and Technology, I.P., within the scope of the project Ref. UIDB/05583/2020. Furthermore, we would like to thank the Research Centre in Digital Services (CISeD) and the Instituto Politécnico de Viseu for their support.

References

1. DataReportal. Digital 2023: Portugal, https://datareportal.com/reports/digital-2023-portugal, last accessed 2023/02/25
2. Bendixen MT (1993) Advertising effects and effectiveness. Eur J Mark 27(10):19–32
3. Janssens W, Pelsmacker P (2005) Emotional or informative? Creative or boring? Int J Advert 24(3):373–394
4. Tsang M, Ho SH, Liang TP (2004) Consumer attitudes toward mobile advertising: an empirical study. Int J Electron Commer 8(3):65–78
5. Naini S, Shafia M, Nazari N (2012) Examining different factors in effectiveness of advertisement. Manag Sci Lett 2(3):811–818
6. Dao W, Le A, Cheng J, Chen D (2014) Social media advertising value. Int J Advert 33(2):271–294
7. Saadeghvaziri F, Hosseini HK (2011) Mobile advertising: an investigation of factors creating positive attitude in iranian customers. Afr J Bus Manage 5(2):394–404
8. Chan K (2015) Chinese viewers' perception of informative and emotional advertising. Int J Advert 15(2):152–166
9. Petrescu M, Korgaonkar P (2011) Viral advertising: definitional review and synthesis. J Internet Commerce 10(3):208–226
10. Taylor DG, Lewin JE, Strutton D (2011) Friends, fans, and followers: do ads work on social networks? How gender and age shape receptivity. J Advert Res 51:258–275
11. Rauwers F, Remmelswaal P, Fransen M, Dahlén M, Noort G (2018) The impact of creative media advertising on consumer responses: two field experiments. Int J Advert 37(5):749–768
12. Rosengren S, Dahlén M, Modig E (2013) Think outside the ad: Can advertising creativity benefit more than the advertiser? J Advert 42(4):320–330
13. Delgadillo, Y., Escalas, J.: Narrative word-of-mouth communication: exploring memory and attitude effects of consumer storytelling. In: Kahn, B., Luce, Valdosta (eds.) GA: Association for Consumer Research, pp. 186–192, Advances in Consumer Research (2004)
14. Lee M, Shen H, Huang JZ, Marron JS (2010) Biclustering via sparse singular value decomposition. Biometrics 66(4):1087–1095

15. Van Laer T, de Ruyter K, Visconti L, Wetzels M (2014) The extended transportation-imagery model: A meta-analysis of the antecedents and consequences of consumers' narrative transportation. J Consum Res 40(5):797–817
16. Brock, Green MC, Brock TC (2000) The role of transportation in the persuasiveness of public narratives. J Person Social Psychol 79(5), 701–721
17. Dunn J, Sinclair J (2000) For better or worse? Examining the effectiveness of envy mitigation strategies. International Association for Conflict Management Conference, Chicago, IL
18. Friestad M, Wright P (1994) The persuasion knowledge model: How people cope with persuasion attempts. J Consumer Res 21:1–31
19. O'Shaughnessy J, O'Shaughnessy NJ (2004) Persuasion in advertising. Routledge, Londres
20. Citelli A (2002) *Linguagem e Persuasão.* São Paulo: Editora Ática
21. Koslow S, Sasser SL, Riordan EA (2003) What is creative to whom and why? Perceptions in advertising agencies. J Advert Res 43:96–110
22. Heiser RS, Sierra JJ, Torres IM (2008) Creativity via cartoon spokespeople in print ads: Capitalising on the distinctiveness effect. J Advert 37:75–84
23. Wang A (2006) Advertising engagement: A driver of message involvement on message effects. J Advert Res 46:355–368
24. Dens N, De Pelsmacker P (2010) How advertising strategy affects brand and USP recall for new brands and extensions. Int J Advert 29:165–194
25. De Pelsmacker P, Geuens M, Van Den Bergh J (2013) Marketing communications: A European perspective (5th ed.). Harlow, UK: Pearson Education
26. Hollebeek LD (2011) Demystifying customer brand engagement: Exploring the loyalty nexus. J Mark Manag 27:785–807
27. Hollebeek LD (2011) Exploring customer brand engagement: Definition and themes. J Strateg Mark 19:555–573
28. Yang S, Lin S, Carlson JR, Ross WT Jr (2016) Brand engagement on social media: will firms' social media efforts influence search engine advertising effectiveness? J Mark Manag 32(5–6):526–557
29. Hollebeek LD, Glynn MS, Brodie RJ (2014) Consumer brand engagement in social media: Conceptualisation, scale development and validation. J Interact Mark 28:149–165
30. Kilger M, Romer E (2007) Do measures of media engagement correlate with product purchase likelihood? J Advert Res 47:313–325
31. Liu X, Shin H, Burns AC (2019) Examining the impact of luxury brand's social media marketing on customer engagement: using big data analytics and natural language processing. J Bus Res 125:815–826
32. Bianchi C, Andrews L (2018) Consumer engagement with retail firms through social media: an empirical study in Chile. Int J Retail Distrib Manag 46(4):364–385
33. Moslehpour M, Dadvari A, Nugroho W, Do BR (2020) The dynamic stimulus of social media marketing on purchase intention of Indonesian airline products and services. Asia Pac J Mark Logist 33(2):561–583
34. Gambetti RC, Graffigna G (2010) The concept of engagement a systematic analysis of the ongoing marketing debate. Int J Mark Res 52:801–826
35. Hollebeek LD, Conduit J, Sweeney J, Soutar G, Karpen IO, Jarvis W, Chen T (2016) Epilogue to the special issue and reflections on the future of engagement research. J Mark Manag 32(5/6):586–594
36. Cheung ML, Pires G, Rosenberger PJ (2020) The influence of perceived social media marketing elements on consumer–brand engagement and brand knowledge. Asia Pac J Mark Logist 32(3):695–720
37. Brodie RJ, Ilic A, Juric B, Hollebeek L (2013) Consumer engagement in a virtual brand community: an exploratory analysis. J Bus Res 66:105–114
38. Chahal H, Wirtz J, Verma A (2019) Social media brand engagement: dimensions, drivers and consequences. J Consum Mark 37(2):191–204
39. Gavilanes JM, Flatten TC, Brettel M (2018) Content strategies for digital consumer engagement in social networks: why advertising is an antecedent of engagement. J Advert 47(1):4–23

40. Lee J, Hong IB (2016) Predicting positive user responses to social media advertising: The roles of emotional appeal, informativeness, and creativity. Int J Inf Manage 36(3):360–373
41. Solja E, Liljander V, Söderlund M (2018) Short brand stories on packaging: an examination of consumer responses. Psychol Mark 35(4):294–306
42. Santos S, Santo P (2021) Brands and stories in ads: the relationship between storytelling and brand distinctiveness. In: Lorena Clara M, Raluca A, Anda D (Eds.), Handbook of research on contemporary storytelling methods across new media and disciplines 116–137
43. Santos S, Santo P, Ferreira S (2021) Storytelling in advertising: from narrative to brand distinctiveness. In: Martins N, Brandão D (Eds.): Advances in design and digital communication. Springer International Publishing 516–527
44. Hair JF, Ringle CM, Sarstedt M (2011) PLS-SEM: Indeed a silver bullet. J Market Theory Practice 19(2):139–152
45. Ringle C, Wende S, Becker J-M (2015) 'SmartPLS 3'
46. Hair JF, Black WC, Babin BJ, Anderson RE (2018) Multivariate data analysis. Pearson Education Limited
47. Chin WW (1998) The partial least squares approach for structural equation modelling: 'Modern methods for business research.' Lawrence Erlbaum Associates Publishers, 295–336
48. Fornell C, Larcker DF (1981) Evaluating structural equation models with unobservable variables and measurement error. J Mark Res 18(1):39–50
49. Henseler J, Ringle CM, Sarstedt M (2015) A new criterion for assessing discriminant validity in variance-based structural equation modelling. J Acad Mark Sci 43(1):115–135

The Role of Design in Online Communication for Higher Education Institutions During the Covid-19 Pandemic

Ana Rute Cardoso, Daniel Brandão, Nuno Martins, and Leonardo Pereira

Abstract This chapter presents a study developed at the Department of Communication Sciences of the University of Minho (Portugal) about the importance of design in the online digital communication of a Higher Education Institution, in a pandemic context. The pandemic scenario, caused by the SARS-CoV-2 coronavirus which in turn is responsible for Covid-19, experienced in Portugal since the beginning of 2020, resulted in rigorous social and human relationship restrictions, forcing individuals, institutions and companies to find alternatives to answer to these limitations. One of the fields that suffered most from these restrictions was the communication one, not only at an interpersonal level, but also at an institutional level. The study presented here follows, therefore, a methodology that crosses literature review with semi-structured interviews and focuses on how communication design can contribute to the improvement of institutional communication in socially dramatic moments like those experienced during the pandemic, which isolated academic communities of teachers, staff and students, reducing their interactions to virtual online environments. Communication design, through its main structuring elements, such as typography and colour, has the potential to "humanise" the messages with information about health and safety, academic procedures, among other formal information usually transmitted by universities to their communities, giving them emotional and appealing attributes. In this way, the communication design contributes to draw the receiver's attention to important information, reinforcing the effectiveness of communication.

A. R. Cardoso
Institute of Social Sciences, University of Minho, Braga, Portugal

D. Brandão (✉)
CECS, Institute of Social Sciences, University of Minho, Braga, Portugal
e-mail: danielbrandao@ics.uminho.pt

N. Martins
Research Institute for Design, Media and Culture, School of Design, Polytechnic Institute of Cavado and Ave, Barcelos, Portugal

L. Pereira
DigiMedia, DeCA, University of Aveiro, Aveiro, Portugal

N. Martins et al. (eds.), *Perspectives on Design and Digital Communication IV*, Springer Series in Design and Innovation 33, https://doi.org/10.1007/978-3-031-41770-2_13

Keywords Visual communication · Communication design · Colour · Typography · Pandemic

1 Introduction

Online presence is fundamental to strengthen relationships between educational institutions and their communities of students, teachers and staff, especially in situations of isolation and confinement, such as those lived in 2020 and 2021, during the pandemic caused by the new coronavirus. In this context, can communication design be seen as a useful tool for the online communication of Higher Education Institutions (HEIs)? This was the motto for this research's starting question: What role can communication design play in the mitigation of problems caused by the pandemic in the communication of and in HEIs?

For this brief study, the following goals were set: to understand the role of communication design in the process of transmitting the messages of organisations; and to understand how the pandemic impacted on communication of HEIs.

The pandemic's emergence completely changed the way the organisational world used to communicate. Its sudden appearance led organisations such as HEIs to rapidly change and adapt their way of communicating with their audiences, which was an expected outcome. Therefore, semi-structured interviews were conducted with both senders and receivers of institutional messages, in order to deepen this phenomenon.

This chapter is therefore divided into three sections. The first one is dedicated to the literature review and to the theoretical framework on the research question to which this chapter is dedicated. The role of communication design in the communication of HEIs is described, highlighting two of its constituent and distinctive elements: colour and typography. In the second section, the data collection methodologies used in this study are presented, in an attempt to find clues that help answering to the starting question previously mentioned. And finally, in the third section, the data collected with the interviews is presented and analysed. But it's analysed with the intent of understanding the point of view of the communicational agents of HEIs about the combined use of design and communication in pandemic contexts.

2 Design and Its Communicational Role

2.1 Visual Communication and Design

Nowadays, the overwhelming amount of information that comes from the most varied media, only becomes communication if it effectively reaches its target audience and influences it [12]. Design can play a decisive role in fulfilling this purpose, due to the differentiating and empathetic properties that it can give to the message. The design must be appealing and attractive, "(…) it must draw attention in crowded environments and implant a seed" [3, p. 129].

So, communication design should not only be seen for its aesthetic function, but also for its functional side, its usability and adaptability to different situations and formats [7, 10, 11].

In times that are so demanding in terms of adaptation to new challenges, such as the ones posed by the pandemic scenario, design can help inform, using elements such as colour and typography so that messages are more immediately and effectively perceived [2, 10]. Because, according to Heller and Ilić [3], by definition, the most important role of a graphic designer is to inform, being even able to change and influence the behaviour of a message's receiver.

From Lupton's [6] perspective "the design process is a mixture of intuitive and intentional actions." (p. 4). By default, designers apply some techniques to develop their projects creatively, in the search for the problem definition and its resolution. After these steps, it is necessary to start developing the solution and, in order to do so, the basic elements of communication design have to be put to use.

2.2 Communication Design and the Awakening of Cultural Associations

In a research conducted by Eva Heller on the Psychology of Colours, this researcher tries to understand the relationship and response of human beings to colours, thus defining the effects that each colour causes in individuals. Colours are therefore responsible for creating and awakening associations in our brain, triggering different interpretations.

We communicate and express ourselves through the use of colours. It is important to understand that the same colour can generate different sensations and interpretations from person to person, since they depend on the personal, social, cultural or situational context of the person to whom it is shown. For example, the colour red can both symbolise love or anger. The same happens with the colour black that in some contexts can symbolise sophistication but in others it might symbolise loss or death.

So, when developing a graphic composition, colour can be a key element to ensure that the message one (designer) wants to convey is understood by the audience in an objective and direct way. As mentioned before, colours are sensory information, that is cognitively interpreted taking into account the experiences stored in personal memory (the conscious and subconscious) and the culture of each individual. Therefore, the interpretation and response to a certain colour may differ from the interpretation and response of another person, as colour can have different effects and experiences from person to person. For Mahnke [8] "To perceive colour means to "experience", to become aware, to be alert. Many factors contribute to this process, partly on a conscious level and partly on an unconscious level" (p. 10).

The effects of colour are essentially the ability to influence our behaviour by provoking sensations and emotions, as stated by Lancaster [4]: "The functions of colour are to attract attention, convey information, create illusions and stimulate emotions" (p. 8). Colours are information, triggering certain actions in human behaviour, and can influence our choices, our mood or state of mind, in a negative or positive way, always depending on the final objective.

Colour is an important element, especially in terms of digital communication, due to its capacity to capture and retain the attention on a given message and the ability to create empathy, by transmitting emotions, which lead to the creation of a closer relationship and involvement of the target audience with the message.

The influence of typography. It is also believed that another of the main tools of communication design is typography. Typography, in turn, has the ability to add emotions, "personality" and meanings to a message transmitted through writing. Looking at the role and at the act of choosing typography, in this context of creating a relationship with the public, typography asserts itself as one of the most important elements of design, "(…) being, as stated by Lupton [5], an essential resource for graphic designers, just as crystal, brick and steel are essential materials for architects." (p. 47).

Each typeface has its own "personality" and unique features and each one needs to be adapted to each problem, always finding the best solution—not always the most minimal or beautiful one works or fits all situations. One has to think about what's behind the message, so that it mirrors the expected emotions. Lupton [7] goes deeper and even states that "Typography is not just about choosing fonts" [7, p. 49].

Similarly, Heller and Ilić [3] add that fonts can transform the way a message is received and perceived. Following this idea, these authors [3] state that typography has an influence on the way information is understood.

Looking now at the opinion of the designer David Carson—"Don't confuse legibility with communication" [1, 13]—, we can state that, for something to be communicated, it must necessarily be understood. Only in this way can the process of transmitting information be carried out successfully and the message that is intended to be disseminated not end up being lost.

Carson [1], UKEssays [13], also states that not everything that is legible communicates, but it is up to the designer to understand this issue and to perform their work in the best way, since their main goal is to communicate messages effectively and efficiently, to the point that the audience has no doubt about what is intended to be transmit. What we believe Carson wanted to explain with his Ted Talk in 2003 is that not everything that is legible is communicable, since it may not expose a coherent message, easy to read and with the necessary transparency so that the public or audience emotionally connects to it.

For this reason, if there is no convergence between what is written and the typographic sensations perceived by the audience, one can say that the message will not be passed on and processed correctly by its receiver, which is clearly a problem.

What is important to retain is that choosing the font to use for a message is not a simple or linear process. The designer has a preponderant role in this choice, due to his/her knowledge and experience on the issue. It is extremely important that the choice falls on the font that best characterizes the emotions that are intended to be aroused or triggered and to pay attention to how the message will be received. Moreover, the "personality" of the chosen font will be a determining factor to convey a message that reaches its target audience "intact" and without different interpretations from those intended by the designers.

2.3 *Communication Design and the Pandemic*

In times like the ones previously faced where we all lived with a global pandemic, combined with a society constantly thirsty for information, we need to be able to filter what is most important from a message [9]. Colours are, therefore, a great aid in the process of disseminating messages, highlighting what we should retain immediately. Not to mention that colours can also help in transmitting more positive and hopeful messages, which help to reduce anxiety. So the way messages are communicated is very important. At a time of constant jitters and anxiety, (textual) messages need to contribute to some calm and not to emotional turmoil. That is why it is so important to pay attention to the typography chosen for the dissemination of those messages and to pay attention to the emotions it may provoke.

What we also tried to do in this chapter was to define, in some way, the area of communication design and two of its basic elements—colour and typography—highlighting their role in the dissemination of messages.

Communication design is, thus, a fundamental tool to correctly put into practice this process of messages dissemination and to reach its main goal, which is to inform the target audience. It is also an important digital communication tool that, when used by HEIs in social networks, can promote greater effectiveness in the dissemination of messages in the context of the Covid-19 pandemic.

3 Applied Methodology

With goals defined, the path was traced in order to reach them. Authors such as Ellen Lupton, Eva Heller, Steven Heller, Paul Rand and David Carson were studied in order to understand the role of communication design and of elements such as colour and typography, in the process of disseminating messages and which led, in this research and chapter, to the discussion about what is online communication in HEIs and how it works in the referred context.

In addition, 5 interviews were applied to people with different profiles in the communication network: the executive coordinator of an innovation and experimentation agency (interviewee A); a student of the Degree in Communication Sciences of the University of Minho (interviewee B); a student of the Master in Communication Sciences of the University of Minho (interviewee C); the person responsible

Table 1 Identification of interviewees and date of each interview

Interviewees	Position/job held	Interview's date
A	CreateLab's executive coordinator	October 11th 2021
B	CC undergraduate student	October 26th 2021
C	CC Master's student	October 28th 2021
D	Responsible for ICS' communication and image	November 9th 2021
E	DCC lecturer	November 10th 2021

for communication and image of the Institute of Social Sciences of the University of Minho (interviewee D); and, finally, a lecturer of the Department of Communication Sciences of the University of Minho (interviewee E). These 5 non-designer profiles were purposely chosen in order to understand points of view from outside design the field and to discover the resonance of design issues in related fields of communication (Table 1).

The questions' script used during the interviews was derived from the readings carried out. The script was divided into two sequential and chained groups of questions which, based on the data collected in the pilot interview, were refined. Since the adjustments were not substantial and did not alter the set goal, it was decided that it could be used and incorporated into the body of analysis of this work. The interviews were conducted via zoom, given that at the time they were applied we were in a pandemic period, with restrictions to on face-to-face contact put in place. The interviews were recorded, as well as the due consent of the interviewees. They were later transcribed so that they could be analysed and the necessary conclusions could be drawn.

4 Results

The hardships of communicating during the pandemic provided greater prominence to communication design, a field with a value that is usually undervalued. Communication design stimulates our interest, directs the audience, attracts/retains our attention, making certain content and messages stand out from similar ones. As mentioned by one of the interviewees of this study:

> [design] is extremely important, especially at a time when we live overwhelmed with information (...) so it is very important to be able to highlight our contents and our information. And, taking this need into account, this idea of personalisation, demarcation and identification is very much entrenched into the field of design (interviewee A).

It not only matters how messages look, but what works. And this is exactly what design is: it combines the aesthetic side of any visual communication medium with its function. This is an idea also mentioned by interviewee A: "Design is not only aesthetic (…) design effectively has a functional side and can define, at the limit, the understanding or not of the information and contents it contains."

According to interviewee D, we have to combine design with communication and image with words. Only then does everything work and the messages are disseminated effectively: "We have to think of this as a duet. A design by itself is not enough, but it works when combined with the creativity that is also provided by the text." (interviewee E). Furthermore, this interviewee also states that "design is fundamental to the idea of trust, credibility and reputation", since it contributes to the creation of a well-executed and coherent visual identity, capable of positioning any institution.

Interviewee A states that "design is an area of communication and that the sooner we start looking at it as a field of communication, the better our communication will be". He even states that "we will reach a time when design, besides being a part of communication, will be communication" (interviewee A).

The arrival of the pandemic brought a greater concern for the field of design. As the popular saying goes, "art makes engineering", and the period of confinement and respective adaptation fits perfectly into this popularly used idea. The need for a meaningful and attractive communication, as well as the need for content with a uniform visual identity, lead to the beginning of the valorisation of this communication field and the growing awareness that design is something that reinforces meaning and is needed in the world of communication. As interviewee A states: "I can assure you that design and communication concerns have been growing".

This happens because, although institutions are starting to understand the negative consequences that the lack of proper communication design brings to their general communication, many still don't: As mentioned by interviewee A:

> We still have a long way to go, but I also think, on the other hand, that we are taking very important steps in the right direction. Also related to the pandemic issue, the "visuality" of the information also ended up being a driving force behind this new look at design (Interviewee A).

Communication designers have a very well developed visual culture and know how to communicate visually, using it to highlight information assertively. Interviewee D even states that "it has to be a design at the service of communication and not communication at the service of design".

Colour and typography are, as discussed above, two basic elements of design, that are crucial to the effective dissemination of messages through the evocation of meanings, associations and emotions in the target audience. However, something that arose from the conducted interviews was that these two elements are still not seen as a set or group that works together to help messages communication.

During the pandemic, when emotions were running high, colour played a very important role in the communication of messages. As mentioned by one of our interviewees, "the messages (…), during the pandemic, became more informal, more emotional. And I think colours are also very much allied to this." (interviewee B).

When asked about the colour that defined the Covid-19 pandemic, all interviewees stated that in all communications blue stood out, a colour closely associated with health, as it conveys a sense of security.

Besides, we have typography, which "immediately transmits states of mind, intentions. (…) typography also carries an intentionality." (interviewee E). In this sense, and also from the perspective of interviewees B and C, the typographic choice influences the way we receive information and messages, determining our interest or not in them: "I think it is very important, especially when there is a lot of information to convey, (…) this is a basic principle of design" (interviewee B). Likewise, it also influences our interpretation of them, as these "have an impact on the way the message is conveyed." (interviewee C).

That said, these are two fundamental elements—colour and typography—that greatly contribute to the interpretation of the messages. However, they are still not perceived as having the same level of importance by everyone and, from the interviewees' perspective, it is somewhat seen as having an unequal importance in messages communication.

From the point of view of interviewee A, the chromatic choice is still more thought out than the typographic one: "we are starting to understand the importance of colours, but regarding fonts they still lack the same level of recognition".

It is important to mention that information only becomes communication when the audience is able to decode the message, i.e. when they attribute meaning to what they are told; that is, "it is only when the end result reaches us and we attribute meaning and decode what was there, that communication happens" (interviewee A). And here, once again, it is important to mention that the appropriate chromatic and typographic choices help in this process of decoding messages.

We face an increasingly lazy society, as stated by interviewee D: "People are getting lazier and lazier. (…) people have stopped reading", meaning that society increasingly prefers to see something rather than to read something. Alongside an important role in fighting this laziness, design will play an increasingly important role in the pending visual future of communication. As expressed by interviewee D:

> I think that design is increasingly going to be king in the message. (...) I think that this is where we are heading towards: an era of aesthetic communication. (...) I think that we are increasingly going to be communicators of images rather than communicators of words (interviewee D).

There are also those who think that "communication without design doesn't even make sense." (interviewee C). And converging with C, interviewee A states that:

> Our processing capacity is getting smaller and smaller and we need to be told everything we want to know, but in a beautiful way and in a way that touches us. And design is the key!

Regarding the design field and its future in communication, interviewee E states that there has been a path that he feels is not yet completely mature, but that already gives us an optimistic sense for the future.

As has been discussed, the pandemic revolutionised the world in many ways. It affected different sectors of society and made the act of communicating imperative to fight misinformation. So the importance of communication and communication design became obvious.

Thus, we entered an era in which there was a need to constantly disseminate messages with a meaning (a purpose), messages that stated a position and fought the misinformation that was felt in a new situation for most people worldwide. In

Interviewee D's opinion, "Image was imperative in this period. And maybe it was the trigger for this revolution, which we are only just now starting in terms of social media presence." (…) "image was indeed king, specifically in social media, during this period."

It is not enough to just communicate. It is necessary that, in this communication, the messages reach their addressees or target audience. This is where design and its role in this digitalised communication comes in. According to interviewee C, "design in social media is fundamental". He adds that: "we can convey a message through an image" (interviewee C), thus explaining that we do not necessarily have to communicate in words. That communication does not necessarily have to be verbal or written, but rather that it can be only visual.

Interviewee D believes that image will be increasingly imperative in the messages and their transmission. Information needs to stand out in the midst of so many other similar information and it needs to be attractive to the targeted audience. The visual part—the image—can indeed strategically assist and reinforce in this highlighting need.

Design gives meaning to the message, helping its decoding in a very effective and assertive way. This, according to the interviewees' perspectives, manages to make the messages more attractive, as evidenced by the following statement made by interviewee B:

> One thing that design helps a lot with is to turn something 'boring', less interesting, into something interesting. (...) I think that in this sense, it was undoubtedly a privileged field and an anchor for communication. (interviewee B).

Interviewee E states that he believes that "design is absolutely structuring, but it is a tool and a means. Therefore, alone it does nothing", adding that "there is a set of tools that we aggregate to achieve goals". This idea is aligned with that of interviewee A who argues that "the role of design is the conjugation and complementarity that it offers to communication in social media".

Design is also pointed out as something "important that obviously helps us capture our audience's attention and helps us to make them receptive to the message." (interviewee E). This implies that design is not seen as a synonym for communication, but rather as a methodology, a tool to help communication take place.

It is also mentioned during the interviews that communication and design complement each other and should form a uniform and coherent pair:

> The creation of imagetic content is also the message. (...) there is a symbiosis here that is very important and that cannot solely be based either in aesthetics alone, or in communication alone. They have to go along together (interviewee D).

Interviewee C believes that institutions preferentially opt for face-to-face communication with their audiences, a reason that leads to a lack of attention and, consequently, a perceived devaluation of design and online digital communication. He reinforces this idea by stressing that:

> People, in general, still do not give much importance to communication and design. They think that it is better to have a face-to-face contact with the person than to communicate

through social media, through a defined communication strategy and an appealing design. (Interviewee C).

Colour and typography are, as previously said, two basic elements of design that are embedded into the creation of visual identities. And these are factors that help create the institution's image and brand in the public's mind. This is where communicative aesthetics becomes relevant.

In communication, as in life, time is short. So communication must be efficient in its purpose of reaching its target audience, and the message needs to be easy to decode, otherwise it will end up lost. At a time when there is so much communicational noise around us, we need to be selective in what we see. And here, both communication design and communication strategy are fundamental to fulfil this requirement.

Everything seems to indicate that, by combining communication design with strategic communication, it is possible to make the delivery of messages to their recipients more efficient. In other words, if, on the one hand, strategy allows us to set goals to achieve objectives, on the other hand, communication design is a factor that helps to achieve those same objectives.

Design, therefore, did not come to change the 'flow' of communication. Rather, it came to help messages to stand out in the midst of the chaos that is competition for the attention of audiences, thus asserting itself as an accelerator and catalyst that can aid the fluidity of the communicative action.

5 Conclusions

This study was based on the views and opinions of Paul Rand, Ellen Lupton, Steven Heller, Eva Heller and David Carson, renowned design authors, who led to the suspicion that design may play a preponderant role in communication. However, there are still few authors and scientific contributions that address design in the context of communication in pandemic situations or risk communication. Therefore, it was important to interview actors who dealt with this type of restrictions, on both sides of the communication—senders and receivers—in order to understand what worked in the pandemic context and, more importantly, to understand what did not work. It was also interesting to understand the perception of these actors about the role of design in the communicative flow of institutional messages of the HEIs in which they work or are part of. So, a set of people connected to institutional communication and students were interviewed, interviews from which it was possible to extract a set of conclusions.

During the pandemic period, among many other problems, one that was most evident was the problem of communication: a lot of information to be communicated and little time, or even resources to do it quickly and with the greatest possible reach. And this is where communication design can make a positive contribution.

Communication design stands out for differentiating the content of the messages according to the theme and the targeted audience. Using elements such as colour and typography, it is possible to capture the attention of the audience to the messages in a more immediate, targeted and even emotional way.

As analysed and discussed above, colours, their meanings and the associations and emotions they evoke, vary according to the context, the culture and even the personal lived experiences of each individual. However, colour is a powerful communicational tool. It is up to the designer to minimise the discrepancies in interpretations and responses, adapting the choice of colours according to the meaning that is intended to give to the message and according to the emotional response that is intended to arouse in the different audiences it is communicated. Typography also evokes different meanings, emotions and associations. And that is what also makes it a strong communication tool, which may give different meanings to the same message, depending on the goal and message previously outlined. However, it was possible to perceive through the interviews that, so far, colour and typography are not seen with the same degree of importance in the communicational process, being typography considered less important than colour. So, there is already a recognition of the importance of colour in a message, but the same does not (yet) apply to typography.

Design can, through its objectivity, be interpreted and understood as a solution to a problem, helping to achieve the defined goals in a clearer and direct way, helping to give a meaning and function to the messages.

In this sense, and after the completion of this research study, we can conclude that communication design helped and continues to help in the mitigation of communication problems caused by the pandemic. It allows communication through the dissemination of clear messages. It also allows these to be understood by any individual (whether they have more or less academic training) through visual communication.

So, there is an interest in the field of design and in the knowledge of designers. But it still is not fully understood. Having said this, we can say that the pandemic caused design to be more taken into account and caused it to have a more relevant role in the process of communicating messages. However, the way forward, for it to be assumed as a communication tool, has to continue to be worked on and studied.

It is important to further conduct studies and research on the importance that design has or can have in institutional communication. So, it is necessary to carry out an evaluation on and of the impact of design on the communication of organisations in crisis contexts.

References

1. Carson D (2003) Design and discovery [Video]. In: TED conferences. http://www.ted.com/talks/david_carson_on_design
2. Heller E (2012) A Psicologia das Cores: Como as cores afetam a emoção e a razão. Editorial Gustavo Gili, SL
3. Heller S, Ilić M (2012)Stop, think, go, do. How typography & graphic design influence behavior. Rockport Publishers
4. Lancaster M (1996) Colourscape. Academy Editions

5. Lupton E (2004/2011) Pensar con tipos. Un guia clave para estudiantes, diseñadores, editores e escritores. Editorial Gustavo Gili, SL
6. Lupton E (2011/2012) Intuição, Ação, Criação: graphic design thinking. Editorial Gustavo Gili, SL
7. Lupton E (2014) Type on screen. A critical guide for designers, writers, developers & students. Princeton Architectural Press
8. Mahnke FH (1996) Color, environment & human response. Wiley
9. Martins N, Brandão D, Penedos-Santiago E, Alvelos H, Lima C, Barreto S, Roberti A (2022) Self-initiated practices in the urban community of Balteiro: design challenges in a post-pandemic setting. In: Amic G.H (ed) Human factors in communication of design; AHFE (2022) International conference. AHFE Open Access, vol 49. AHFE International, USA. https://doi.org/10.54941/ahfe1002029
10. Rand P (1947/2014) Thoughts on design. Chronicle Books
11. Santos B (2013) A importância do design para tornar as redes sociais mais interativas. Texto Livre: Linguagem e Tecnol 6(1):150–165. https://periodicos.ufmg.br/index.php/textolivre/article/view/16637/13396
12. Silva J, Raposo D, Neves J, Almendra R, Brandão D, Martins N (2021) Narrative benchmarking in scenario assessments for digital design. In: Advances in ergonomics in design: proceedings of the AHFE 2021 conference on human dynamics for the development of contemporary societies, Lecture notes in networks and systems. Springer, Cham, New York, USA, pp 3–9. https://doi.org/10.1007/978-3-030-79760-7_1
13. UKEssays (2017) Don't mistake legibility for communication—Analysis. https://www.ukessays.com/essays/design/dont-mistake-legibility-communication-2998.php?vref=1

Game Design and Virtual Environments

Friction Firestarter: A Toolkit for Designing Meaningful Friction in Game User Interfaces

Isabella Silva, Pedro Cardoso, and Bruno Giesteira

Abstract Despite the prevailing paradigm of user-friendliness and enjoyment in mainstream game design and user interface design, intentional friction in game user interfaces that can be used to create meaningful experiences and to encourage reflection in players. This work aims to explore such use of intentional friction, providing designers with a valuable resource to generate unconventional game interfaces. As a starting point, we previously identified six strategies for intentional friction: (1) exploit memory shortcomings; (2) faulty feedback; (3) mismatched mental models; (4) impairment of ability; (5) deliberate inefficiency; and (6) oppressive constraints. Afterwards, to help operationalise these strategies and identify others, we ran co-creation workshops with game and UI designers, which lead to the development of a tool composed of three decks of cards, combining additional friction strategies, intentions, emotions, and ideation triggers, and enabling designers to create expressive game interfaces that intentionally incorporate friction as a design strategy. The *Friction Firestarter* toolkit is intended to inspire designers to explore various options and think creatively about friction in UI design.

Keywords User interface (UI) · Digital games · Game design · Friction · Card deck

I. Silva (✉)
Faculty of Engineering, University of Porto, Porto, Portugal
e-mail: Isabella.barbosa.s@gmail.com; up201902811@up.pt

P. Cardoso
University of Aveiro/DigiMedia, Aveiro, Portugal
e-mail: pedroccardoso@ua.pt

B. Giesteira
University of Porto/ID+, HEAD Health + Design Lab, Porto, Portugal
e-mail: bgiesteira@fba.up.pt

N. Martins et al. (eds.), *Perspectives on Design and Digital Communication IV*, Springer Series in Design and Innovation 33, https://doi.org/10.1007/978-3-031-41770-2_14

1 Introduction

Despite the potential of game user interfaces (UI) to challenge players' beliefs and encourage questioning of the status quo, UI practices are guided by an overarching paradigm of usability and enjoyment [24]. Although interface designers typically aim to limit or eliminate friction in their designs, many authors [6, 12, p. 139, 19, p. 100, 33, 38] indicate that games should also cater for friction and negative experiences. Therefore, designers of game UI can challenge the design rules of UI to go beyond the conventional paradigm of making user-friendly and pleasant interfaces and instead consider the emotions and behaviours evoked by interactions with the end user.

In this way, we explore the tension between the expectation of interface usability and the potential benefits of incorporating friction and challenge alongside the mechanical aspects of games for artistic, expressive, or esthetic reasons. The use of intentional friction strategies in game interfaces could help achieve several goals: Building on the natural antagonism of games [37], disrupting flow [33], avoiding dominance through positive psychology [15], promoting critical reflection [14], exploring a broader range of human emotions than just pleasure [11, 12, 25, 38], and expressing empathy, disobedience, deception, uncertainty, and uncontrollability [6]. We shouldn't hesitate to explore the full range of human emotions because, as stated by Norman [28, p. 49], both relaxed and anxious emotional states are effective and influential tools for human creativity and action. Moreover, in this way, we can broaden the range of representations of game interfaces and make them a more diverse medium.

This chapter expands the work presented in the paper *Strategies of Intentional Friction in the User Interface of Digital Games* [32] by introducing the card tool developed in the context of the dissertation titled *Intentional friction in the User Interface of Digital Games* [31]. The research began with a literature review to establish the context for using friction as a strategy. Next, we explored game instances where friction is employed to express a point of view, challenge current systems, or promote critical reflection. We drew from Norman's [28] seven design principles and Nielsen's [27] usability heuristics to identify six strategies for intentional friction [32]:

1. ***Exploit memory shortcomings***: The strategy is to exploit human memory limitations by hiding information and forcing users to recall it from memory. The game *Bomb the Right Place* [4] is an example of this, as it deliberately hides labels on a map, making it difficult for players to locate and strike targets. This violates Nielsen's [27] usability heuristic #6 of recognition rather than recall and aims to make players feel their lack of knowledge and sensitivity toward geopolitical conflicts.
2. ***Faulty feedback***: The strategy is to employ faulty feedback to mislead or overbear players, creating anxiety and tension. The game *Hellblade: Senua's Sacrifice* [17] uses a bluff feedback system where the player is threatened with permadeath if they fail, but in reality, they can always continue from the last saved point. This is designed to provoke fear and anxiety in players to simulate the fear of death

experienced by those with psychosis. The game also uses sound to simulate auditory hallucinations. It does not display health levels, violating usability heuristic #1 of *visibility of system status* and contributing to creating anxious emotional states in players.

3. ***Mismatched mental model***: It involves exploiting players' assumptions and expectations about game genres and terminology to create confusion and mislead them, transgressing Nielsen's [27] usability heuristics #4 of *consistency and standards* and #5 of *error prevention*. For example, *Undertale* [36] uses typical Role-Playing Game acronyms for game stats but misleads players by using EXP to stand for "EXecution Points" and LV to stand for "Level of Violence." *September 12th: A Toy World* [30] intentionally uses a crosshair as the mouse pointer to make it look like a shooting game but makes it impossible to win if players shoot terrorists, as it causes more civilians to become terrorists. These designs make players contemplate their actions and the consequences of unreflected violence.
4. ***Impairment of ability***: It involves intentionally impairing a user's abilities to develop empathy or provoke reflection and understanding through experience. This can be done by simulating sensory or perception impairments. *Hellblade: Senua's Sacrifice* [17] uses an audio and visual interface to simulate a character's mental illness experiences, aiming to create empathy and understanding. *Hypnospace Outlaw* [20] intentionally frustrates players with a difficult-to-read interface, mimicking the challenges faced by those with low vision or difficulty navigating confusing interfaces to create empathy for their struggles. These examples allow users to experience first-hand the daily frustrations and challenges that individuals with impairments face.
5. ***Deliberate inefficiency***: The deliberate inefficiency strategy involves intentionally making the UI of a game less efficient, which can increase the challenge and convey a message or feeling to the player. Games *Papers, Please* [29] uses UI inefficiencies to increase immersion and convey the natural antagonistic nature of the game. In *Papers, Please*, the UI mimics the frustrating experience of going through paperwork in real life, which serves to immerse the player in the role of a border-crossing immigration officer and reflect on the impact of bureaucracy imposed by institutions of power.
6. ***Oppressive constraints***: The oppressive constraints strategy involves limiting users' actions through physical, logical, semantic, or cultural curbs in the interface design. Games such as *Depression Quest* [9] and *Before Your Eyes* [2] use interface constraints to limit users' choices, express hardship, diminish the sense of freedom in players, and enhance the game's expressive argument. The constraints used in these games include lockouts and lock-ins [28] and transgressing Nielsen's [27] usability heuristics #3 of *user control and freedom*.

This chapter presents the objectives and methods employed in the next of steps of the research, a summary of the insights collected that contributed to the UI *Friction Firestarter* toolkit development, describes the toolkit usage, present its card decks' contents, briefly presents the first formative evaluation of the tool, and finally our concluding thoughts.

2 Designing the Toolkit

Considering the aforementioned UI design strategies, the next steps of the research aimed to identify further strategies and proposed a tool to aid designers and the community in putting these strategies into practice. The objectives were as follows:

1. To discuss the potential of intentional friction on UI as a way for communicating ideas, building arguments, and encouraging reflection;
2. To develop a toolkit for designing expressive video game interfaces using friction.

2.1 Methodology

To attain these objectives, we conducted workshops where we collaborated with other designers, gathered their viewpoints, and explored additional tactics. We aimed to acquire a variety of perspectives to discover novel ways of introducing intentional friction in video game user interfaces. Hence, these workshops fostered conversations about intentional friction in design, solicited feedback, and generated strategies for its incorporation.

The activities of the workshop were divided into stages featuring distinct methods. Stage 1 was designed as a focus group [16, p. 277] to discuss the theme with the participants and gather their viewpoints on the subject matter. Firstly, the participants were onboarded to the session, ensuring everyone was familiar with the tools and processes. Secondly, the group discussed the problem landscape serving as a focus group stage and paving the way for the idea generation to follow. Thirdly, a lightning demonstration and sharing of examples were conducted, highlighting relevant examples of intentional friction in game interfaces.

Stage 2 consisted of an ideation session with individual brainstorming and collaborative affinity diagramming [16, p. 24]. The group aligned on design principles, establishing a mutual understanding of the terms and concepts used in the ideation process. Later, participants engaged in an ideation session to generate as many design principles as possible to incorporate intentional friction into game interfaces. Afterwards, the group discussed the generated ideas, seeking to understand and explore the different perspectives and considerations presented. Finally, the session was wrapped up, with any outstanding questions or issues addressed.

We used the How Might We (HMW) statement as an alternative problem-framing method to introduce the research problem to the participants.[1] This method is helpful in brainstorming and is commonly used to explore various creative solutions, from insights to ideas. According to Hanington and Martin [16, p. 320], the HMW approach enables a creative exploration of multiple solutions by providing an alternative framing.

[1] The statement presented was "How might we, game and UI designers, leverage friction thought the interface to communicate ideas, build arguments and promote reflection".

The sessions were conducted remotely using a communication platform that enabled video and audio communication and recording. A Miro[2] board was set up in advance with relevant information and workspaces to facilitate collaboration and interaction with the participants. For qualitative analysis, transcription, and documentation of insights, we used Dovetail.[3]

2.2 *Characterization of the Sample of Participants*

To determine the participants for the workshop we followed the recommendations of ACET [1] to consider which individuals could provide valuable insights into the workshop's theme. We therefore sought insights from knowledgeable students and professionals in either game design or common UI design principles like usability heuristics. We limited the number of participants to 6 per session, and at least 2 were required to possess some game design experience. The workshop was held over two separate sessions, each lasting approximately two hours, between March 29th and April 1st, 2022. Three distinct participants attended the first session, and four attended the second.

As a safety measure during the COVID-19 pandemic, we conducted the workshops remotely, allowing us to reach participants from different nationalities and cultural contexts, including Portuguese, Brazilian, Slovenian, American, and Russian participants.[4]

2.3 *Stage 1: Focus Group*

The planning and development of the script for the focus group were based on the advice provided by ACET [1] for facilitating focus groups. The elaborated script consisted of an introduction session to explain the purpose and context of the workshop, an ice-breaker activity between participants, core questions to prompt conversation and a space for sharing examples of design friction.

The core questions were:

- When you think of design frictions, what are the three words or phrases that immediately come to your mind?
- What is your understanding of design friction in the user interface (UI)?
- Please share your thoughts on the relevance of design friction to UI design, both in general and for games.

---

[2] Visual collaboration whiteboarding platform—https://miro.com (Retrieved July, 2022).

[3] User research solution that assists in analyzing, synthesizing, storing, and sharing customer research—https://dovetailapp.com (Retrieved July, 2022).

[4] To ensure effective communication, all workshop materials were written in English.

After discussing what was understood as design friction in the UI, participants were presented with two different definitions of friction to stimulate the discussion further. These definitions were:

- Definition 1: "points of difficulty occurring during interaction with technology" [7, p. 2]
- Definition 2: "anything that prevents users from accomplishing a task" [34].

2.4 *Analysis and Insights*

The data from the workshop were analysed qualitatively. Dovetail[5] was used for data. The transcriptions were target of thematic analysis with main citations highlighted and grouped into codes. These groups were further divided into semantic categories to identify subtleties in how the participants discussed the topics. After coding and analysing each highlighted unit within the group, the findings were examined to reveal insights.

This section will discuss key insights that emerged from the two workshop focus groups stage after the qualitative data analysis.

2.4.1 Entertainment Games and Art Games Have Distinct Characteristics and Priorities

Participants discussed using friction as a tool for designers to express ideas and experiment in both entertainment and art games. Some argued that friction could create unique art pieces but negatively impact revenue and player experience, while others desired to experiment with unconventional game design to express emotions and perspectives.

It was interesting to hear the participants' takes on their views regarding games as an art form. This debate has been previously explored by other authors such as in a Koster's keynote [23], that suggest that entertainment games tend to be conservative and reinforce societal norms, while art games take risks and promote empathy. He also suggests that entertainment games prioritize delight and have a low cognitive load, while art games take risks and promote empathy and hard fun, potentially making intentional friction better suited for the latter.

2.4.2 Games Are a Safe Environment for Friction

Participants pointed out that games provide a secure space for players to try, fail, and learn from their mistakes. A game designer mentioned that he expects and accepts

[5] The software performed automatic audio transcription to speed up the process, but a manual review was still necessary to correct errors.

some level of friction when playing games. For example, a "virus" downloaded in a game will not harm a player's real-life computer, so any negative consequences are temporary and do not pose a real risk. This allows players to experiment and take risks that they may not take in real life. Safety is a crucial aspect of games, according to Crawford [8], and this discussion in the session highlighted the idea that games offer a way to experience psychological conflict and danger without any physical harm.

2.4.3 Friction May Create Exclusion

While participants regarded friction as a tool to reflect on people who experience accessibility barriers or other undesirable situations, it can also create accessibility and usability barriers, frustration, or negatively impact business and commercial considerations. Participants discussed the potential of friction as a tool for designers to express ideas and experiment, but some expressed concerns about alienating and excluding players. A balance between friction and accessibility or an awareness of these consequences was raised by participants as an important aspect to consider when designing intentional friction.

2.4.4 Friction Can Have a Close Relationship with Everyday Life

During the session, participants discussed examples of everyday friction experiences encountered while interacting with government services online, learning new software, buying tickets on websites, consuming media, or even interacting with other people. One participant mentioned the idea of "friendly design" as UI that nags the user for attention, subverting the usual meaning of the term, and added that "there is more friction in friendship than in utility." One participant commented on reclaiming the word "friendly" in a more interpersonal context instead of the corporate context of making things easy to use.

The discussion also touched on how intentional friction in the interface can serve to explore mundane frustrations and struggles in interpersonal relationships or daily life. Although newsgames [3] examine and discuss significant global news events, incorporating friction into the user interface can also allow for exploring mundane annoyances and difficulties. This kind of friction could help ideate in-game experiences and offer a unique perspective on everyday life.

2.4.5 Game UI is More Ambiguous Than Traditional UI

Video games can be classified as software systems comprising a multitude of interdependent components that interact to form a cohesive whole [13, p. 5]. Crawford [8] also noted that games exhibit a fundamental characteristic: they are systems comprised of numerous components that frequently interact in intricate and complex ways. This observation underscores a crucial aspect of games we observed to have caused some confusion among some participants, as they were often unable to clearly distinguish between various design elements such as mechanics, interface, and the blueprint outlined in the DDE framework developed by Walk et al. [37].

This was particularly true for diegetic interfaces. Since these are more tightly integrated with the game's spatial and narrative elements in comparison to non-diegetic interfaces, the confusion among participants was more conspicuous when the user interface was intricately linked to the world of the game's story. During discussions of the game *Hellblade: Senua's Sacrifice* [17], despite initial disagreement from other participants, one participant highlighted that the character's arm in the game could be interpreted as a diegetic representation of a progress bar. This example illustrates the interconnectedness of numerous design elements such as the world rule set, permadeath mechanics, diegetic interface report system, narrative, character design, and others within this game. As a result, it serves as an instance that underscores the intricate relationships among various design components that form a complex whole, thereby emphasizing the ambiguous nature of game UI compared to traditional non-game digital products and more work-oriented systems.

2.4.6 UI Designers Less Familiar with Game Design May Have Difficulty Grasping the Concept of Intentional Friction

UI designers lacking Game Design experience may find intentionally building friction in the user interface to be a foreign concept, to the point where they may not initially consider it a potential strategy. One participant emphasized this point. They expressed difficulty designing user interfaces that intentionally elicit emotional responses, as their typical work is guided by the goal of avoiding frustration and misleading users. However, they recognized the value of intentionally considering how to provoke emotions in the user interface. On the other hand, UI designers with more familiarity with Game Design may be more amenable to subverting established UI principles. For instance, despite working in traditional UI design, one participant had previously conducted heuristic analyses of games. They offered an alternative viewpoint that designers should distinguish between intentional design friction with a purpose and unintentional design mistakes. By being aware of both types of friction, designers can use them as they would other design patterns or heuristics to create "clean" interfaces. Additionally, this participant suggested that intentional friction could be used as an opportunity to teach or create something new. This suggests that, by recognizing the purpose and value of certain types of design friction, UI designers can use them to create innovative solutions.

2.4.7 Interface as the Dominant Actor and Autonomous from the Needs of the User

During the discussion, participants talked about how friction in user interfaces can shift power dynamics from the user to the technology. One participant noted that users typically see themselves as the dominant actor, but technology has more control in some situations. Examples during the workshop included companies monitoring remote workers and devices shutting off for updates without user input. Additionally, participants discussed that unfriendly interfaces might not be trying to be aggressive but rather focusing on the internal logic of the world it inhabits instead of catering to the user's wants and needs.

2.4.8 Definition and Vocabulary of Friction in the UI

During the discussion on the definition of friction, some participants felt that the definitions[6] we provided were too broad or too narrow. One participant's use of the term *abrasiveness* brought up interesting conversations on the metaphorical erosion of trust and patience in UI design. A reflection on the physics concept of friction and its counterpart in game and UI design led to the realization that friction can hinder and aid in achieving goals, and its absence can lead to a lack of control and precision. During the workshop, there were discussions about situations where friction and resistance are necessary to prevent slips in critical situations with grave consequences. One participant cited the example of shooting nuclear weapons, emphasizing the need for complexity and difficulty in such operations. These real-life friction characteristics were later mapped to usability efficiency, efficacy, and satisfaction [22]:

- Friction introduces resistance, which requires extra effort and energy from users and can reduce efficiency.
- Friction affects the ability to achieve a goal by providing too much resistance or removing it altogether, which can impact efficacy.
- Friction has the potential to gradually wear down and degrade a person's happiness, comfort, or enjoyment, which can have an impact on satisfaction.

2.5 *Stage 2: Design Workshop*

During this stage, participants had the chance to generate ideas for design principles that could be used to incorporate intentional friction into game interfaces. Design principles are guiding principles that help organizations, teams, or individuals make design decisions. Designers use these principles to integrate generally applicable laws, guidelines, biases, and design considerations into their work [21]. However,

[6] Definitions presented to participants: "points of difficulty occurring during interaction with technology" [7, p. 2] and "anything that prevents users from accomplishing a task" [34].

using them carefully is crucial, as they may need to be adapted depending on the project's goals and restrictions. Design principles aren't prescriptive because they're intended to be universally applicable, and their interpretation may vary depending on the designer. Nonetheless, they can be useful in game design, as they provide a framework for decision-making that can be adapted to various game genres.

During this stage, we provided examples of design principles and explained how they could be used in the context of game design, especially under a subversive lens of intentionally inserting friction. We also offered a cheat sheet with common design principles to help familiarize participants with the concept. Even where a lack of usability and understandability is deliberate, it's still essential to know the principles of good design since they can state in reverse how to move forward with our designs [28]. Therefore, the aim of providing the cheat sheet was to assist participants in generating ideas for intentionally breaking common design principles. The cheat sheet contained a variety of design principles, including Nielsen's 10 Usability Heuristics [27], Norman's seven fundamental design principles [28], Federoff's Game Interface heuristics [10], and Whitney Hess's [18] guiding principles for UX designers.

Participants engaged in an individual brainstorming activity to generate ideas to produce as many design principles as possible. After that, they shared their ideas in an affinity mapping exercise, where they grouped similar ideas together to better understand the connections and patterns between them, providing a basis for further discussion and exploration.

We used the ideas generated by participants as the basis for the tool that combines different friction strategies and intentions, emotions, or messages to communicate through friction. Combining the insights from the workshops with previously identified friction strategies [32], the tool offers designers a wide range of friction techniques to create unconventional game interfaces. The tool and its usage are presented in the next sections.

2.6 Tool Development

Workshop participants were asked to create an affinity map of their own and others' ideas after individual brainstorming. This approach was found to be effective in collecting final thoughts and identifying where ideas converged. After analysing the ideas for design principles, we needed to decide on a venue for practical usage. At first, we considered creating posters or a website to disseminate the strategies, but we chose to develop a card tool instead due to the considerable number of strategies. A card-based tool has several advantages, such as starting design discussions, enabling knowledge transfer, clarifying concepts, guiding the design process, facilitating shared understanding and communication, and providing a playful way to engage people [35].

It was also important to consider how the tool would be used in a practical context, and we took inspiration from Brandalise's [5] Infiltration opening process to structure the card deck and tool usage activities. Designing artefacts that challenge existing systems, even design itself, may be difficult. Brandalise [5] suggested a process that could help infiltrate dominant systems of power and institutions that seem unyielding and impenetrable. Her approach involves appropriating the characteristics of the system structure to question and subvert it using unconventional interventions. The Infiltration-opening process proposed by Brandalise provides a valuable perspective on how design can promote dissent and celebrate otherness. In Game Design, this approach can be leveraged to create a reflective and expressive UI that can convey powerful social and political commentary. By incorporating unconventional elements in UI design, designers can challenge existing systems and even the rules of UI design itself, which can serve as a powerful tool for promoting change and addressing societal issues.

To create the tool, we considered the insights and strategies collected during the workshop, information gathered from previous steps, and strategies identified in prior stages [31]. Therefore, we organized the strategies, insights, and knowledge in a spreadsheet to inform the cards' content. We then sorted the content based on three categories: emotions, strategies to incorporate friction, and reasons where friction could be used as a strategy. During this process, we ended up with three card decks that made up the tool: the intention deck, the expression deck, and the trigger deck. The cards in the trigger and intention decks have a description to aid their understanding and are identified by a number. In addition, some titles and text descriptions were adapted from the workshops, and playful explanations and examples provided by participants were kept.

Once the content was compiled, a layout was designed. Each deck was assigned a specific colour to aid quick identification, and a visual element was added to the back of each deck to differentiate them. The basic card layout includes the card number, title, and type (expression, intention, or trigger). The intention and trigger cards also contain descriptions to help with the card's understandability.

The Syne typeface[7] family was used for titles. This typeface explores unconventional weight and style associations, and as it becomes bolder, it gets wider, necessitating daring visual design choices. Because of its experimental and rebellious nature, it was deemed suitable for representing the card deck titles and the tool style. The DM Sans[8] typeface family, which has a geometric sans-serif style with low contrast, was used for the body text of the cards. This choice complements Syne and maintains the legibility of the text descriptions (Fig. 1).

[7] Syne was conceptualized by Bonjour Monde and designed by Lucas Descroix with the help of Arman Mohtadji. https://fonts.google.com/specimen/Syne.

[8] The DM Sans project was commissoned by Google from Colophon. https://fonts.google.com/specimen/DM+Sans/about?query=dm+sans.

Fig. 1 The game UI *Friction Firestarter* cards

3 The Game UI Friction Firestarter

We established a five-step process to use the *Friction Firestarter* tool. Step 1 defines the larger context, challenge, or issue to be explored. Step 2 focuses on defining the intention and reason for using friction as a strategy. Step 3 addresses the intended emotion or message to be suggested through friction should be defined. Step 4 is about selecting the friction strategy to be used as an ideation trigger. And step 5 is when brainstorming takes place to find ways to solve the brief based on the suggestions of the cards (Fig. 2).

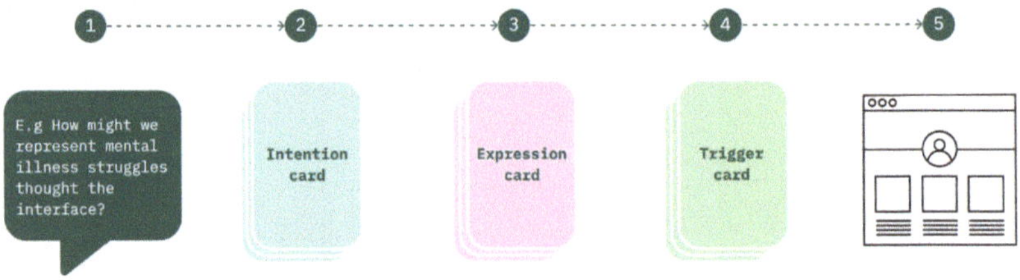

Fig. 2 Five steps for using the tool

3.1 Step 1—Defining the Big Picture: Exploring Context, Challenges, and Issues to be Addressed

The tool's first step is ensuring participants are aligned and focused on the same problem. It induces participants into a thought-provoking mindset and ensures that the ideation process with the aid of the card deck tool is productive. To frame the design challenge, participants should individually brainstorm potential design challenges and phrase them as questions beginning with "How might we...". Afterwards, they should share their statements and vote for the most appropriate scope and context. It is important to strike a balance between a statement that is not too broad or too narrow. Alternatively, if there's not enough time to complete all the tool steps in one session, the team can decide on the HMW statement beforehand.

Instructions

1. Reflect on a challenge the team wish to explore in the context of interface design and frame it with a "How might we..." statement. Example: "How might we represent mental illness struggles thought the interface"?
2. Write the challenge on a sticky note. This exercise is conducted silently and without discussion.
3. Continue writing sticky notes until the 10 min of allotted time is up.
4. After this exercise, participants should discuss and group the ideas generated during the ideation exercise by their similarity. Allocate 15 min for this activity.
5. Next, participants should vote on the most fitting challenge for further exploration. Allocate 5 min for this activity.

Activity duration: Approximately 30 min.

3.2 Step 2—Defining the Intention: Articulating the Purpose and Benefits of Using Friction as a Strategy

This step focuses on determining why friction could be an appropriate design strategy for the given challenge. This step involves using the intention card deck, which consists of different scenarios where friction could help address interface design challenges. After defining the design challenge in the first step, participants are instructed to review each intention card, discuss the ones that resonate with them, and vote for the one they believe is most relevant to the challenge. The card with the highest number of votes is then chosen (Fig. 3).

Instructions

- Pick up the intention card deck, which contains various scenarios where friction can be used to address interface design challenges.
- After defining the design challenge in the first step, review each intention card.
- Discuss with the group the intention cards that resonate with the design challenge.

Fig. 3 The intention card deck

- Vote for the intention card that is believed to be the most relevant to the challenge.
- The card with the highest number of votes is chosen for further exploration.

Activity duration: Approximately 10 min (Table 1).

3.3 Step 3—Defining the Expression: Clarifying the Intended Emotion or Message to Suggest Through Friction

In step 3, the focus shifts to the expressive component of the solution, and participants are asked to consider the potential emotions, experiences, and messages that could be conveyed through friction. This involves the Expression card deck. This card deck offers multiple choices that suggest concepts for the desired player experience. Participants should review the cards individually, discuss the ones that stand out, and vote for the most suitable option for the challenge. This step aligns participants on the desired experience for the players and the message to convey to them (Fig. 4).

Instructions

1. Pick up the expression card deck, which features diverse options for potential intended emotional expression.
2. Review each expression card individually.

Table 1 The intention card deck's contents

#	Intention	Observation
1	Create empathy between the player and a situation	Use the interface to develop empathy or provoke reflection and understanding through experience
2	Make the UI a tool of self-expression for the designer	The interface can be a self-expression tool where the designer can communicate their own worldview and struggles with players or use it for aesthetic purposes
3	Subvert the power dynamic between the user and the technology	Create situations when users are not truly in control, and the interface is not actually there to serve them
4	Challenge dominating structures and bias	The interface can challenge game and society conventions, player expectations, and the plethora of entities and connections that make up digital games
5	Explore human emotions other than enjoyment	Negative emotions are a powerful and valid way to foster action and reflection
6	Build upon the game natural antagonistic nature	Understanding and interacting with a frictional interface can be a formal challenge in the game
7	Make a point, critique, or capture real life/everyday struggles	Get inspired by all the friction real life already has to reflect critically on the triviality of everyday life
8	Teach something to players	Committing errors and failing is part of learning
9	Reinforce the game narrative	Challenges, difficulty and tension are driving forces in narrative. A frictional interface can serve as a metaphor, allowing the player to feel tension as an in-game character would

3. Discuss with the group the expression cards that stand out and have the potential to convey the intended emotions, experiences, and messages to communicate through friction.
4. Vote for the most suitable option to the challenge at hand.

Activity duration: Approximately 10 min (Table 2).

3.4 *Step 4—Define the Friction Strategy: Using Friction as an Ideation Trigger*

This step introduces participants to strategies and principles for intentionally designing friction in the interface through the trigger card deck. This deck contains a range of strategies for creating friction without being overly specific. The aim is to stimulate ideas and facilitate decision-making in the next step. Participants should individually read each trigger card, discuss the ones that stand out to them,

Fig. 4 The expression card deck

and then vote on the most relevant card based on all previous choices. As the most content-heavy deck, participants should be given more time to review and discuss it in comparison with previous steps. Alternatively, this step can be done by randomly choosing a trigger card to avoid potential decision fatigue (Fig. 5).

Instructions

1. Pick up the trigger card deck, which contains a range of strategies for designing friction.
2. Read each trigger card individually and take note of the ones that stand out by physically separating them from the rest or placing a dot sticker on them.
3. After approximately 15 min, the group should be encouraged to discuss the trigger cards with each other and vote on the most relevant card based on all previous definitions.
4. Alternatively, the group can randomly choose a trigger card to avoid potential decision fatigue.

Activity duration: Approximately 20 min (Table 3).

Table 2 The expression card deck's contents

#	Expression
1	Disobedience
2	Oppression
3	Deception
4	Betrayal
5	Uncertainty
6	Uncontrollableness
7	Unforgiveness
8	Interpersonal hardships
9	Overwhelmingness
10	Vulnerability
11	Powerlessness
12	Apathy
13	Annoyance
14	Shame
15	Regret
16	Inscrutability
17	Hostility
18	Fatigue
19	Sluggishness
20	Slyness
21	Isolation
22	Perplexity
23	Boredom
24	Sadness

4 Preliminary Formative Evaluation and Concluding Thoughts

4.1 Goal, Sample, and Procedure

After creating a prototype, we initiated an evaluation process to assess how the *Friction Firestarter* tool was able to aid designers in exploring intentional friction in gaming interfaces. To observe the practical application of the tool and collect qualitative feedback on its capability to achieve its intended purpose, we conducted a workshop with a new set of 4 participants, who were UI or Game Design students and professionals.

Fig. 5 The trigger card deck

During the formative evaluation session, we explained the tool's usage and its three card decks, and participants were guided through each of the five-step process,[9] highlighting the playful nature of the decks. To streamline the workshop, a predefined design challenge was presented to participants in step 1: "How might we remix *Monopoly* [26] to explore equity, inclusion, and transparency in the real estate/housing system?" Participants were free to reimagine how the board game would be represented in a digital game interface.

4.2 Results

During the ideation step, a participant expressed concerns about going too far with friction ideas and potentially breaking the game. Participants found trigger deck cards challenging due to the abundance of suitable options and the fear of missing out on potential ideas. They proposed using multiple trigger deck cards to create unique combinations and challenges to address this.

Despite such challenges, the group was able to generate ideas that met the challenge, with the cards 5 from the intention deck—explore emotions beyond enjoyment; 11 from the expression deck—*powerlessness*; and 29 from the trigger deck—*keep users in the dark*. These cards opened up various avenues for exploring social

[9] Described in the previous section.

Table 3 The trigger card deck's contents

#	Trigger	Description
1	Explore diegetic interfaces	Make everyone aware of your interface, even the characters in the game. The information may become hidden or difficult to parse, increasing the cognitive load of some tasks or making actions take longer
2	Use deliberate inefficiency	If users want to move a file to the trash, have they close their garbage bag, replace it with a new one, and throw it out in the dumpster. I hope they know when the garbage collectors are coming
3	Do a golden detour	People tend to choose the easy way out. Give a generally unfavorable outcome on the path of least resistance. Allow the player to take some roads less traveled by adding a minimally functional UI
4	Slow the player down	Give them time to think: If it can be one click, make it take two clicks. Use long delays or display too many options
5	Drop random inputs	Give the interface power over the user. Not all buttons want to work all the time; it's hard work being pressed. If the button wants to be nice, it can tell you that it is not feeling like doing it now
6	Lie or lie about lying	Use the interface to display completely untrue information. Who knows? Eventually, they may start using their critical thinking and skepticism
7	Make content not understandable	Sometimes, it is tiresome to explain everything. Remain loyal to the material of your design language even at the cost of user comprehension. Exploit readability: use jargon and unfamiliar terms
8	Use mismatched visual mimicry	Not everything is what it seems. When a design copies the visual appearance of a known object, it indicates the way it will work or be utilized (due to its familiar look). Exploit this assumption
9	Use faulty feedback	Give useless information like it is feedback or fake errors as a part of the experience. Make people wonder what is part of the show and what is not
10	Create perception or ability impairment	Put the players in other people's shoes. Make players experience accessibility barriers first-hand to get some perspective. Consider making the interface less perceptible, operable, simple, and forgiven

(continued)

Table 3 (continued)

#	Trigger	Description
11	Exploit signal-to-noise ratio	Manipulate the proportion of important to irrelevant information. Consider diluting useful information with useless information. Clutter the UI with distractions
12	Give too much feedback	A lot, really, like, an uncomfortable amount, I'm not kidding. Think about visual, auditorial, and haptic feedback. Every interaction could have a fun sound, making that mute button useful
13	Stop trying to make the UI invisible	Stop trying to hide it. If it's in there, I bet it's important. If it's important, put it in the middle of the screen. Please don't be shy; make it enormous
14	Give too much control	Do not constrain players to do anything—and don't warn them about the consequences of doing so. So let players do things very easily, even if they are not sure what they are doing
15	Create a bad first impression	Make players judge you by the cover. Foster a negative attitude and emotional reaction from the get-go
16	Bait players	Build trust in the system by fostering internal consistency. After the user trusts the system, double-cross them
17	Make errors unrecoverable	Don't let the user take it back. Don't provide any way for players to reverse their actions, any safety nets, confirmation, warnings, or help
18	Use minimalistic design	That is it! Simplify interfaces to the point of abstraction
19	Create aporias	Intentional use of gaps, lacks, and omission in information delivery leaves room for interpretation and experimentation. Let them wander a bit
20	Design for slips	Skateboarding is fun because you may slip at any moment. Therefore, design the interface for users to commit actions they did not intend to do
21	Give lazy feedback	Delay feedback, so the player needs to wait to understand the result of their actions. Make it difficult for them to course-correct their mistakes immediately
22	Exploit memory shortcomings	Make users recall information from memory as much as possible. Maybe interrupt users while they are in the middle of it. Is the cognitive load too high? It doesn't matter; Make them do it faster

(continued)

Table 3 (continued)

#	Trigger	Description
23	Reclaim "friendly" design	There is a lot more friction in friendship than there is in utility. So, make the interface mimic human relationships and behaviors, even the annoying ones
24	Provoke the user with oppressive constraints	Tease the user with options they can't use or are missing, drawing their focus away from their task or play and onto the interface. Where is that mute button?
25	Use a mismatched mental model	If a thing is expected to work in some way, break it. Make them have to relearn it and remind the user things only work a certain way because some human beings decided that one day
26	Require precision	Make the players thread the needle by designing the interface to demand players to be precise, accurate, or fast
27	Consider the world outside of the game	Make the real-world part of the game through the interface. Get inspired by how unpleasant the real world is to create moments of humor, discomfort, or insight
28	Use confusing mapping	Map the controls to be unlike the player's mental map. Make it difficult for players to understand the layout of the controls and the devices being controlled
29	Keep users in the dark	What they don't know won't hurt them. Refrain from giving players critical information on their performance and status of the system

issues within the game, such as financial literacy, contracts, housing conditions, and unexpected expenses, e.g. their ideas introduced financial literacy challenges where players could not effectively make informed decisions regarding investments, budgeting, and resource management. The combination of card 11 (*powerlessness*) and card 29 (*keep users in the dark*) generated ideas for situations where players faced unexpected expenses or contractual obligations that they were unaware of, adding a layer of complexity and realism to the gameplay.

5 Concluding Thoughts and Future Work

While requiring more research, it is promising to note that expressiveness played a significant role in the participants' discussions and ideation process. By incorporating social issues and emotional exploration into the game, it may be possible to create a more thought-provoking experience. Through iterative playtesting and feedback, designers can refine the implementation of these ideas and ultimately create a unique and socially conscious gaming experience.

Also, as a tool with the potential to reflect on different everyday difficulties, it can empower non-designers to contribute to a game interface design that effectively conveys a message or emotion of a diverse group. For example, this tool could be useful in pedagogical contexts associated with inclusion. Teachers could use the tool to design educational games that promote empathy and understanding towards diverse groups by introducing friction in the form of challenges that require the player to think critically about social issues and consider different perspectives. In the field of environmental sustainability, the tool could be used to design games that raise awareness about environmental issues and promote sustainable behaviors. In the field of social justice and activism, the tool could be used to create games that raise awareness about social injustices and promote activism. Overall, the potential applicability of the Friction Firestarter tool is broad and varied, and further testing and exploration in different contexts could lead to new insights and possibilities for its use.

While the workshops provided relevant information, subsequent testing with different procedures and samples may yield different results. With a bigger sample, new strategies and conclusions may emerge. Future work could also focus on assessing the tool in a real ongoing game project to see if it can still provide value when faced with constraints such as system impact, feasibility, and effort. Additionally, while this research primarily focused on game UI Design, it also explored processes and concepts that could potentially be useful in other areas of game development. This idea should be further researched.

Friction Firestarter is intended to inspire designers to explore various options and think creatively about friction in UI design. It should not be viewed as a definitive guide but rather a tool to encourage designers to approach their work with an open mind and consider new possibilities. That is the direction this tool is heading and where future work will reside.

References

1. ACET (2011) The art of facilitating focus groups. http://www.interpretereducation.org/wpcontent/uploads/2015/02/NCIEC_focus_group_manual_07-20112.pdf
2. *Before Your Eyes* (2021) GoodbyeWorld games. https://www.beforeyoureyesgame.com
3. Bogost I, Ferrari S, Schweizer B (2010) Newsgames: journalists at play. MIT Press. https://doi.org/10.7551/mitpress/8854.001.0001
4. *Bomb the Right Place* (2016) Everyday arcade. https://everydayarcade.com/games/bomb-the-right-place
5. Brandalise I (2016) Eventual everydays: infiltrating and opening systems through design [Master's thesis, Parsons School of Design]. ISSUU. https://issuu.com/isabellabrandalise/docs/isa_thesisbook_issuu

6. Cardoso P, Melo R, Carvalhais M (2019) Breaking the hedonistic loop: meaning before fun in videogames. In: Artech 2019: proceedings of the 9th international conference on digital and interactive arts, no 52, Portugal, pp 1–4. https://doi.org/10.1145/3359852.3359902
7. Cox AL, Gould SJJ, Cecchinato ME, Iacovides I, Renfree I (2016) Design frictions for mindful interactions: the case for microboundaries. In: CHI EA '16: proceedings of the 2016 CHI conference extended abstracts on human factors in computing systems, USA, pp 1389–1397. https://doi.org/10.1145/2851581.2892410
8. Crawford C (2011) The art of computer game design (n.p.)
9. *Depression Quest* (2013) Zoë Quinn. http://www.depressionquest.com
10. Federoff MA (2002) Heuristics and usability guidelines for the creation and evaluation of fun in video games [Master's thesis, Indiana University Bloomington]. CiteSeerX. https://citeseerx.ist.psu.edu/viewdoc/summary?doi=10.1.1.89.8294
11. Forgas JP (2017) Mood effects on cognition: affective influences on the content and process of information processing and behavior. In: Jeon M (ed) Emotions and affect in human factors and human-computer interaction. Academic Press, pp 89–122. https://doi.org/10.1016/b978-0-12-801851-4.00003-3
12. Frasca G (2007) Play the message: play, game and videogame rhetoric [Unpublished Ph.D. dissertation]. IT University of Copenhagen
13. Galloway AR (2006) Gaming: essays on algorithmic culture. University of Minnesota Press.
14. Grace, L. D. (2014). Critical games: critical design in independent games. In: Proceedings of 2014 digital games research association (DiGRA), USA. https://doi.org/10.13140/2.1.2607.3603
15. Han BC (2021) Sociedade paliativa: a dor hoje. Editora Vozes
16. Hanington B, Martin B (2019) Universal methods of design: 125 ways to research complex problems, develop innovative ideas, and design effective solutions. Rockport Publishers
17. *Hellblade: Senua's Sacrifice* (2017) Ninja theory. https://www.hellblade.com
18. Hess W (2010) Guiding principles for UX designers. UX Magazine. https://uxmag.com/articles/guiding-principles-for-ux-designers
19. Hodent C (2018) The gamer's brain: how neuroscience and UX can impact video game design. CRC Press
20. *Hypnospace Outlaw* (2019) Tendershoot. https://www.hypnospace.net
21. Interaction Design Foundation (2017) *Design principles*. https://www.interactiondesign.org/literature/topics/design-principles
22. International Organization for Standardization (2018) Ergonomics of human-system interaction—Part 11: usability: definitions and concepts (ISO Standard No. 9241-11:2018). https://www.iso.org/obp/ui/#iso:std:iso:9241:-11
23. Koster R (2012) A theory of fun 10 years later [Video]. GDC vault. https://gdcvault.com/play/1016632/A-Theory-of-Fun-10
24. Kuang C, Fabricant R (2020) User friendly: how the hidden rules of design are changing the way we live, work, and play. WH Allen.
25. Lawhead N (2019) On UI design & using UI as a means to tell a story, convey emotion, create personality… An in-depth look…. The Candybox Blog. http://www.nathalielawhead.com/candybox/on-ui-design-using-ui-as-a-means-to-tell-astory-convey-emotion-create-personality-an-in-depth-look
26. *Monopoly* (1935) Elizabeth J. Maggie, & Charles Darrow. http://www.monopoly.com
27. Nielsen J (1994) Enhancing the explanatory power of usability heuristics. In: CHI '94: proceedings of the SIGCHI conference on human factors in computing systems, USA, pp 152–158. https://doi.org/10.1145/191666.191729
28. Norman D (2013) The design of everyday things: revised and expanded edition. Basic Books
29. *Papers, Please* (2013) Lucas pope. https://papersplea.se
30. *September 12th: A Toy World* (2003) Gonzalo Frasca. http://newsgaming.com/newsgames.html
31. Silva I (2022) Intentional friction in the user interface of digital games [Master Dissertation]. University of Porto. https://hdl.handle.net/10216/145082

32. Silva IP, Cardoso P, Giesteira B (2022) Strategies of intentional friction in the user interface of digital games. In: Advances in design and digital communication III. DIGICOM 2022. Springer Series in Design and Innovation, vol 27. Springer International Publishing, pp 49–62. https://doi.org/10.1007/978-3-031-20364-0_5
33. Soderman B (2021) Against flow: video games and the flowing subject. MIT Press
34. Swallow D (2018) A web of anxiety: accessibility for people with anxiety and panic disorders [Part 2]. TPGi. https://www.tpgi.com/a-web-of-anxiety-accessibility-forpeople-with-anxiety-and-panic-disorders-part-2/100
35. Tahir R, Wang AI (2020) Transforming a theoretical framework to design cards: LEAGUE ideation toolkit for game-based learning design. Sustainability 12(20):8487. https://doi.org/10.3390/su12208487
36. *Undertale* (2015) Toby fox. https://undertale.com
37. Walk W, Görlich D, Barrett M (2017) Design, dynamics, experience (DDE): an advancement of the MDA framework for game design. In: Korn O, Lee N (eds) Game dynamics. Springer, pp 27–45. https://doi.org/10.1007/978-3-319-53088-8_3
38. Wilson D, Sicart M (2010) Now it's personal: On abusive game design. In: Futureplay '10: proceedings of the international academic conference on the future of game design and technology, Canada, pp 40–47. https://doi.org/10.1145/1920778.1920785

In[the Hate Booth]: A Case Study on How to Deal with Online Hate Speech

Susana Costa, Mirian Tavares, José Bidarra, and Bruno Mendes da Silva

Abstract In today's hyper-connected world, digital games and online gaming communities occupy a prominent place in the communication system, in social media, forums or Internet communities, where online hate speech (OHS) takes place, frequently and publicly, triggering toxic environments. In this chapter we present a case study based on interviews, distributed in two sessions, to ten participants with 12 and 13 years old, and an experience over the SG *In[The Hate Booth]*, as a counterproposal to address OHS. The qualitative data approaches three aspects: the experiences with OHS, the perspectives about OHS and the possible solutions to counteract OHS. We conclude that OHS is a common complaint from players and a characteristic behavior in game communities. Data shows that even users who don't identify themselves with this behavior accept it as part of online environments and agree that this toxicity continues outside the in-game screen with effects in everyday life. The pedagogical approach, namely through SG, is perceived as a possible measure to counteract the OHS.

Keywords Games · Serious games · Online hate speech

1 The Game as a Generator of Learning and Meaning

The works of Huizinga and Caillois have broadly shown the relevance of games in culture. Huizinga [1] describes the game as a free and meaningful activity, spatially and temporally separated from the demands of practical life and limited by a self-contained system of freely consented, but mandatory rules, immersing the player in an intense and total mode.

S. Costa · M. Tavares · B. M. da Silva (✉)
Research Centre for Arts and Communication, Universidade Do Algarve, Faro, Portugal
e-mail: srsilva@ualg.pt

J. Bidarra
Research Centre for Arts and Communication, Universidade Aberta, Lisboa, Portugal

N. Martins et al. (eds.), *Perspectives on Design and Digital Communication IV*, Springer Series in Design and Innovation 33, https://doi.org/10.1007/978-3-031-41770-2_15

Salen and Zimermman [4] applied and recreated the magic circle concept, initially mentioned by Huizinga, to digital games. According to the authors, the magic circle is the space where the game takes place: playing means entering this magic circle, a limited space with specific characteristics, but with infinite possibilities [4].

Caillois [5] proposes four fundamental categories to describe the complexity of games: competition, chance, simulation, and vertigo. These categories can intersect and combine in different ways.

Huizinga [1] also highlights the construction of game communities that tend to remain beyond the duration of the game. The members of these communities share the feeling of being in an exceptional situation of sharing something important, of distancing themselves together from the rest of the world and suspending the usual norms. This sense of belonging, manifestly present in contemporary platforms, has allowed for the aggregation and gathering of communities around digital games.

The researcher Celia Pearce [6] observed how the voluntary nature of the game, an aspect already pointed by Huizinga [1], can contribute to lead players to establish a participatory group of collective sharing around an online game, as happens on platforms such as *Reddit*, *Twitch* or *Discord*. Some of the main motivations for the formation of these communities include a sense of pleasure and happiness arising from sharing and collaborating with peers [6].

The emergence of SG brought a new pedagogical approach. SG are specifically developed for educational purposes, to help acquiring some specific concept, skill, or technique, beyond play and entertainment [7] [8]. These games are mainly based on the constructivist theory, emphasizing the experience, the construction of knowledge through argumentation, collaborative work, discussion, and debate. Since the 1980s, when the first SG appeared for military training, the variety of SG has been increasing significantly, not only regarding the type of game, but also regarding the type of skill that is concerned by the game [7].

During a game, the player is an active participant, that makes interpretations of experiences, elaborates, and tests these interpretations, appropriating the given information. In this perspective, the construction of knowledge is not spontaneous, it needs to be activated, in a provocative and challenging practice [9].

Non-governmental organizations, consortia benefiting from public funding, researchers, educators, and politicians have been promoting the use of videogames in education to address complex subjects. Through the narrative and interaction with the game, players are expected to understand that they are active agents, influencing the environment around them. Climate change, water saving, waste separation, health and financial education have been topics of games, through local, national, and European programs that support projects promoting the development of videogames that promote sustainable behaviours.

For its part, the videogame industry has been attentive to the potential of games in education and has presented proposals in various scopes of training intervention [3].

According to Gee [2] immersing an individual in a virtual environment with characteristics of the physical world is one of the most effective forms of learning. Gee highlights how games encourage players to think about different game variables to solve problems and accomplish goals. When playing, the engagement in the process of problem solving is high, because of the internal mechanics of the game.

Engagement and immersion in the game are achieved through the interaction with game mechanics: the login bonus; levels, rewards, rankings, and scores, as well as immediate feedback.

In this chapter, we analyze the importance of games and the possibilities SG to address a contemporary problem: the spread of OHS on the online communities and its effect on youth. The case study presented analyzes the collection of qualitative data from two sessions carried out with 10 volunteer participants who were interviewed in two moments: before and after trying a SG tackling OHS.

2 The Problem: OHS

The use and dissemination of OHS pervades online platforms in a ubiquitous way. It flows instantly and massively. As a result, OHS has come to be recognized as a serious problem by democratic governments, giving rise to several international initiatives as countermeasures to the problem.

OHS may occur in a spontaneously way, as a reaction to the stress of the game, or in a programmed and strategic way, as a path to extremism and radicalization often as a resource to program attacks within and out of the virtual world [10, 11]. If, on one hand, the individual right to freedom of expression is inalienable and indisputable, it is no less important to underline that the exercise of this right implies responsibility and respect for the Other, ensuring the difficult balance between fundamental human rights.

The committee of Ministers of the Council of Europe issued in May 2022 [12] a Recommendation on hate speech, including OHS, to assist European States in preventing and combating hate speech within the framework of human rights, democracy, and the law. This Recommendation also highlights the need to ensure legislation addressing OHS and foreseeable provisions for the swift and effective removal of OHS that is prohibited under criminal, civil or administrative law, as well as the mechanisms for reporting the cases of OHS to public authorities and private actors, including internet intermediaries, and rules for the processing of these reports [12].

Misogyny, racism, antisemitism, homophobia, xenophobia, and other forms of alterophobia have various mechanisms for producing victims and causing harm. The concern of democratic governments is precisely the solution to this problem, without harming the values of freedom of expression, seeking a sensitive balance between freedom and equality or inclusion. Wachs et al. [13], distinguished six categories of motivations to perform OHS: (1) revenge, (2) power, (3) joy, (4) ideology, (5) group conformity, and (6) status enhancement.

OHS is based on the use of hostile and malicious speech, directed at an individual or a group of people, motivated by a discriminatory, intimidating, disapproving, antagonistic and/or harmful attitude towards one or more characteristics such as gender, race, religion, ethnicity, skin color, national origin, sexual orientation and/

or disability. This phenomenon has gained greater proportion and visibility through the Internet, due to the rapid haste with which it can be disseminated.

The study of the state of the art in this field shows that many youngers play and use social platforms, from earliest ages to communicate with each other within communities, often without any adult supervision [11].

Sellars [14] surveyed a set of common traits that help defining and identifying hate speech: (1) the fact that it is addressed to a group or an individual, as a member of a group; the presence of content that expresses hate and may cause harm; (2) intent to harm; (3) the public nature of the discourse and, finally, (4) the existence of a context that makes a violent response possible. Citron and Norton [20] defined four forms of response to online hate messages: (1) inaction (2) deletion/suspension of the message and user (3) education (4) counter-narratives.

To define OHS, it is necessary to address human rights. According to the United Nations, human rights are inherent to all human beings, regardless of race, sex, nationality, ethnicity, language, religion, or any other status, including the right to freedom of opinion and expression, which sometimes seems to clash with the definition of hate speech.

In accordance with the Universal Declaration of Human Rights, [21] every human being is entitled to enjoy the rights and freedoms set forth in this Declaration, without distinction of any kind, such as race, color, sex, language, religion, political opinion or of another nature, national or social origin, wealth, birth, or any other status. In these definitions, human rights are guarantees for all individuals, regardless of their singularities. This, in turn, goes against OHS, which preaches prejudice against human beings who are part of some social minority. Hate speech violates the guarantees and fundamental rights of every citizen. The Recommendation on hate speech [12] underlines its deep commitment to the protection of the right to freedom of expression, seen as the essential foundations of a democratic and pluralistic society.

3 The Methodology: A Case Study

In this study, we intend, in a first phase, to analyze the experience of a group of youngers on the Internet, regarding the contact with OHS. In a second phase, we propose an experience of a SG about OHS to provoke a reflection and discussion on the effects of hate speech and the possible means to contain it, prevent it or cope with it.

The methodology was based on a case study [15], articulated with a narrative approach that departs from the experiences of youth and, at the same time, requests for participatory action within the community of participants in the case study.

Within the narrative approach, the fiction-based research, using the metaphor, can cause changes in the way individuals relate to themselves and others [16, 18], through a reflective, participatory, and aesthetic process, since the research developed is more truthful, meaningful, useful, accessible, and human [17]. In the SG, this methodology allows the description of the participants' reactions in the face of events and actions.

This is a case study descriptive and instrumental [15], that aims to contribute to the understanding of a broader problem, identifying its roots and wider societal context, as well as its various expressions and different impacts on those involved in OHS experiences. In this case, the experiences described are instruments to understand the effect of OHS on youth and the possibilities of education, namely the use of counter and alternative narratives to understand, face and cope OHS.

This experience is also the basis for the development of a questionnaire to be applied on a large scale, with closed answers, in order to obtain quantitative data from a larger sample.

3.1 The Sample and the Context of the Case Study

Olhão is a city located in the South of Portugal in the region of Algarve. It is a multicultural municipality, with 33% foreign inhabitants. Its contribution has been important in demographic terms, namely for maintaining a positive migratory balance and for generational renewal, as well as for boosting the regional economy, bridging the labor shortage in certain periods and/or sectors, bringing ideas and investment and adding experience. Of the 17 main nationalities living in Olhão, 10 correspond to European Union countries, 2 to Eastern European countries, 3 to Asian countries, 1 to an African country and 1 to country from Latin American [19].

The community of Olhão is often in the news for issues relating cases of violence. Last January, the community was surprised by a video that circulated on the social network *Instagram* in which a group of 4 students from the school Agrupamento de Escolas Francisco Fernandes Lopes physically assaulted, with extreme violence, a Nepalese immigrant that was returning home from work.[1] Further investigation showed that it was not an isolated case. The circumstance was so outrageous that the President of the Republic decided to intervene by visiting the school, where he did an open class on hate speech and immigration.

This multiculturality and specific characteristics of the region are important aspects to contextualize the sample, since the participants of this case study are a group of youngers, living in Olhão and attending this same school, Agrupamento de Escolas Francisco Fernandes Lopes. Schools in Olhão are a mirror of the municipality's multiculturalism, as we can find students of various nationalities. The sample is composed of a group of ten volunteer students aged between 12 and 13 years old of the following nationalities: 5 Portuguese students, 1 Russian student, 2 Brazilian students, 1 Ukrainian student, 1 Indian student. Six of the students identified themselves as boys and four identifies themselves as girls.

Our intervention in this community looks for a change of behavior and empowerment of the participants and the institutions involved. This type of research

[1] https://www.dn.pt/sociedade/tres-jovens-de-16-anos-detidos-por-agressoes-A-imigrante-nepales-em-olhao---psp-15831311.html.

also presupposes that the researcher will proceed collaboratively with the participants (in constructing content, formulating questions, collecting data, and analyzing information).

3.2 The Process and the Implementation

To implement this case study, we followed Yin's research design and methods [15]. We defined the stages to implement the study: development of a plan of action, definition of the sample, the strategies to promote data collection, analysis, and the report of the experience.

The planning stage focused on identifying the research questions of the study for the in-depth interview with attributes of validity and reliability. It is a time-intensive method in which two interviews are conducted, each one week apart, to allow participants to reflect on what they have shared and to help them reconstructing meaning of their experiences.

The contact with the participants was organized in two sessions and interviews: Session 1 (S1) and Session 2 (S2). The first interview intended to understand the experiences and events leading to the phenomenon under study, such as the experiences with OHS and the strategies used by the participants to cope with the phenomena. After this first interviews, participants were invited to play the SG *In[The Hate Booth]*.

In[The Hate Booth] is SG inspired in several examples of OHS. This game is based on a fictional blog signed by two fictional characters and commented by fictional followers. Following the blog entries, it is possible to experience the escalation of hatred.

This interactive fiction game gives the player a written position or situation about OHS and allows the player to enter text-based commands that the computer will respond to, progressing to the next post. In this game, the word takes all the emphasis. The aim is to highlight the power of words, as a starting point of the discussion and reflection on OHS.

Throughout the blog authors' posts, an atmosphere of hatred can be felt in the comments that focuses on various stereotypes: issues of gender, sexuality, nationality, freedom of expression and disrespect for the difference and for the Other.

The levels are the different posts and comments. To advance to the next post, the player should discover a keyword hidden in the text. The hate escalation also progresses with each level. The game is a starting point for the reflection on OHS and its effects in everyday life.

The S2 intended to analyze the experience of the SG and the perspectives for possible solutions to address OHS on games and communities. After S1, we outlined the profile of the gamers in order to extract the essence of their experiences for data analysis. For the case presented here, the phenomenon of the research was to understand the lived experiences with OHS in online communities.

The first interview was based on two main questions divided in groups of sub questions to ease the description of the experiences (Table 1). In this phase, the participants had staggered entry times in the room and the interviews were conducted individually.

After the interview, the experimentation of the game *In[The Hate Booth]* was made in small groups. The group activity provided a space for reflection and collaborative critical thinking, fostering a constructivist learning and thinking environment with potential to build a perspective on OHS and its effects on digital and non-digital life. The pedagogical approach, through the SG, was a facilitator, influencing engagement and mapping the group comments and interactions.

A second approach to the group of participants was scheduled a week after the first session. In this S2, the goal was to promote meaningful discussion and critical thinking around democratic values and participation on online spaces. To this effect a second main question and four sub questions were launched to the groups (Table 2).

Table 1 The questions and sub questions of S1

Main question	Sub questions
What is your experience with online hate speech?	Where do you experience it? Can you tell a specific episode? How do you react when you see or hear hate speech episodes? Do you follow youtubers or gamers that use OHS? Can you name same examples?
What is your perspective on hate speech?	How do you see haters? Do you have an idea of when hate speech is more likely to occur? Do you perceive a "climate" of hate speech in games or social communities? Where? Which are the consequences of hate speech in everyday life?

Table 2 The questions and sub questions of S2

Main question	Sub questions
What is the possible solution to counteract hate speech?	How do you evaluate the possibility of using SG to address OHS? What would you suggest to game designers to prevent hate speech? What would you suggest to community managers? What can video gamers (and communities of gamers) do to prevent hate speech?

3.3 Results and Discussion

The answers to the first question "What is your experience with online hate speech?" were wide-ranging. All participants reported experiences of OHS, as presented on Table 3.

The reported experiences occurred in different environments, with an emphasis on *Discord*, *Free Fire* and *GTA*. In six cases, the Participants (P2, P5, P6, P7 and P8) revealed to be, simultaneously, victims and aggressors, because when they are attacked, they respond aggressively. The reactions of P2, P4, P5 and P7, shows that toxic environment increases the use of OHS within the community.

The participants' reports about their online gaming practices consisted of behavioral complaints arising from interactions between players in a particular instance of gaming. OHS is merged with common personal attacks against other gamers, ranging from offences to the performance of the player, as reported by P1, P2, P5, P6 and P8. Some narratives also exposed complaints on racism (P9), sexist speeches (P1), and nationality discrimination (P9). We can find, in these episodes, the 4 common traits established by Sellars [14] to identify hate speech.

Participants reported some names of YouTubers with aggressive, racist, and sexist content in their videos or streaming's. Live streams can create stars with the ability to influence other players to act on certain types of attitudes. Taspio (5 occurrences), one of the leading figures in the *Epic Games Battle Royale* community, currently averaging 12,000 concurrent viewers on *Twitch*, is in the top of the list. The Participants mentioned the name of YouTubers which are often blocked at some point of their streaming' sessions, due to the use of OHS which doesn't comply with the terms and policies of use of the platforms.

The participants reported their perspective on OHS, as represented in Table 4, recognizing communities and online games as places for the dissemination of discrimination speeches and toxicity. The most referred online spaces were *Tik Tok*, *GTA*, *Roblox* and *Fortnite*. Participants P1 and P2 underlined the fact that OHS addresses the avatar and not the person behind it. Participants P1, P2, P5, P7 and P10 find a correlation between the stress of the game, the performance of the gamers and the occurrence of OHS, and this is the most used argument to tolerate and accept OHS as part of the game. There are also two occurrences indicating that OHS can help improvements in gamers performance (P9) and that it can add fun to the game (P8). One participant (P3) indicated a correlation between violent games (like *GTA*) and the occurrence of OHS. Participant P8 related OHS to the number of followers (more OHS generates more followers to the gamers account).

Finally, all Participants (excepted Participants P1 and P10) recognized the effect of OHS in everyday life, highlighting the relation with low self-esteem, isolation, and anxiety. Participant P2 and P10 correlates the effect on everyday life with the familiarity of the aggressor, indicating that if the aggressor is a strange, then OHS has no consequences.

In S1, the experimentation of *In[The Hate Booth]* was the last part of the session. S2 took place a week later with the same group. In Table 5, we summarize the

Table 3 Answers from S1: Experiences with OHS

Experience with OHS	
P1	
Where	*GTA* e *Discord*
Description	"I was insulted when I died. The insults were about the fact that I was a girl, and I didn't know how to play."
Reaction to OHS	"I ignore"
Youtubers using OHS	Taspio
P2	
Where	*Free Fire*
Description	"Someone told me I was a bot"
Reaction to OHS	"I reply back"
Youtubers using OHS	T3ddy e Taspio
P3	
Where	*Roblox* e *Rocket League*
Description	"They called me gay and said I didn't know how to play. I was attacked by a group."
Reaction to OHS	"I try to beat the aggressors in the game. It is the best answer"
Youtubers using OHS	Pew Die Pie
P4	
Where	*Gartic*, *Roblox*, *Fortnite*
Description	"I was insulted in Gartic, by Brazilian users, because of my nationality (Portuguese)."
Reaction to OHS	"I react with anger"
Youtubers using OHS	–
P5	
Where	*Discord* e *Instragram*
Description	"I was insulted because of my performance in the game."
Reaction to OHS	"I reply back"
Youtubers using OHS	T3ddy e Taspio
P6	
Where	*GTA* e *Fortnite*, *Free Fire*

(continued)

Table 3 (continued)

Experience with OHS	
Description	"I was playing *Free Fire* and a boy said that I didn't know how to play and that my avatar looked like crap."
Reaction to OHS	"At first, I would get upset and respond badly back. After some time, I gained a certain tolerance for this type of people, so whenever it happens, I simply mock the situation in a sarcastic way. Nowadays, I only answer if the hate speech is against someone else, if it is against me, I just make fun of the situation."
Youtubers using OHS	Soph and Taspio
P7	
Where	*GTA*, *Free Fire*
Description	"In *GTA* I am frequently insulted. It is the game."
Reaction to OHS	"I reply in the same tone."
Youtubers using OHS	MOE
P8	
Where	*Discord*, Fortnite, *Free Fire*
Description	"I saw a famous *Free Fire* YouTuber disguised as a beginner player who was the target of hate speech by another player just because he didn't have a skin and the player said as much nonsense as possible."
Reaction to OHS	"OHS is normal because it happens a lot. I think that a person who plays a lot is already used to these offenses and doesn't care about them."
Youtubers using OHS	Destiny and FaZe Tfue
P9	
Where	*GTA* e *Overwatch*
Description	"Insults against the color of my avatar or my nationality."
Reaction to OHS	"It makes me want to play more and to be well ranked."
Youtubers using OHS	Taspio e Zigueira
P10	
Where	*Roblox*, *Discord*, *Gartic*
Description	"Often, my cousin (who has depression because he doesn't have friends) tried to establish these relationships in online games and he was often attacked. As it we can imagine, for a person with this type of problem it becomes a very complicated situation"
Reaction to OHS	"I didn't say anything, but I was more careful the next time I stayed there in the same game. I took hate speech as a learning situation to improve in the game"
Youtubers using OHS	Xbox Mil Grau

Table 4 Answers of the S1: perspectives on OHS

Perspectives on OHS	
P1	
Perspectives on OHS	"The OHS can be part of the game. It is part of the learning on the game. When we insult someone based on their skin, we are using the characteristics of the avatar and not the player."
Occurrences of OHS	"It depends a lot on the type of game and on the player, but it's common to offend new players who don't know the game well and judge them for having a low level."
Perception of a climate of hate	*Overwatch*, *GTA*, *Roblox*, *Fortnite*
Consequences in everyday life	"OHS shouldn't be taken seriously because it's just a game and it's just haters."
P2	
Perspectives on OHS	"OHS happens in moments of tension. When we turn the game off, we forget."
Occurrences of OHS	"The moments of the game that generate more OHS are those that involve more adrenaline or when there are teams."
Perception of a climate of hate	*Roblox*
Consequences in everyday life	"An offense will only really offend the person if the person is familiar or if the offending person is a friend of the offended person. It is almost impossible in a game with a large or medium community to be offended because the improbability of knowing who the player is behind the avatar."
P3	
Perspectives on OHS	"OHS may just be a reaction to moments in the game. You insult someone, but you don't really want to affect anyone."
Occurrences of OHS	"More violent games, like GTA, have more OHS."
Perception of a climate of hate	*GTA*, *Roblox*
Consequences in everyday life	"Not everyone has the same level of sensitivity, but there are people who can be truly affected."
P4	
Perspectives on OHS	"Sometimes games are invaded by groups of haters who act as a group against a certain person. This can affect who is attacked."
Occurrences of OHS	"When the attack is in groups, there is sometimes the intention to insult certain types of people."
Perception of a climate of hate	All online games

(continued)

Table 4 (continued)

Perspectives on OHS	
Consequences in everyday life	"Many gamers use games as a way to escape their reality. If they have a very bad day, it is possible that they will play when they get home, hate speech can, in this way, worsen the emotional state of a player who is not used to it. This can be extreme and can lead the player who has been the victim of this type of speech to do things that are detrimental to their mental or physical health."
P5	
Perspectives on OHS	"OHS is part of the internet and gaming. It doesn't affect me, and I try to ignore it. Unfortunately, there are bad people everywhere."
Occurrences of OHS	"Aggressive language is a reaction to the game."
Perception of a climate of hate	*League of Legends* (*LoL*)
Consequences in everyday life	"It can affect the player both psychologically and physically, it can lower the victim's self-esteem and lead to mental problems."
P6	
Perspectives on OHS	"For some people being a hater is cool, it's like being a leader. When I'm in a game with a toxic environment, I try to ignore it."
Occurrences of OHS	"Hate speech is inevitable, it is present on the internet, on social networks. People like to say things like that to make themselves feel superior."
Perception of a climate of hate	*Tik Tok* e *Gartic*
Consequences in everyday life	"It can affect the player both physically and psychologically, as it can lower the victim's self-esteem and make them feel bad about themselves."
P7	
Perspectives on OHS	"All online game communities always have people who practice hate speech."
Occurrences of OHS	"The insults and aggressive language come at the worst moments of the game or when you loose."
Perception of a climate of hate	In all online games
Consequences in everyday life	"Gamers transfer these actions to everyday life."
P8	
Perspectives on OHS	"OHS can add more excitement to the game."
Occurrences of OHS	"Some gamers use hate speech to get more followers."
Perception of a climate of hate	*Fortnite* and *Free Fire*
Consequences in everyday life	"Being the target of hate speech can be anxiety-provoking during the game and beyond. It can affect who plays and who is attacked."

(continued)

Table 4 (continued)

Perspectives on OHS	
P9	
Perspectives on OHS	"Sometimes, OHS help us understanding a mistake we did in the game."
Occurrences of OHS	"Insults come as an instinctive response to the stress of the game and one's performance"
Perception of a climate of hate	*GTA*
Consequences in everyday life	"It causes low self-esteem"
P10	
Perspectives on OHS	"OHS is an Internet problem and it affects people. In games there is no moderation, often when you are attacked you have to leave the game."
Occurrences of OHS	"Most of the time it happens when gamers are losing, and they start accusing the others of hackers, or insulting because they are nervous about losing the game."
Perception of a climate of hate	*Tik Tok*
Consequences in everyday life	"In offline life, I don't think there are many consequences, it depends on the person, I don't take this in a wrong way, because those people who make fun of me have never even seen me in real life, but other people get upset!"

feedback on the experience of the SG, which was mainly described as useful and fun.

All participants were unanimous in recognizing the power of games to learn and the possibilities of SG to address OHS and propose a reflection on the subject. Participant P2 has highlighted the importance of being able to talk about such an important topic in a school environment.

As measures for counteract OHS, there was a wide range of suggestions. Participants considered game designers as the less responsible for OHS. Participant P1 suggested games should be more inclusive, with representative characters; P5 suggested the elimination of expressions of hate from the games; P8 suggested the creation of more SG to address OHS.

Most Participants (P2, P5, P7, P8, P9, P10) indicated that managers should block and ban gamers reported for OHS. Participant P6 suggests that gamers should not be banned, but muted. Managers are perceived as a key figure for controlling OHS, because of their power to block and punish users and also because they are seen as an example of behaviour.

Gamers are seen as the main triggers of OHS. Participants suggest respect, inclusion, and calm in the reactions, as well as the punishment of being expelled of the community and of the game. Participants P3 and P10 were unable to express suggestions to counteract OHS, indicating that they don't believe that OHS will ever disappear.

Table 5 Answers of the S2: Solutions for OHS

Solutions for OHS	
P1	
Feedback on the SG experience	"We can learn a lot from games. It was interesting to see the examples of OHS"
Suggestions to game designers	"Make games more inclusive and representative of all."
Suggestions to community managers	"Be more active in the moderation, being an example to the gamers."
Suggestions to gamers	"Be more respectful, calmer and inclusive"
P2	
Feedback on the SG experience	"It loved that we were able to talk about games in the classroom. The game was very amusing. It is a good way to start talking about OHS"
Suggestions to game designers	"I think the designers should be responsible, because they are the ones who have to make a game that doesn't allow that kind of speech from the outset, not having to take corrective measures after its launch."
Suggestions to community managers	"Block certain words or phrases from comments."
Suggestions to gamers	"Leave the game."
P3	
Feedback on the SG experience	"The game helped us to learn how to manage situations of hate. I missed some images in the game"
Suggestions to game designers	"I don't know. I think it is part of the game and we cannot eliminate it."
Suggestions to community managers	"I don't know. I think it is part of the game and we cannot eliminate it."
Suggestions to gamers	"I don't know. I think it is part of the game and we cannot eliminate it."
P4	
Feedback on the SG experience	"It was fun to pass the levels as a team and only by reading and searching words. I think we can learn how to deal with situations."
Suggestions to game designers	–
Suggestions to community managers	"I think managers are the most important. They usually play and enjoy the game, like gamers, but they have the power to silence OHS. I think what they usually do is enough."
Suggestions to gamers	"Being banned from gaming, it's like an addiction."
P5	
Feedback on the SG experience	"I've learned a lot from games, keeping calm under stress, paying attention to details without too much effort, like being a team player, historical episodes, scientific phenomena, geography and much more. This game is a good example for talking about hate."
Suggestions to game designers	"They can eliminate expressions or language from the game that could be used to offend, not allowing the game to send any message from the player that contains any of these expressions or language."

(continued)

Table 5 (continued)

Solutions for OHS	
Suggestions to community managers	"The punishments go directly through the platforms where the games are, and therefore through their managers. The player base is on the platform, hence the accounts (and inventories, as in the case of Steam)."
Suggestions to gamers	"Respect the others."
P6	
Feedback on the SG experience	"It was fun, I think it is a good idea to use games for education"
Suggestions to game designers	"I think it has nothing to do with the games."
Suggestions to community managers	"Banning the person from the game is not the best way to end the game, as the person can simply create another account to play, but if the player is muted, he would continue with the false impression that he manages to hurt other players, however no one can hear or read what he writes in the chat."
Suggestions to gamers	"Gamers should take it easy."
P7	
Feedback on the SG e experience	"I really enjoyed it; this game is a fun way to put someone in someone else's shoes."
Suggestions to game designers	"I don't think it's a problem with games, but with the people who play."
Suggestions to community managers	"In cases of hate situations in these groups, haters should be permanently banned and could no longer access the platforms."
Suggestions to gamers	"It's the players who start to incite and start these groups. Those are the ones who should be banned."
P8	
Feedback on the SG experience	"It was fun and a good moment to talk and experience a subject that is important. I would suggest adding sounds to the game"
Suggestions to game designers	"Create educational games."
Suggestions to community managers	"Kick out those persons."
Suggestions to gamers	"Those who create and manage the communities are not to blame for the game having a toxic community, it is the players who make the community toxic with OHS towards noobs, newbies, players who criticize, even if constructively, the game."
P9	
Feedback on the SG experience	"It is a serious problem, and it is important to talk about it in a funny way."
Suggestions to game designers	"It is difficult because there is OHS even in non-violent games."
Suggestions to community managers	"Block trolls and haters."

(continued)

Table 5 (continued)

Solutions for OHS	
Suggestions to gamers	"If players don't use hate speech it will disappear. I think designers and managers already do everything they can to avoid it."
P10	
Feedback on the SG experience	"It was a good session and make me think about how OHS affects people."
Suggestions to game designers	–
Suggestions to community managers	"Ban the haters."
Suggestions to gamers	"Players have to try not to take the game too seriously."

4 Final Considerations

During an online game, a player's performance can cause a defeat of a team. Players can react aggressively through common and personal insults that are based in some kind of discrimination related to color, gender, ethnicity, and location.

The interviews revealed the acceptance by most participants of toxic environments, even though they claim not to appreciate this kind interaction. This attitude is reflected in the act of ignoring hateful speech rather than confronting it.

The responses of the participants allowed us to perceive that the reactions to occurrences of OHS are mostly inaction or denunciation. However, we also found a large number of responses, not indicated by Citron and Norton [20], who describe an equally aggressive behavior as a reaction leading to an increase in toxicity.

The answers of the Participants allowed to distinguish different motivations to perform OHS. We could distinguish power, joy, group conformity, and status enhancement. There is, however, an attitude of reaction that is not described in Wachs' scale for perpetration of OHS [13]. Participants confessed to perform OHS as a reaction to the toxic environment of the game or the community.

Almost all Participants recognize that the speeches voiced by the players have a real effect on everyday life and may be impregnated with prejudices that arise in heated moments or frustration. This gives rise to racist, sexist, xenophobic speeches, among others. Toxic comments don't translate in physical aggression and harm, but they disturb the activities of games and communities. Even those who do not support OHS and suggest that it will never be eliminated.

Conclusions on how players perceive the responsibility for OHS are divided. Few Participants blame the game designers, or the communities' managers, but most of the elements point to the gamers as the responsible for online toxicity.

The experience of the SG *In[The Hate Booth]* brought a new perspective on the possibilities of education and a pedagogical approach to the problem, within the school, which received the approval of the majority of the participants and was pointed out as a solution to address the phenomenon. While working in small groups, it was possible to create a sense of community, united by a common goal, based on the SG. This strategy appears to be a useful tool and a starting point in the mobilization against OHS.

After hearing and analyzing the Participants testimonials, it was possible to conclude that there is a need for awareness, in order to curb OHS, since the game and the experiences in the game are proven to be, as pointed by Huizinga, Caillois and Pearce, a constructor of meaning.

This case study intended to analyze the online interactions of a group of young participants and the contact and perspectives they have in relation to OHS. From this first approach, we can understand that this is an issue with effects on youngers offline and online experiences. The data collected is important to further the research, namely creating a survey by questionnaire, with closed questions allowing to collect data on a larger scale.

Although this study proposes a pedagogical approach, based on the experience of a serious game, further research also needs to analyze more systematically the issue of supervision, parental support and educator's monitoring of youth online experiences.

Acknowledgements This publication is financed by national funds through the projects "UIBD/04019/2020 CIAC" and "UI/BD/150850/2021" of the Foundation for Science and Technology.

References

1. Huizinga J (2003) Homo Ludens: um estudo sobre o elemento lúdico da cultura. Edições 70, Lisboa
2. Gee JP (2003) What video games have to teach us about learning and literacy. Palgrave Macmillan, New York
3. Martinez K, Menéndez-Menéndez MI, Bustillo A (2022) A new measure for serious games evaluation: gaming educational balanced (GEB) model. Appl Sci 12:11757. https://doi.org/10.3390/app122211757
4. Salen K, Zimmerman E (2003) Rules of play: game design fundamentals. MIT Press, Cambridge
5. Caillois R (2017) Os jogos e os homens: a máscara e a vertigem. Editora Vozes, Lisboa
6. Pearce C (2009) Communities of play. The MIT Press, Cambridge
7. Marsh T (2011) Serious games continuum: between games for purpose and experiential environments for purpose. Entertain Comput 2:61–68. https://doi.org/10.1016/j.entcom.2010.12.004
8. Bayart C, Bertezene S, Vallat D, Martin J (2014) Serious games: leverage for knowledge management. TQM J 26. https://doi.org /https://doi.org/10.1108/TQM-12-2013-0143
9. Zin, N. A. M., W. S. Y.: Design and evaluation of history digital game-based learning (DGBL) software. Journal of Next Generation Information Technology 4.4 (2013) 9.
10. Uyheng J, Carley K (2021) Characterizing network dynamics of online hate communities around the COVID-19 pandemic. Appl Netw Sci 20(6). https://doi.org/10.1007/s41109-021-00362-x
11. Gagliardone I, Gal D, Alves T, Martinez G (2015) Countering online hate Speech. United Nations Educational, Scientific and Cultural Organization, Paris (2015). https://bit.ly/3pFWTjp

12. Council of Europe: Recommendation CM/Rec (2022) 16 of the Committee of Ministers to member States on combating hate speech. Strasbourg (May 2022) bit.ly/3Zf3cJS last accessed 2023/03/25.
13. Wachs S, Wettstein A, Bilz L, Gámez-Guadix M (2022) Adolescents' motivations to perpetrate hate speech and links with social norms. Comunicar 71:9–20. https://doi.org/10.3916/C71-2022-01
14. Sellars A (2016) Defining hate speech. Berkman klein center research publication, 2016–20, pp. 16–48 Boston University School of Law (2016). https://doi.org/10.2139/ssrn.2882244
15. Yin RK (2014) Case study research design and methods, 5th edn. Sage, Thousand Oaks
16. Clandinin DJ, Rosiek J (2007) Mapping a landscape of narrative inquiry: borderland spaces and tensions. In Clandinin DJ (Ed.) Handbook of narrative inquiry: Mapping a methodology, pp. 35–75. Sage Publishing. https://bit.ly/3sjNXSP
17. Leavy P (2009) Method meets art: Arts-based research practice. The Guilford Press, New York
18. Riessman CK (1993) Narrative analysis. Sage, New York
19. Comissão de Coordenação e Desenvolvimento Regional do Algarve: Números em Destaque: População estrangeira residente. CCDR Algarve, Faro (2019).
20. Citron, D. & Norton, H.: Intermediaries and Hate Speech: Fostering Digital Citizenship for Our Information Age. Boston University Law Review, V. 91 (2011). https://ssrn.com/abstract=1764004
21. Organized Nations General Assembly: Universal Declaration of Human Rights (UDHR). New York: 1948 https://www.un.org/en/about-us/universal-declaration-of-human-rights

Endogenous Asymmetry in Games: Expanding the Typology

Abel Neto, Pedro Cardoso, and Miguel Carvalhais

Abstract Asymmetry is a prevalent feature in modern games. It allows each player to have a unique gameplay experience, which improves a game's replayability and makes in-game interactions more analogous to those we encounter in real life. Yet, there has been little discussion about the phenomena of asymmetry, how they occur during play, as well as their repercussions on player experience. This chapter begins with a definition for asymmetry, followed by its deconstruction into two major categories: endogenous and exogenous. We analyse six types of endogenous asymmetry described in the literature—(1) Ability, (2) Challenge, (3) Goal, (4) Responsibility, (5) Information, and (6) Interface—and present six other types: (7) Operation, (8) Location, (9) Time Frame, (10) Interdependence, (11) Outcomes, and (12) Feedback, discussing how they may interact with one another.

Keywords Asymmetric gameplay · Gameplay mechanics · Games · Game design

1 Introduction

The term 'asymmetric games' generally describes games in which players have sets of mechanics that are different to the point of seemingly belonging to different games. This usually entails different in-game perspectives, goals, and actions that

A. Neto (✉)
Faculty of Engineering of the University of Porto, Rua Dr. Roberto Frias, 4200-465 Porto, Portugal
e-mail: abelneto@live.com

INESC TEC—Institute for Systems and Computer Engineering, Technology and Science, Campus da Faculdade de Engenharia da Universidade Do Porto, Rua Dr. Roberto Frias, 4200-465 Porto, Portugal

P. Cardoso
University of Aveiro/DigiMedia, Campus Universitário de Santiago, 3810-193 Aveiro, Portugal

M. Carvalhais
i2ADS/Faculty of Fine Arts of the University of Porto, Avenida Rodrigues de Freitas, 265, 4049-021 Porto, Portugal

N. Martins et al. (eds.), *Perspectives on Design and Digital Communication IV*, Springer Series in Design and Innovation 33, https://doi.org/10.1007/978-3-031-41770-2_16

lead each player to have a set of mechanics and play style so unique that they may specialise in just one of the roles. This employment of the term has some utility, as it allows players to find games that aim to provide such disparities. However, we find that this terminology can almost be a misnomer, since most—if not all—games are asymmetric to some degree, allowing players to experience them in different ways. Even the game of Chess, so apparently equitable that it might be mistaken for a perfectly symmetrical game, has an asymmetry: White plays first. This unavoidable imbalance, while seeming inconsequential at first, has trickle-down repercussions over the course of a match, resulting in up to a 64% win rate for White in tournaments [4]. This shapes players' expectations and compels them to adopt different strategies depending on the colour they are playing.

If Chess, with this single asymmetry, evokes these differences in gameplay experience, what is there to say about other games in general? In this text, we point to the fact that asymmetry is a pervasive feature that can be observed across genres, and for good reasons: asymmetry boosts replayability by increasing the number of novel experiences a game can provide; it also improves strategic decisions by adding complexity to the process of reaching the optimal strategic response [14]. Asymmetries present in games often emulate the asymmetries present in real life [1], offering unique perspectives and experiences [15].

However, designing asymmetric play is not without challenges. Keith Burgun levies several criticisms towards asymmetric play, addressing how it "generally causes games to be vastly harder to balance than they should be", since adding asymmetric elements brings about the need to balance them fairly [5]. Furthermore, asymmetry can be "a smokescreen" that makes it harder to assess the quality of games and to find problematic elements that aren't working as intended. The added complexity brought by asymmetry can be an obstacle towards the design and development of games. As Burgun points out: "[i]f you make a fighting game with just 4 characters, what you've actually done is create ten different games. Each matchup is a distinct game".

The term asymmetry itself seems to be undefined in the literature. As we will discuss later, we were only able to find definitions for the term asymmetric game, and even those were found to be incomplete and unreliable. The aforementioned design difficulties, coupled with the fact that we weren't able to find an overarching categorization for asymmetry—or even a concise definition for it—indicate that it should be better understood if we aim to enjoy its advantages while minimising its challenges. The definitions and the categorization presented in this text seek to contribute to establishing a vocabulary for scholars, game designers, and developers. Our goal is for this to represent a step towards a greater understanding of asymmetry, as well as its effects on a game's dynamics and aesthetics.

This chapter revises and expands the contents of "Asymmetric Gameplay: Types and perspectives" [13], including new contributions and an updated perspective on video game asymmetry.

2 Asymmetry

2.1 General Definition of Asymmetry

Before we delve into the different forms asymmetry may assume in a game, we must define it. Despite its pervasiveness in games, we were not able to find a concise definition of *asymmetry* in the literature. However, we uncovered several definitions for the term *asymmetric game*. While this term has some utility for players, we believe that for designers and scholars alike, it is more productive to define *asymmetry* itself, because even games that aren't often regarded as asymmetric contain some degree of asymmetry. Addressing this issue, we could aim to reduce the colloquial use of this term by expanding the category of *asymmetric games* to include all those games (most games, if not all), but this could also increase the confusion that already exists when discussing the phenomena of asymmetric play. Alternatively, we chose to adopt the nomenclature that aims to describe what is at the core of our analysis: *asymmetry* itself. While such terminology can be hard to adopt by a not so specialised audience, we believe that scholars and designers will benefit greatly from this approach: *a definition that addresses what is at the root of asymmetric play, and not one that describes the games that feature it.*

We used current definitions of asymmetric game as a starting point for a definition for asymmetry, but we found them lacking in several aspects. Tracy Fullerton ties a definition of asymmetric game to the existence of variations in "abilities, resources, rules, or objectives" [7], a definition that misses a number of other ways in which games can offer asymmetric experiences. Also, by enumerating a closed set of possibilities, this definition cannot possibly encompass future developments in game design, becoming increasingly outdated as designers find new ways to create asymmetric experiences, and as such merely reflecting those that existed at the time of its writing. Most definitions we found share this issue. Christophe Bortolaso et al. state that in asymmetric games, "players take on different roles, sometimes using different forms of interaction, different hardware, and different views of the game world" [3]. This definition lists some ways for a game to achieve asymmetry, but like Fullerton's, it focuses solely on some of the expressions of asymmetry that already exist. A better definition must, therefore, be capable of including all current forms of asymmetry, but also be as adequately prepared as possible to encompass forms yet undiscovered.

Ernest Adams describes asymmetric games as those in which players play by different rules and chase different victory conditions [9]. Here, we see rules being used as a catch-all term that encompasses all the ways that asymmetry can originate from a game's formal rules. However, as we will see later, asymmetry can occur even when a game's formal rules do not account for or encourage it. Furthermore, there are many examples of games that can be considered asymmetric in which players do not chase different victory conditions. In *It Takes Two* (2021), two players play collaboratively and make use of their distinct abilities to progress through the game and achieve their shared goal.

These definitions of asymmetric games point towards the idea that the distinguishing feature of these games is the emphasis on disparities in how different players act and experience the game world. With this in mind, we can propose a definition of asymmetry to be used throughout this text as *variations in gameplay caused by how differently players can act in a game.* This definition is worded very deliberately to allow us to observe differences in how players are allowed to act within the context of the game, but also in how they may choose to act. Furthermore, by establishing our focus on "variations in gameplay", we mean to hone our analysis in how these different actions impact a player's experience and interpretation of the game.

Our analysis distances itself from the colloquial understanding of asymmetry because it does not exclude single player games. While the term *asymmetry* may suggest that we are comparing two different experiences, there's no reason why those experiences need to belong to different players playing concurrently. Asymmetry can exist between two solo playthroughs of the same game. When two different players engage with the same game, they inevitably make different choices that differentiate their experience with the game. Such disparity can exist even if those two playthroughs were experienced by the same player. When a player revisits a game, their experience will be conditioned by the fact that, for instance, they know which strategies work best for each of the game's challenges. Therefore, we believe that single player games should not be exempt from being examined in regard to how they can provide asymmetric gameplay experiences.

2.2 Endogenous and Exogenous Asymmetry

It is evident that variations in game mechanics can fundamentally alter the different players' experiences within a single game. However, there is also another category of asymmetry: one that concerns aspects that are external to the game, but nevertheless impact it in one way or another. These aspects can include differences between the players themselves, or the context in which the game is taking place. Let us return to our quasi-symmetrical example of a game of Chess. An amicable match between friends will invariably feel very distinct from one in a tournament, as much as a match between two similarly skilled opponents will be dissimilar to one between a teacher and their pupil.

In these examples, asymmetry at the level of game mechanics will surely emerge throughout the match, as players trade pieces or establish different positions on the board. However, at the same time, the context the game is taking place in, as well as the players' respective skill levels, will also play a role in shaping expectations and goals for each player, therefore fundamentally influencing their experiences with the game. Figure 1 illustrates the influence that differences in player profiles have in each of the three design components described in Robin Hunicke et al.'s MDA framework [11].

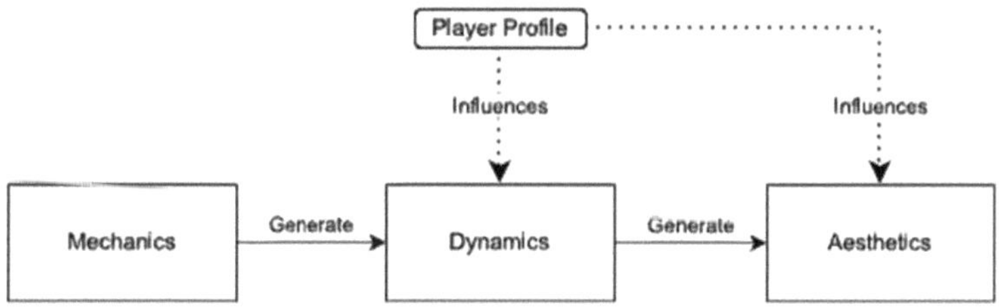

Fig. 1 Player profile's influence on a game's dynamics and aesthetics, based on the MDA framework [11]

Here, *player profile* refers to any and all aspects that will affect a player's input in the game, as well as their gameplay experience. These aspects can include factors such as:

- The player's level of skill and proficiency with the game;
- The player's motivations and expectations;
- The player's current frame of mind; and
- The play context[1] in which the player is interacting with the game.

While different player profiles cannot exert influence in a game's mechanics, they can have a substantial impact in how its dynamics and aesthetics unfold during play.

We will refer to these two categories as endogenous and exogenous asymmetry. Their nomenclature is a reflection of where the origin of those asymmetries lies: either inside, or outside the explicit rules of the game, respectively. As seen on Fig. 2, endogenous asymmetries exist at the level of a game's mechanics, while exogenous asymmetries are the way a game's dynamics and aesthetics can be influenced by the player. Of course, these components still exert influence on one another, as described in the MDA framework. That is to say, different mechanics will generate different dynamics, which will in turn generate different aesthetics.

The inclusion of a category for asymmetries originating from aspects outside of the game's rules may seem fruitless—after all, these asymmetries are beyond pervasive: they are inevitable—but in fact, games often take exogenous asymmetries into consideration. Skill-based matchmaking systems can be described as simply methods of diminishing asymmetries related to the players' skill levels. Games also often include accessibility options meant to facilitate their use by people with impairments. We acknowledge exogenous asymmetry not with the impossible goal of avoiding it, but for game designers to better understand its role in shaping a gameplay experience and accommodate for it.

It then stands to reason that when exogenous asymmetry is taken into account, no game with human players can ever be totally symmetrical.[2] We are aware that

[1] As defined by Jurie Horneman, play context relates to "the when and the where" a game is taking place, and "how those factors influence how people feel and which devices they use" [10].

[2] Note that games can still have perfect endogenous symmetry in their starting conditions. For instance, in *Tetris* [43], both players receive the same sequence of pieces.

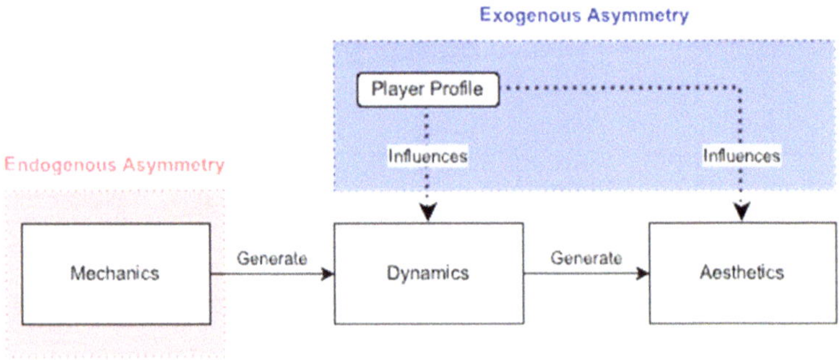

Fig. 2 Endogenous and Exogenous asymmetry

this fact defies current discourse, which tends to consider asymmetry only in its most pronounced and endogenous expressions. However, our categorisation is not concerned with adhering to pre-existing conceptions of what an 'asymmetric game' is or is not, and rather focuses on the phenomena of asymmetry itself, and how it influences player experience.

The inclusion of these two broad categories in our definition of asymmetry will allow us to gain a deeper understanding of how it might be used for the design of gameplay experiences that more accurately capture the diversity of human experiences. That being said, the focus of this particular article will lie on the mechanics of asymmetry and how they can interact with one another, rather than exogenous asymmetry and the specifics of player experience and interpretation of a game.

2.3 *Asymmetry Beyond Starting Conditions*

The colloquial understanding of asymmetric play seems to be that for a game to be asymmetric, it must offer players distinct experiences right from the start. But we disagree, since asymmetry fluctuates over the course of play, often in response to player choice. Some games have player characters start from a blank-slate state. In *Runescape* [38], all players' characters start with their skills at level 1 and are all able to do the same actions. Only by engaging with the levelling process do players unlock different actions by progressing through the different skills.

Other games feature asymmetry from the start. In class-based role-playing games such as *Diablo III* [21], the player character's chosen class will impact the actions they have available at the start. However, even in this case, asymmetry only increases further and further as the game progresses, as each class will unlock more and more actions that distinguish it from the others.

It might seem, from the previous examples, that asymmetry can be seen as an entropy-like feature in games, always increasing over time. However, there are instances where asymmetry can also decrease. For instance, two players in *Battlefield V* [18] may unlock guns at different rates, depending on the amount of time they can dedicate to the game. Nevertheless, at some point, one of them will reach a point where they have unlocked every weapon available. From this point onwards, the asymmetry between the two players (in respect to which weapons they can use in the game) will decrease over time.

Not all games follow these simple patterns of increase and subsequent decrease of asymmetry. Some allow for asymmetry to fluctuate more freely and in interesting ways. In *OBEY* [31] all players spawn as bunnies. However, the first player to enter the giant robot in the level gets to control it. There is an extreme power differential between the robot and the bunnies that the robot player uses to coerce the other players. However, the catch is that the bunny players may decide to covertly conspire against their overlord, with ultimately one of the bunnies hijacking control of the robot. In this example, the great power asymmetry associated with being in control of the robot is constantly shifting from player to player, evolving over time in interesting ways.

Any of the types of asymmetry discussed in this text could be designed to be shaped over time by the actions of players. By rejecting the idea that asymmetric play presupposes static differences between players, we hope to encourage designers to create innovative ways in which this feature evolves over time, fostering interesting and unique gameplay.

3 Types of Asymmetry

3.1 Literature Review

Despite asymmetric gameplay's appeal, there is little written on the subject. Anyhow, Harris et al.'s work features a brief discussion regarding the mechanics of asymmetry: "potential mechanical manipulations that designers can utilize in order to give rise to asymmetric player experiences" [9]. The list of mechanics featured in this discussion is, as acknowledged by the authors, not a complete account of all the types of asymmetry. Nevertheless, it served as an excellent starting point for a categorization of the different methods through which a game's mechanics might generate asymmetry in play.

What follows is our analysis of Harris et al.'s classification. We will present their formulation of each type of asymmetry, followed by our examples found in case studies, as well as additional considerations and potential repercussions of each type of asymmetry. Unless otherwise noted, we maintained the original authors' nomenclature.

Ability. *Asymmetry of ability* occurs in games in which one player can do things that another player cannot [9]. That is to say, each player has their own unique repertoire of actions they can perform within the game.[3] This is a prevalent type of asymmetry, allowing for each player to engage with different tools to complete their goals. It can be present at the beginning of the game or develop as the game progresses as a result of player action.

In some games, asymmetry of ability results in sets of mechanics so dissimilar that they might be seen as belonging to different game genres—only they exist within the context of a single game. This is the case with *Resident Evil Resistance* [37], in which four *Survivor* players face a *Mastermind* player. The *Survivor* players experience the game through mechanics from the survival genre, while the *Mastermind* is acting through strategy and resource management game mechanics. When games take this approach, they are in effect appealing to players with different gameplay preferences. This could result in different types of players engaging with the same game, whereas otherwise they would be engaging only with players with the same preferences.

This type of asymmetry also brings opportunities for more inclusive game designs with greater accessibility, by diversifying the game mechanics the players engage with. For instance, a round-based first-person shooter akin to *Valorant* [45] could accommodate older players with slower reaction times by allowing them to participate through the role of a strategist in charge of devising the strategy for their teammates to follow. By allowing different kinds of players to act differently, games could become more accessible, and therefore more attractive, to demographics that are typically underrepresented in gaming spaces.

Challenge. *Asymmetry of challenge* occurs when players face different kinds of challenges. It does not refer to instances where players face different degrees of the same challenges, e.g. by opting for different difficulty settings [9].

An example of this asymmetry would be a game in which one player engages with puzzles while the other engages in combat. Different challenges are usually coupled with different abilities, to adequately equip players with the tools to face those challenges. However, this does not always have to be the case. Players may be prompted to tackle different challenges even when they have access to the same mechanics. Take for instance a game in which two players have access to the same ability set, one typical of a first-person-shooter game. If one player faces the challenge of defeating enemies while the other must shoot at targets in the correct sequence to solve a puzzle, these players are engaging with asymmetric challenges, despite interacting with the game through the same set of actions.

An example of asymmetry of challenge occurs in *Panoptic* [33], where the *Challenger* player must stealthily sneak through the level while avoiding the gaze of the *Overseer* player. The types of challenges present in this game are *Stealth* for the *Challenger*, and *Finding Hidden Object* for the *Overseer*.[4]

---

[3] This does not mean that each action is exclusive to one player. In most games, there is some overlap between the different player's sets of actions they can perform. For instance, even in games where each playable character has different abilities, all of them are often given the shared ability of moving in 3D space.

[4] The terminology used to describe these challenges is the one delineated by [2].

Different demographics of players may be more inclined to face different kinds of challenges. As such, games that allow players with different preferences to coexist in a multiplayer environment may result in a greater level of heterogeneity in their communities.

Goal. *Asymmetry of goal* exists whenever players seek to achieve different outcomes [9]. This does not include instances where players have the same goals, but diametrically opposed, as would happen, for example, in a game with two players playing opposite sides in a virtual football match, trying to score in the opposite goal. We make this distinction because, in the perspective of the players, their goals are the same: 'I must score more goals than my opponent'.

Vast: The Crystal Caverns [46] presents a prime example of this asymmetry as each one of up to five players is chasing a different goal. The *Knight* must kill the *Dragon*, the *Goblins* must kill the *Knight*, the *Dragon* must wake up and escape the *Cave*, the *Cave* must fully expand and then collapse to bury everyone, and the *Thief* must collect and stash treasures.

Different goals are often coupled with different sets of available actions, to allow players to pursue those unique objectives. Furthermore, different goals often entail different perspectives of the same game. Once players switch roles, they gain a new perspective over the game, gaining insight into each other's roles' strengths and difficulties, as well as how it feels to enact that role. This approach may entice players to experience the same game in new ways, increasing its replayability and longevity.

Asymmetry of goal does not always entail asymmetry of ability. In *Counter-Strike: Global Offensive* [20], one team must plant a bomb while the other team attempts to stop them. Despite those asymmetric goals, players engage with the game through the same mechanics (those of a first-person shooter), with only minimal asymmetry of ability.[5] This means that the skillset that players build playing one role will easily transfer to the other.

Responsibility. *Asymmetry of responsibility*[6] exists whenever players are tasked with different responsibilities [9]. In this context, the term *responsibility* refers to a player's perceived role to play in the pursuit of their in-game objectives. This type of asymmetry can sometimes overlap with asymmetry of goal, i.e. it is often the case that players with different goals will also be tasked with different in-game responsibilities. But this isn't always the case. Players can share a single overarching goal while having different responsibilities. This often occurs as consequence of the existence of asymmetry of ability: each player's responsibility is informed not only by their goals, but also by the actions they can execute in pursuit of those goals. In *Keep Talking and Nobody Explodes* [26], the two players share the goal of defusing the bomb. However, the tasks they must perform to achieve this objective are different: one is allowed to see and interact with the bomb, while the other has

---

[5] The asymmetry is minimal and not inexistent due to the fact that some of the weapons each team has access to are team exclusive. For instance, only members of the attacking team may purchase an AK-47.

[6] In the literature, this type of asymmetry was originally bundled together with asymmetry of goal, as "asymmetry of goal/responsibility". However, we believe these to be sufficiently different types of asymmetry to warrant separate discussions.

access to instructions on how to defuse said bomb. In this example, players have a shared goal but asymmetric responsibilities.

In some games, players can have the same abilities and goals and yet be encouraged to adopt asymmetric responsibilities. This is the case of *Overcooked* [32], a game in which players act as cooks. They share the same abilities and the goal of cooking meals within a short amount of time. However, the game's tight time constraints make it so that players are encouraged to delegate responsibilities. For instance, one player may be in charge of chopping and cooking meat, while the other serves the finished meals and washes the dishes.

Information. *Asymmetry of information* exists when one player knows something other players do not [9]. This type of asymmetry is present in games in which players are given different information regarding the game state, be that information in accordance with reality or not.

Games in the social deduction genre would simply not work if players were all given the same information. In games such as *Spyfall* [40], players are assigned hidden roles that are not meant to be revealed to other players. This allows players to bluff and deceive other players, adding a lot of uncertainty and suspense to an intended gameplay experience that requires it.

Cooperative games often give one of the players exclusive information, while disallowing them from directly acting on it. Instead, they must relay that information to another player so that it can be interpreted and acted upon. In *We Were Here* [48] this relationship is bidirectional, as each player frequently requires information that the game presents to the other player. This approach makes it so that players must rely on one another to better tackle the game's challenges.

We argue that, in cooperative games, it is important to ensure that each player does not act solely as a conveyor of information. Instead, they should also be in charge of processing and filtering that information. If not, games run the risk of nullifying their asymmetry of information and allow themselves to be solved by the most proficient player in the group. Features such as strict time limits[7] can be used to limit the amount of shared information, so that players can still retain their own agency, while also relying on others to reach the overarching goal.

Interface. *Asymmetry of interface* exists when the means by which players engage with the game differ; both in terms of input and output [9]. Different interfaces will have their own advantages and pitfalls, and be more suited for some kinds of experiences and gameplay mechanics than others.

An example of this asymmetry occurs in *Acron: Attack of the Squirrels!* [17], in which one player is equipped with a virtual reality (VR) headset, and acts as a giant tree that must protect its acorns. In contrast, the other one to eight players interact with the game through their phones, controlling squirrels that aim to steal the acorns.

Other VR games follow this same structure, with one VR player interacting with multiple players on their phones. *Reiko's Fragments* [36] is a horror game in which the VR player must escape a haunted house, while the other players trigger game events using their phones. These games would be fundamentally different if this

[7] E.g., as implemented in *Keep Talking and Nobody Explodes* [26], *Captain Sonar* [19].

asymmetry were to be inverted (one phone player versus eight VR players). The immersive nature of VR is better suited for a horror experience, while the phones are an interface that's comparatively cheap and ubiquitous enough that the players will already have access to one and can play the game with no added monetary investment.

Some games allow players to participate in the same in-game roles through different interfaces. For instance, in *Phasmophobia* [34], players can either play using a typical keyboard, mouse, and monitor setup, or they can play through a VR headset. Likewise, *Marvel Snap* [28] players can play on their computer or through the mobile app. Even though these different interfaces will affect the way players perceive and interact with the game, developers often take steps to guarantee that no particular interface brings competitive advantages to those who use it.

Asymmetry of interface can have a great impact on how the game is experienced, as well as how logistically feasible a game is. However, it may also bring about some design challenges for developers trying to make a game equitable despite each interface's idiosyncrasies. Designers must be aware of factors such as these when deciding which interface to design for to arrive at their intended gameplay experiences.

3.2 *Our Contributions*

Harris et al. state that their list of types of asymmetry is not exhaustive [9]. We analysed current games in search of more types of endogenous asymmetry. As a result, we found six more types of endogenous asymmetry: *Operation*, *Location*, *Time Frame*, *Interdependence*, *Outcomes*, and *Feedback*. We describe each of them in the following sections, while also providing examples of how they occur in games and additional considerations about how they may interact with one another.

Operation. *Asymmetry of operation* takes place whenever players' gameplay experience is affected by their own actions and choices within the context of the game. For this to take place, a game's mechanics must be designed in a way that allows these differences in input to be translated into meaningfully different gameplay experiences. This may be one of the most common types of asymmetry, as games are an inherently interactive medium in which player choice often results in gameplay variations. As such, this type of asymmetry can present itself in numerous different ways. This section will describe some different expressions that this type of asymmetry may take.

In *Kingdom Come: Deliverance* [27] the player character's *Skills* improve over time according to player action. The more the player performs an action associated with a *Skill*, the more proficient their character will become at all actions related to that *Skill*.

Differences in player proficiency with the game can also trigger asymmetry of operation, so long as the game's formal rules allow for those differences in proficiency to express themselves during play. This is often the case with games in general, but in some game genres, such as Visual Novel games, skill expression[8] is not as prevalent. Games in which skill expression is highly prevalent exacerbate the differences in player skill level. For instance, a more skilled player can more easily access specific content within the game, while less skilled players may even be faced with an obstacle that prevents them from progressing past a certain point in the game.

Harris et al. mention *asymmetry of investment* as existing in games in which the amount of time players dedicate to the game differs [9]. This is particularly prevalent in Massively Multiplayer Online Role-Playing Games such as *World of Warcraft* [49], Which reward players for completing daily missions and participating in weekly raids. Since this kind of asymmetry emerges as a result of player action, we included it under the umbrella of asymmetry of operation. Player investment does not have to be related only to time. It may also be monetary investment, in the form of microtransactions through which players can acquire in-game advantages. These microtransactions influence the player's expectation of how the game should play out, as they may feel entitled to having better in-game outcomes. They also often allow players who engage in microtransactions access to privileged game mechanics, creating other kinds of asymmetry. For instance, they may allow players to craft special items, or gain access to exclusive equipment, thus generating asymmetry of ability.

Another way that games can include asymmetry of operation is by allowing players to customize their experience with the game. This can happen very explicitly through a settings menu,[9] but also through other in-game systems that allow players to make decisions regarding their gameplay experience. For instance, in *Metal Gear Solid V: The Phantom Pain* [29], enemies adapt to the strategies that the player relies on the most by acquiring equipment that counters them. When met with this, the player can either adopt different strategies to defeat these enemies, complete missions that reset the enemies to their base state, or simply continue playing in the same way and facing the harder version of the enemies. Regardless of their choice, the player is, in a sense, adjusting the game's difficulty and/or choosing to play the game in a way that uses a greater variety of strategies, effectively customizing their gameplay experience.

Location. This type of asymmetry occurs when different players inhabit different locations within the game. These locations might give players access to different resources and might bring specific advantages and disadvantages to each player.

An interesting way in which this asymmetry can present itself is when player space, the physical space where the player resides, overlaps with game space, which relates to the actual game world[10] [6]. In *Pokémon GO* [35], players must visit local real-world landmarks and monuments to get useful in-game resources. Depending on

[8] By *skill expression*, we are referring to the ability for a player's skill level to be reflected in their in-game actions. Games that require precise inputs from the player can be said to have more skill expression than games where skill does not play such a big role.

[9] Games like *Fell Seal: Arbiter's Mark* [24] even give players the option to change how the game's systems work (e.g. how the game handles random number generation in combat calculations).

[10] These categories are based on Jesper Juul's *3D space*, *screen space*, and *player space* [12]. However, the term game space is agnostic of whether the game is two- or three-dimensional, or

whether a player lives in an urban or rural area, they can have access to an abundance of game resources, or face scarcity, respectively [16].

A more common expression of asymmetry of location occurs when player characters simply occupy different in-game locations. In games with *Battle Royale* game mode, such as *Fortnite* [25], players can choose which location they want to move to during the start of the game. Different locations provide different opportunities and challenges, as locations that offer more chances of finding good equipment are sought after, and thus will have more players competing for the resources.

Time Frame. *Asymmetry of time frame* occurs when a game offers different gameplay experiences depending on when players engage with the game. A prime example of this occurs in *Death Stranding* [22], in which players can build structures to assist themselves and others in overcoming the game's challenges. These structures decay over time and must be maintained by players. One of the most helpful structures are the roads, which aid in a faster and easier traversal of the game world. However, roads are expensive to build, which meant that at the game's commercial launch, they were a rare sight, becoming more common over time. The availability of the structures fluctuates over time, which means that a player may or may not find those structures depending on when they are engaging with the game. The fact that these structures heavily affect gameplay and the difficulty of the game, means that this asymmetry has a great impact on the players' experience. Games that strive to explore this type of asymmetry may be able to create the conditions for players with different schedules to interact asynchronously in meaningful ways, creating interesting gameplay experiences.

Interdependence. In multiplayer games, the player is inevitably reliant on other players to assume supportive roles to achieve cooperative goals, or antagonist roles that opposes their progress. Sometimes these relationships can become asymmetrical, meaning that this reliance on other players isn't reciprocal [9]. This occurs in *The Isle* [44], in which players can play either as carnivorous or herbivorous dinosaurs. The herbivorous dinosaurs can subsist by feeding on the world's vegetation, while carnivorous dinosaurs must feed on herbivorous dinosaurs to survive. This means that the relationship of dependence between these players isn't reciprocal: herbivorous dinosaurs can survive without the presence of carnivorous ones, while the opposite isn't true.

In other games, in particular those where one player competes against multiple players, the solo player is less dependent on others in order to accomplish their goals, while the rest of the players must be efficient in their cooperation in order to win. In *Secret Neighbor* [39], up to six players play as *Kids* sneaking into their neighbour's house, trying to collect six keys to get to the basement. However, one of the players is secretly the titular *Neighbor*, who must catch and eliminate the *Kids* in order to prevent them from achieving their goal. Both the *Neighbor* and the *Kids* need each other's input in the game for the competition to take place. However, the *Kids* have the

---

whether it is viewed through a large or small screen, or even if it is entirely represented through the use of audio. Game space is "where the game actually happens" [6].

added interdependence of having to rely on one another to win, while the *Neighbor* acts more independently.

Games featuring audience participation are an extraordinary example of asymmetry of interdependence. In *Warhammer: Vermintide 2* [47], *Twitch.tv* audience members can influence the game through a voting system. They can elect to provide support or hinder the players' progress. However, the nature of the voting system means that audience members that didn't vote for the same option as the majority of the audience are prevented from acting in the game world. Since only the most voted option choice is put into effect, the members of the audience are extremely interdependent on one another, as they cannot even act in the game without a consensus.

Outcomes. Some games present players with different outcomes even when they perform the same action. This can occur when games deal with randomness, such that the players are not given full assurance of the consequences of their actions. This can be used in a purposeful way to increase a game's uncertainty, or to simulate the complexity that is present in real life scenarios. For instance, in *Dungeons and Dragons* [23], players roll dice to determine the outcome of their actions. This, among other things, allows the game to simulate combat scenarios in which attacks may hit their target, or be successfully defended against. Whereas in a real setting this would depend on a multitude of factors, in this game it is simplified and abstracted through a simple roll of the dice. The uncertainty encourages players to adapt to changing circumstances and heightens the tension of playing the game. Two different groups of players engaging with the same combat scenario may have very different experiences, caused by this asymmetry of outcomes.

The difference in outcomes does not need to originate from randomness. Outcomes triggered by the same action can be deliberately decided by the game's internal rules. For instance, the Handicap system in *Super Smash Bros. Melee* [41] aims to balance matches between players by giving the less-skilled player's attacks more knockback, as well as making them more resistant to getting knocked back. Players can adjust the level of *Handicap* themselves, or set the system to *Auto* mode, and have the game dynamically change it in response to the matches' results. In this sense, this endogenous asymmetry can assist in diminishing the impact of exogenous asymmetries relating to each player's individual skill. This results in closer matches, even if players have different proficiency levels with the game.

Feedback. *Asymmetry of feedback* exists when games provide different types of feedback to their players. This can occur, for instance, when one player experiences the game exclusively through visual feedback, while another experiences it through audio cues.[11] Despite being exceedingly rare, this type of asymmetry has a great impact in how players perceive the game world. Furthermore, it may be the only way how some pairings of players are able to participate in the same game together. For

---

[11] To disambiguate the terms *information* and *feedback*: for our purposes, *information* refers to the content of what is being transmitted to the players (e.g. the board state in a game of chess), whereas *feedback* refers to the 'vehicle' through which that information is communicated (e.g. visual or auditory cues).

instance, in the case of games for mixed-visual-ability pairs, this type of asymmetry allows for a visually impaired person to participate in the same game as a sighted person. Gonçalves [8] designed games for mixed-visual-ability pairs, with one role dependent on visual abilities, while the other relied on auditory cues. This resulted in balanced collaborative play, with both types of participants feeling they played an important role in the game.

This type of asymmetry may also be used simply to create different experiences and challenges for each player. Although we weren't yet able to find any commercial videogames that do this, there are some examples specific to board games. In *Nyctophobia* [30], four *Hunted* players wear black-out glasses that deprive them of sight during the game. Thus, to navigate the game board, they must rely on touch and memory to escape the *Hunter* player, who can freely look at the board during the game. In *Team3 Green* [42], a team of three players must transfer information from one to the next through different means. The first communicates by gestures, the second one by speech, and the third one, who wears a blindfold, communicates through their in-game actions.

In both examples, asymmetry of feedback is paired with asymmetry of information: the flow of information is heavily informed by the means through which each player perceives the game state. We believe it would be interesting to see asymmetry of feedback take on a bigger role in videogames, as their board game counterparts have proven that there are interesting ways to implement this type of asymmetry in gameplay.

4 Conclusion

In this article, we presented a new way of examining *asymmetry* in games. We evidenced it to be a lot more widespread than the term 'asymmetric game' may lead us to believe. In fact, from our analysis stems the conclusion that asymmetry is quite pervasive in games, especially when exogenous asymmetries are considered. We acknowledge that this fact alone may originate some confusion regarding our definition and categorization. However, we believe our approach may lead to more nuanced and productive discussions regarding asymmetry.

A designer can only inject asymmetry into a game through its mechanics. Thus, we found it important to dedicate this chapter to endogenous asymmetry, as these are the mechanisms through which a game designer can directly influence a game's asymmetry. Nonetheless, exogenous asymmetries are also important to be considered, and in some ways they already are. Games often employ systems for skill-based matchmaking, so that asymmetry in player skill does not play such a great role in how the game unfolds. We hope that developers continue to find ways to implement game mechanics that cater to players with different preferences, but also players with different faculties, so that videogames become a more democratised medium, accessible by anyone willing to engage with them.

As far as we've been able to tell, asymmetry is not discussed in the literature through the lens of exogenous and endogenous asymmetry. It seems to us that both categories of asymmetry should be accounted for when designing games. With this insight, designers may be equipped to more effectively design experiences that consider not only how asymmetries at the level of game mechanics can impact how a game unfolds, but also how asymmetries in player profiles can influence how a player interacts with and experiences a game, and design mechanics that aim for that.

The types of asymmetry we described can be used as lenses through which we can analyse asymmetry in games and how they affect the players' gameplay experiences. As reiterated throughout the text, each type of asymmetry can be—and usually is—coupled with other types of asymmetry. Our categorization will allow these couplings to be deconstructed and better analysed. One might be tempted to assign links of causality to these couplings. For instance, take the idea that in a particular game, two players have different responsibilities because they have different abilities. This link of causality could easily be inverted from the perspective of the game designer, who would claim that the two players have different abilities because they were meant to have different responsibilities. Therefore, we prefer to look at these couplings in a more agnostic manner, as we cannot truly access the designer's intent when examining case studies.

These types of asymmetry are not final. There's no telling which new types of asymmetry will be designed in the future. However, our categorization can serve as a palette of options from which game designers may look to for inspiration on how to instil asymmetry in their games. For instance, they could combine different types of asymmetry in novel ways, searching for interesting combinations that result in engaging gameplay experiences.

While an in-depth analysis of exogenous asymmetry was not in our scope for this chapter, we do believe that it may be worthwhile to explore the different ways in which a player's profile may influence the way they play, as well as how they experience the game. This kind of insight into the endogenous/exogenous dichotomy should give us a more holistic view of how asymmetry is generated during gameplay as well as its effects on player experience.

Finally, asymmetry seems to be lacking attention in scientific literature. Occasionally throughout this text we relied on the perspectives of game designers, who currently are leading the discussion on this topic. It would be beneficial to have more academic authors approach this complex topic and offer their viewpoints as well. By doing this, both groups stand to benefit from a greater understanding of asymmetry and the opportunities it offers.

References

1. Adams E (1998) Designer's Notebook: A Symmetry Lesson. Game Developer. Retrieved May 3, 2023, from https://www.gamedeveloper.com/design/designer-s-notebook-a-symmetry-lesson
2. Adams E (2014) Fundamentals of Game Design (Third Edition), pp. 11. United States: Pearson
3. Bortolaso C, Bourdiol J, Graham TN (2019) Enhancing communication and awareness in asymmetric games. Joint International Conference on Entertainment Computing and Serious Games
4. Brams S, Ismail M (2021) "Fairer Chess: A Reversal of Two Opening Moves in Chess Creates Balance Between White and Black" MPRA Paper 111075. University Library of Munich, Germany
5. Burgun K (2013) "Debunking Asymmetry". Retrieved March 27, 2023, from https://keithburgun.net/debunking_asymmetry/
6. Cardoso P (2016) "Playing in 7D: An Action-Oriented Framework for Video Games". Ph.D. Thesis. Faculty of Fine Arts, University of Porto
7. Fullerton T (2018) Game Design Workshop: A Playcentric Approach to Creating Innovative Games (4th ed.), pp. 326. A K Peters/CRC Press
8. Gonçalves D, Rodrigues A, Richardson ML, de Sousa AA, Proulx MJ, Guerreiro T (2021) Exploring asymmetric roles in mixed-ability gaming. In: Proceedings of the 2021 CHI Conference on Human Factors in Computing Systems
9. Harris J, Hancock M, Scott SD (2016) Leveraging asymmetries in multiplayer games: Investigating design elements of interdependent play. Proceedings of the 2016 Annual Symposium on Computer-Human Interaction in Play
10. Horneman J (2015) "Play Context". Retrieved March 27, 2023, from https://www.gamesindustry.biz/play-context
11. Hunicke R, Leblanc M, Zubek R (2004) MDA: A Formal Approach to Game Design and Game Research. AAAI Workshop – Technical Report, 1
12. Juul J (2012) A casual revolution: reinventing video games and their players. The MIT Press. ISBN: 978-0-262-01337-6
13. Neto A, Cardoso P, Carvalhais M (2022). Asymmetric gameplay: types and perspectives. https://doi.org/10.1007/978-3-031-20364-0_64
14. Shafer, J. (2013). Asymmetry. Game Developer. https://www.gamedeveloper.com/design/asymmetry.
15. Snyder J (2014) "Asymmetrical Gameplay: Gimmick or Revolution? Retrieved May 3, 2023 from https://web.archive.org/web/20190405215117/; http://www.theoryofgaming.com/asymmetrical-gameplay-gimmick-or-revolution/
16. Weinberger M (2019) "Pokémon Go is finally fixing a problem that's plagued it since launch, and players in rural areas should be thrilled". Business Insider. Retrieved January 15, 2022, from https://www.businessinsider.com/pokemon-go-rural-remote-players-migration-2019-12

Ludography

17. *Acron: Attack of the Squirrels!* (2019) Resolution Games
18. *Battlefield V* (2018) DICE
19. *Captain Sonar* (2016) Roberto Fraga, Yohan Lemonnier
20. *Counter-Strike*: Global Offensive (2012) Valve
21. *Diablo III* (2012) Blizzard Entertainment
22. *Death Stranding* (2019) Kojima Productions
23. *Dungeons and Dragons* (1974) Gary Gygax & Dave Arneson
24. *Fell Seal: Arbiter's Mark* (2018) 6 Eyes Studio

25. *Fortnite* (2017) Epic Games
26. *Keep Talking and Nobody Explodes* (2015) Steel Crate Games
27. *Kingdom Come: Deliverance* (2018) Warhorse Studios
28. *Marvel Snap* (2022) Second Dinner Studios, Inc.
29. *Metal Gear Solid V*: The Phantom Pain (2015) Kojima Productions
30. *Nyctophobia* (2018) Catherine Stippell
31. *OBEY* (2015) Dez, The Lo-Fi Apocalypse, Inc
32. *Overcooked* (2016) Ghost Town Games Ltd
33. *Panoptic* (2020) Team Panoptes
34. *Phasmophobia* (2020) Kinetic Games
35. *Pokémon GO* (2016) Niantic
36. *Reiko's Fragments* (2019). Pixel Canvas Studios
37. *Resident Evil Resistance* (2020) NeoBards Entertainment
38. *Runescape* (2001) Jagex
39. *Secret Neighbor* (2018) Hologryph, Eerie Guest Studios
40. *Spyfall* (2014) Alexandr Ushan
41. *Super Smash Bros. Melee* (2001) HAL Laboratory
42. *Team3 Green* (2019) Alex Cutler, Matt Fantastic
43. *Tetris* (1989) Tengen
44. *The Isle* (2015) Afterthought LLC
45. *Valorant* (2020) Riot Games
46. *Vast: The Crystal Caverns* (2016) Patrick Leder, David Somerville
47. *Warhammer: Vermintide 2* (2018) Fatshark
48. *We Were Here* (2017) TMG Studios B.V
49. *World of Warcraft* (2004) Blizzard Entertainment

Text-Image-Movement-Interaction: Written Words and Their Artistic Paradigms in Video and Virtual Systems

Alexandre Martins and Bruno Mendes da Silva

Abstract This study analyzes the written word and typography in a modern and postmodern context, regarding its view and usage in audiovisual and digital systems. Analyzing different periods, works and artists, we aim to understand the potential of typography in the contemporary artistic discourse, mainly exploring the visual, dynamic and interactive components of typography. The theoretical study explored here serves as the basis for a project that foresees the production of a video installation, constructed from the documental and textual collection of Fernando Gonçalves Lavrador. The objective is to disseminate this collection through one exhibition, where a selection of his texts will be appropriated and transformed into artifacts that will converge textual, imagetic and animated elements. Until this public presentation, which is expected to happen in 2024, a process of experimentation with different styles, techniques and materials is being prepared. The results obtained so far—four short films and an interactive program—will also be presented in this paper. The preparatory act that precedes the appropriation of Lavrador's heritage aims to improve the artistic processes around the themes explored in this paper and study new possibilities of using typography as an artistic component in a computational and videographic system.

Keywords Typography · Written Word · Digital media-arts · Fernando Gonçalves Lavrador

A. Martins (✉) · B. M. da Silva
Centro de Investigação Em Artes E Comunicação, Universidade do Algarve, Faro, Portugal
e-mail: acmartins@ualg.pt

B. M. da Silva
e-mail: bsilva@ualg.pt

N. Martins et al. (eds.), *Perspectives on Design and Digital Communication IV*, Springer Series in Design and Innovation 33, https://doi.org/10.1007/978-3-031-41770-2_17

1 Typography: Invisible Craft or a New Form of Artistic Expression

An important quality of the written word is its capacity to preserve oral speech and to transmit memories, information, and ideas through time and space [1, 25]. In the 1440 s, Gutenberg, using his knowledge of metallurgy and a wine press, created a modular composition and printing method [15]. With this creation a new discipline was formed: typography, i.e., the composition of letters, words, and sentences in a certain order. It is a repeatable system that can be combined in interminable forms, and the means through which it is possible to translate a mental concept into a visual shape. By introducing a system of movable characters, Gutenberg might have been responsible for one of the most important technological and social breakthroughs in human history—we say might because there is now a significant consensus that Gutenberg was not only not the first to produce the movable type, that invention is attributed to Bi Sheng during the eleventh century in China [6, 23], he might also not have been the architect of "his own invention" [22].

Although it is still debatable the true origins of this technological development, it is undoubtable that knowledge spread swiftly and literacy expanded after this event, thus altering the way humans communicate [19]. But even if this technological revolution changed the pace and capacity for communication, the essence of writing was still to perpetuate speech through time. According to Warde [26] typography should be understood as an "invisible craft." Contrary to calligraphy [7], a genre of writing that has deep ties with the visual arts and where communicative and aesthetic components exist side-by-side, the main purpose of typography is legibility: above all, the author 's ideas should be transmitted to the reader in a clear manner. The reader, when immersed in the contents of a book, is unable to simultaneously concentrate on reading and viewing the text. Because of this, typography must be unintrusive, and should avoid interfering with our inner voices, which traverse the page producing meaning while they interpret the text's story. Warde [26] argues that if a type interferes with our mental picture of the text by, for example, deforming the design or adding excessive color, it should be understood as a bad type.

Bachfischer and Robertson [1] explain that the reason for the arguments cited by Warde may have manifested as a reaction to the distinct cultural and artistic movements that flourished in the beginning of the 1900s. The Dadaists and Futurists, for example, "played" freely with formal and spatial aspects of texts, introducing a new grammar that would constitute a layer over the literal and contextual meaning that we learn to read in Western culture [1]. The futurists contributed to reform printing styles, triggering a new conception of typographic design.

By transitioning from handwriting to machine writing we fundamentally changed the way typographic communication is done. Confident of the strong optical presence of shape, size and color of type on the page, Futurists set out to modernize the visual quality of printing. Filippo Tommaso Marinetti, founder of the Futurist movement, theorizes, in manifestos written between 1912 and 1914, that a book should not be seen as a neutral receiver for the written word, but as a medium where the use of colors and different fonts, the use of numbers, mathematical symbols and onomatopoeias, and the organization of the text in multiple and varied directions, could increase the expressive power of letters, words, sentences and texts: *italics* for a series of similar and rapid sensations, **bold** for violent onomatopoeia, etc. [10]. Futuristic typography proposed new reading codes: the eyes do not read a text continuously following a linear syntactic structure, from left to right, top to bottom, but, as Toschi [24] refers, the meaning of the text can be naturally understood and enriched by networks of analogies born from visual montages of words which work as self-illustrations.

Marinetti also made use of the white space, i.e., the non-print space. He would resort to the contrast between the "materiality" of type and the "immateriality" of white space, to achieve a visual depth in the text that would prove to be innovative. He attempted to explore a new physicality of writing and to express the essence of matter [9, 24]. Marinetti's concepts were in line with some notions presented by Jan Tschichold about what should constitute modern typography. Tschichold declared in *Elementare Typographie* (1925), that the page should assume a more effective visual language where the negative values build a contrast with positive values, which is to say that the unprinted space must balance the printed space. White spaces should also be used to highlight the content, demarcate the text and provide a more incisive effectiveness to the material. Furthermore, different types and sizes of fonts or geometric shapes, such as the black line, could be used to underline or frame a certain content, giving it greater prominence.

Jan Tschichold thus outlines some of the elements that would constitute The New Typography, identified in the homonymous work *Die neue Typographie* (1928). In one of the chapters, Tschichold alludes to the fact that the first steps towards an innovative typography took place in the literary field when certain writers, influenced by the sense of urgency brought by cinema, began to lose interest in description. Bypassing the sonic qualities of words, these authors adopted the visual tools of modern prose [20]. Futurists and The New Typography created a new way to reflect typography; as a way of creating a balance of contrasts and structural relationships. They were also fundamental to free the characters from traditional rules and a fixed grid system. Through modern artistic expression, the written word began to traverse freely through the page [1].

2 Experiencing Words and Texts in Audiovisual and Digital Systems

For millennia poets and writers have tried to experiment with textual elements and typographic properties to add another type of expressiveness to their texts. Almost 2000 years ago, Greek poets—namely Simmias of Rhodes and Theocritus—created works with patterns illustrating eggs and flutes. In modern times other examples emerged that intended to confer writing with original syntactic or semantic qualities by composing text more visually. In the last century we have seen these practices in places like experimental poetry or typewriter art, genres that give pictorial characteristics to text [19].

In the beginning of the twentieth century, with the advent of film, the strict rules which dictated how traditional western typography should operate were gradually rescued, and the convergence with motion helped to create new forms of meaning. Text would gain an important role in the silent movie era, complementing images in conveying meaning, e.g., time, place, and dialogues. Moreover, new works appeared which focused on text as a central element, something that can be attested, for example, in Marcel Duchamp's 1926 short film, *Anémic Cinéma*. Sitney [21] explains that *Anémic Cinéma* was the first moving picture to give equal claim to title and image within the avant-garde tradition (see Fig. 1).

Experiencing *Anémic Cinéma* is like a process that merges the semiotic systems of reading text and seeing images, resulting in a hybrid experience. It's worth noting that other works, such as the constructivist films of the 1960s, 1970s, and 1980s, like *T,O,U,C,H,I,N,G* (1968) by Paul Sharits, *Zorn's Lemma* (1970) by Hollis Frampton, and *So Is This* (1982) by Michael Snow, also aimed to incorporate and merge text and motion. Pethö (2011), argues that in these works, text and moving pictures operate as a mirror of each other, forming a dual reflection of the two mediums [5].

In the later stage of the twentieth century, typography and its relationship with visual forms became more prominent with the appearance and proliferation of

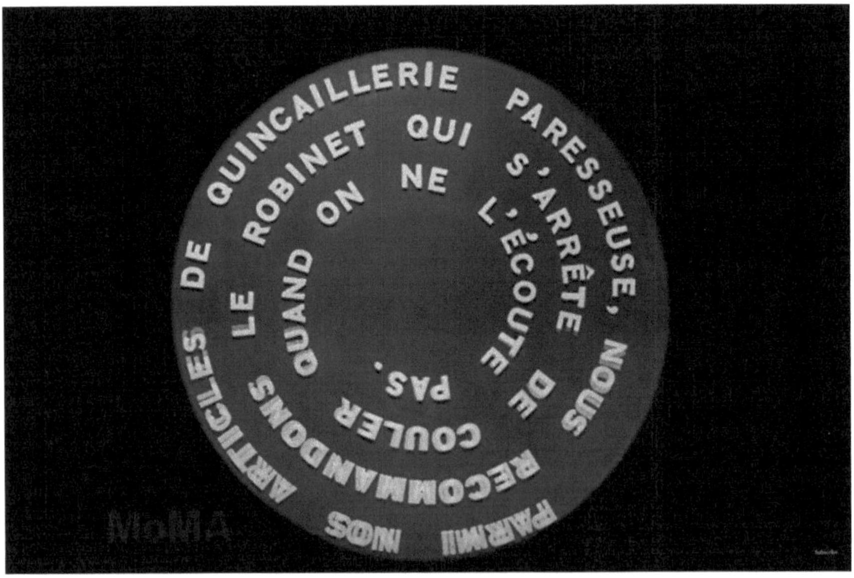

Fig. 1 *Anémic Cinéma*, Marcel Duchamp, 1926. *Source* Duchamp, 2002

personal computers and the consolidation of digital platforms [17, 19]. Contemporary software was able to perform a broad range of animations using typography: for example, the configuration of characters in animorphs or the superimposition of typographic visual effects on images played in real time. And because of the continuous reduction of production costs, moving typography is increasingly seen on computer screens, televisions and e-books [2]. According to Helfand [8] the advances in technology will make us reconsider screen-based typography as a new form of language, one that has a new grammar, syntax and set of rules [18]. Computer systems allow us to portray words in original styles, in an environment which enhances the production of creative works exploring the motion and interactivity of typography.

As for the poetic text it finds new forms of manifesting itself via visual poetry, a genre that links words by using space in a nonlinear fashion, producing a whole new aesthetic that culminates in the emergence of digital poetry [16, 17]. In turn, digital poetry amplifies debates about interaction between reader and poem. It is important to note that it is not realistic to say that these works are a transposition from print to screen. Instead, they embody certain artistic objects which developed and are intrinsic to digital programs. When we talk about the interactive elements in digital poetry, they can materialize, for example, through the reader's movements, creating a sort of dialogue where gestures can establish distinct interpretations [17]. This direct agency readers can achieve constitutes a new way of interpreting poems inviting multiple and different readings of these creations. There is a greater freedom for readers to form their own meanings of the object they consume [16]. In this paradigm, the reader embodies the part of the reader-actor who keeps a gestural relationship with typography, imprinting his gaze on the poems.

Memória (2002) by Alckmar Dos Santos and Gilbertto Prado is a digital poem which demonstrates the potential of user participation and interactivity (see Fig. 2). Composed of nine (at first) indiscernible word illustrations, they become intelligible when the reader-actor moves its cursor above them. Simultaneously we can hear a voice-over saying the word which correlates to each figure. Despite this creative model, it is relevant to say that interactivity is not an imperative prerequisite of digital poetry, as some creations ignore or reject this aspect. For example, certain works delve into the variety of readings and approaches the digital dimension enables as sound and motion give a bigger freedom for intertextual relations [16]. In Joesér Alvarez' *Scalpoema* (2001), the poem is built from a verse from *Memórias Póstumas de Brás Cubas* (1881), by Machado de Assis. Starting from a passage from the original text, words blend; an amalgam which gives rise to a new work, an intertextual product born from another text (see Fig. 3).

Another artistic/technological field that seeks to gain new textual interpretations is that of experimental poetry in immersive environments such as Virtual Reality (VR). According to Uyan Dur [25], works dealing with text in VR do not have the possibility to build an ideal setting for reading. This occurrence is particularly true when the interface experiences are animated and interactive. Therefore, the reading component will lose some relevance. VR communication systems take place through a process of interaction, visualization and other sensory *stimuli*, and the most important effect of these mechanisms emerges from an immersive and multisensory experience. Thus,

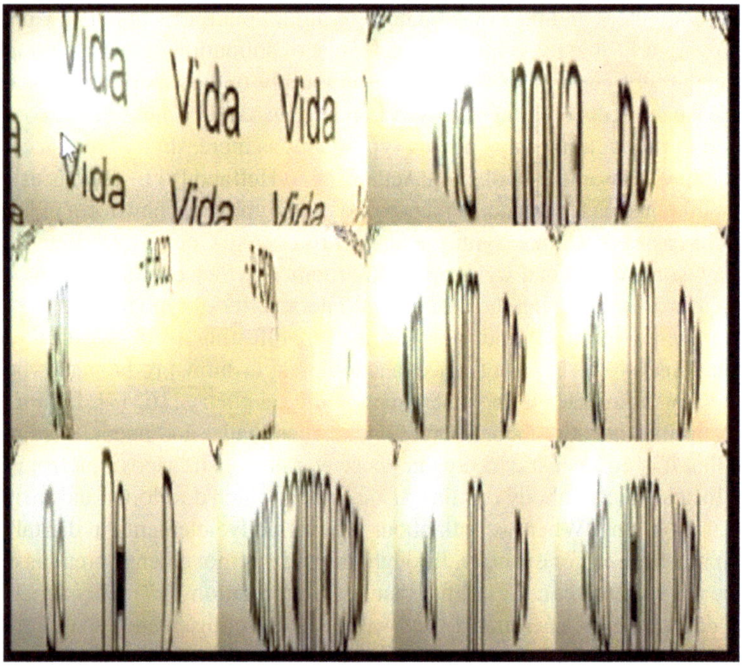

Fig. 2 *Memória*, Alckmar Luiz Dos Santos and Gilbertto Prado, 2002. *Source* Dos Santos & Prado, 2002

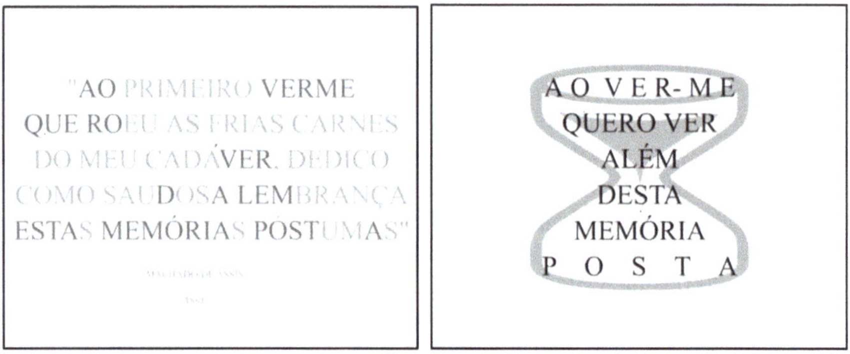

Fig. 3 *Scalpoema*, Joesér Alvarez, 2001. *Source* Joesér Alvarez, 2001

artistic artifacts in VR that make use of text, instead of focusing on the legibility of typography in three-dimensional virtual environments, should rather focus on the exploration of the qualities and new possibilities that result from the use of these systems. In this scenario, experimental typography, a practical approach based on an exploration and interpretation that differs from traditional patterns, provides, for example, the distortion or even abstraction of letters. In this environment, letters,

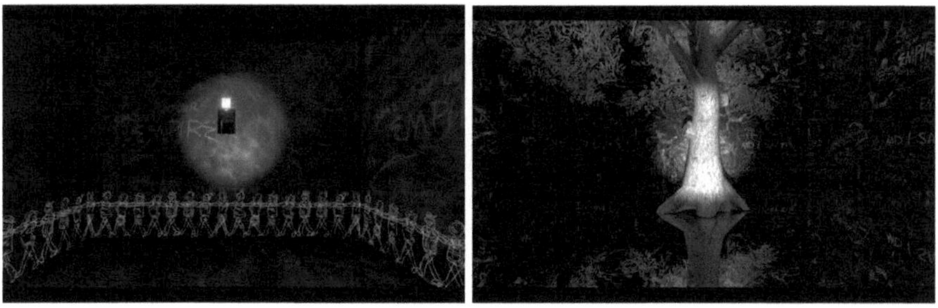

Fig. 4 *Chalkroom*, Laurie Anderson and Hsin-Chien Huang, 2017. *Source* Laurie Anderson

words and sentences also have the possibility of transforming into images or being complemented by sounds, giving a more visual and/or aural dimension to the work. Experimental typography makes room for a freer and more intuitive exploration of the word. In this ecosystem, the objective is based on conceptual solutions that operate on creativity and not, as much, on functionality. Bearing in mind that in this context legibility is not one of the priorities, experimental typography works seek new forms of visual expression and find an original language.

Through Laurie Anderson and Hsin-Chien Huang's virtual exhibition *Chalkroom* (2017), where ambient sounds and narration contribute to the construction of immersiveness, the reader can traverse a structure covered by words, drawings and illustrations. Upon entering this immersive typographic universe, there is freedom for the reader-actor to go through it, while words form, modify, fall and hover around him (see Fig. 4). This work demonstrates how the use of VR provides totally differentiated conceptual and intellectual results. By merging writing, art and VR, new perspectives and creation standards are born [25].

3 Memories in Motion: An Audiovisual and Digital Project

3.1 Fernando Gonçalves Lavrador

In 2021, Cine-Clube de Avanca (CCAVANCA) signed a protocol with the family of Eng. Fernando Gonçalves Lavrador (Porto, 1928–2005), an engineer by profession who was also an important portuguese essayist in the areas of Semiotics, Film Studies and Aesthetics. In this document, Fernando Lavrador's family ceded his estate to CCAVANCA, which is now in charge of studying, organizing and disseminating this cultural heritage. With this began a project to create an archive that would present his memories and works to the public.

To create an archive, whether physical or digital, it is first necessary to develop an extensive study of the materials handed to us. By the time the protocol was signed, during a ceremonial event that took place at the 25th Avanca Film Festival (2021), one of the authors of this text was appointed to coordinate the project. However, we soon realized that to carry out a systematic study of all the documentation—consisting of thousands of objects—digitizing part of it and creating a physical and digital archive would be a job that would take several years to complete. This is a consequence of the massive size and diversity of the collection and the lack of human resources capable of working continuously with these documents. Thus, the strategy of presenting this legacy to a general audience had to be reformulated into a model that would include different small-scale initiatives.

One of these initiatives is an exhibition, which will take place at the 28th Avanca Film Festival, in the summer of 2024. The exhibition will consist in the presentation of distinct audiovisual and digital artifacts constructed from the documentation of the collection. These works will bring together the raw material—for example a poem written by Lavrador himself entitled *Meditação Sobre a Terra com Sede* (1951) (see Fig. 5)—and some of the different fields and artistic practices associated with Digital Media-Arts, specifically, but not exclusively, audiovisual communication, thus maintaining close relations with the works and interests of Fernando Lavrador.

We intend, in line with the precepts of New Typography, digital poetry and works such as *Scalpoema*, that these artistic interventions culminate in artifacts that give shape to Lavrador's words, so that they are read, seen and touched. In short, this

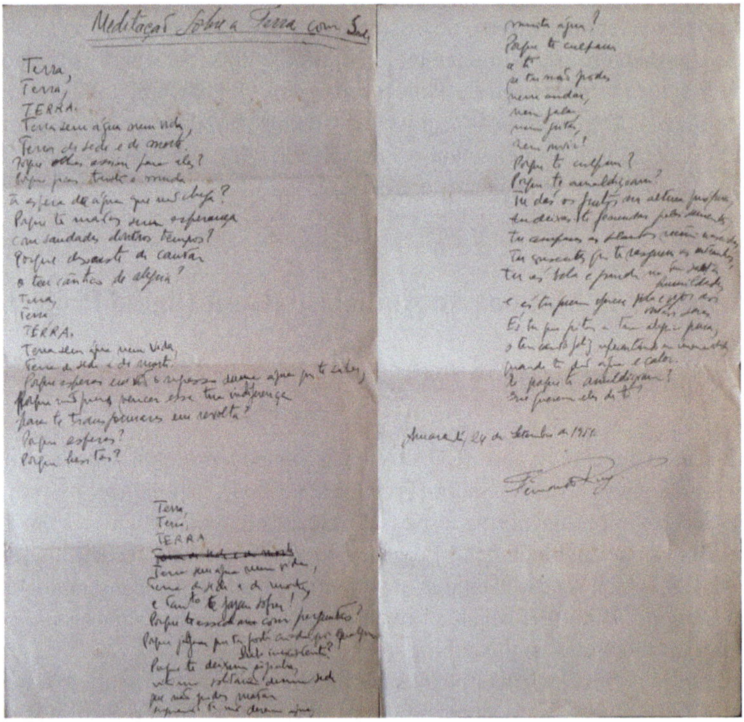
Meditação Sobre a Terra com Sede

Terra,
Terra,
TERRA.

Fig. 5 *Meditação Sobre a Terra com Sede*, Fernando Gonçalves Lavrador, 1951. *Source* authors

proposal seeks to approximate different physical and static objects, composed of paper and text, with virtual and kinetic images and interactive elements, exploring themes alluding to the moving image and the written word, and highlighting the different facets that made up this important figure.

3.2 Experimenting with Words, Images, Movements and Interactions

Until the exhibition takes place—in the summer of 2024 during the 27th edition of the Avanca Film Festival—it is important to develop a set of experiments—even if we don't use the materials from the collection yet—to understand better the artistic practices that surround the use of typography. This is a process that occurs in parallel with the theoretical study of this discipline. From these experiences we will be more familiar with the different styles, methods and tools available. This process is already underway, and so far, we produced four animated short films and an interactive work which converged the written word and typography with image, movement and interactivity.

The first work, *Muted Words* [14], is a machinima—a type of animation that uses real time computer graphics—which explores the concept of subtitles. Subtitles are a topic very important to Portuguese people. Unlike Spain, France, Germany or Italy, which dub the original material [4], in Portugal, watching a film involves reading text and images in a quasi-simultaneously fashion. In this black and white silent short (see Fig. 6)—an allusion to the beginnings of cinema -, subtitles are the primary means of communication between the characters. Even if they don't speak, they still understand each other through text. Such a scenario reveals the role that the text and subtitles have in breaking linguistic and idiomatic barriers that arise from dialogues in a foreign language film.

In addition to the thematic exploration of typography and writing, another important goal was to adopt, early on, different tools and understand the possibilities that arise from their use. *como se constroí uma casa* [13] started with the desire to appropriate a program with a well-defined purpose and subvert it. In this case we used a word processor, namely Google Docs. The premise for this work emerged from the conversion of the characters's illustrations into images formed entirely out of text. These textual drawings were then copied to the word processor, which became the setting for the story. In this environment, each textual symbol was manipulated, either by changing its color, shape, size or position. These instances were then captured through a triple click, a "Command-Shift-5". Finally, all the registered frames were treated in a video editor which gave them movement and coherence. The result would be a work of animated visual poetry using a stop-motion style where the theme in focus was that of the evolution of the alphabet (see Fig. 7).

The third short film originated from new experiments, this time with the P5.js. This open-source online platform that makes use of javascript language for creativity,

Fig. 6 *Muted Words*, Alexandre [18, 20, 22]. *Source* Martins [5] https://vimeo.com/764464638

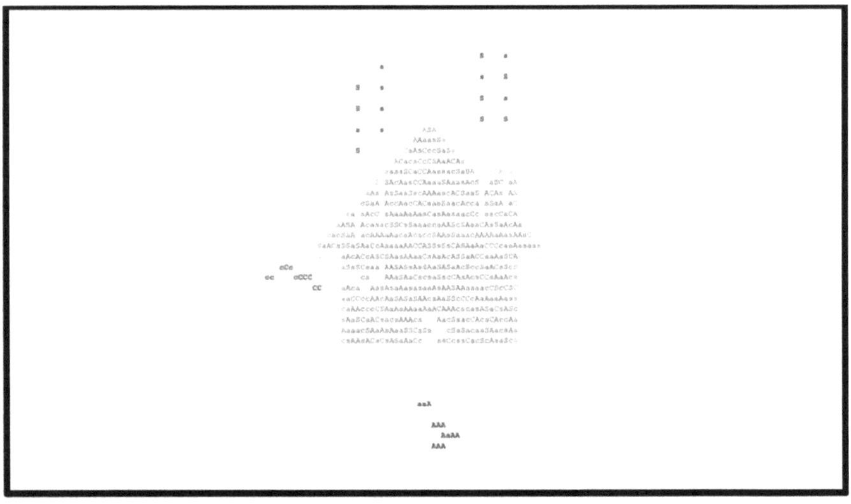

Fig. 7 *como se constrói uma casa*, Alexandre [18, 20, 22]. *Source* Martins [19]. https://vimeo.com/764465418

allowed us to add interactivity elements to a work we had done previously. Firstly, we started by making a stop-motion video clip that portrayed, in an accelerated and chaotic way, the figure of the three wise monkeys, a pictorial proverb that refers to the tradition of "looking away from something". In the second phase, this video was uploaded to Processing, where we added a layer of text over the original images (see Fig. 8). After converting the original video into a video layered with text on top,

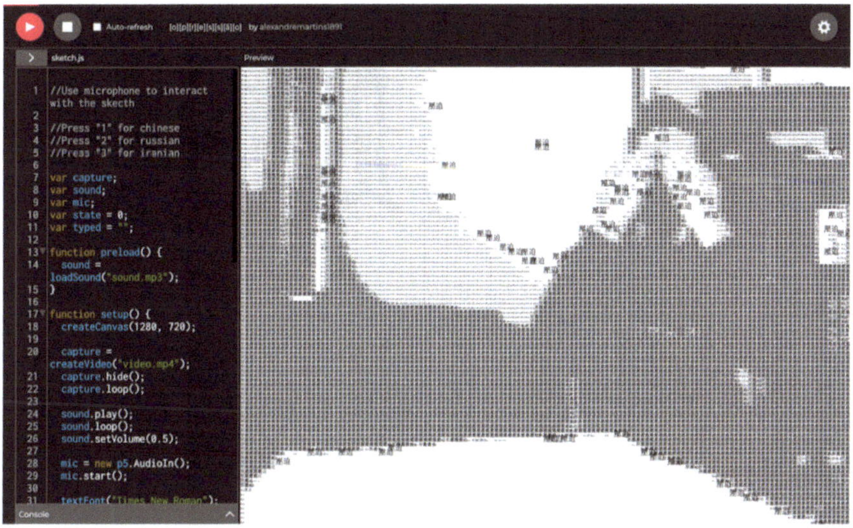

Fig. 8 *[o][p][r][e][s][s][ã][o]*, Alexandre [18, 20, 22]. *Source* Martins [19] http://bit.ly/3JUVQXj

the interaction element was worked on. For this, we used the computer's internal microphone as a volume sensor. As the noise around the computer increased, so did the letters on the screen.

Through the continuous experimentation with this tool, a concept for the work was conceived. The intention was to allude and to talk about the oppression of authoritarian regimes. The example chosen was the Chinese government due to the alarming occurrences in recent years, on problems such as the imprisonment of ethnic minorities, the massive monitoring system of the Chinese population, or the implementation of a system of social credits. Because of this choice we transformed the original video clip into a moving image composed of the Chinese characters that spelled *oppression* (later we implemented a function that could change this word into other languages). The work thus became a metaphor for freedom of expression and its pernicious weaknesses. This P5.js sketch would later result in *[o][p][r][e][s][s][ã][o]* [12], a short film that replicates the work already done in P5.js, but, of course, without the interaction function (see Fig. 9).

Since the recent appearance of A.I. image generators such as DALL.E, MidJourney or Stable Diffusion, we have also been testing these technologies to understand original ways of exploring the combination of text with moving images. After several trials and numerous short clips, we were recently able to produce our first short film using this new language. From the film *Anémic Cinéma*, we built new imagetic layers that established a dialogue between the original and new sequences. The process started by using Deforum Stable Diffusion, a fast image synthesizer, to recreate all 19 original sequences of *Anémic Cinéma*. For the original elliptical/circular sequences we used prompts like: "kanji, japanese language, japanese people". For the original textual sequences, we combined the prompts used

Fig. 9 *[o][p][r][e][s][s][ã][o]*, Alexandre [18, 20, 22]. *Source* Martins [19] https://vimeo.com/771213662

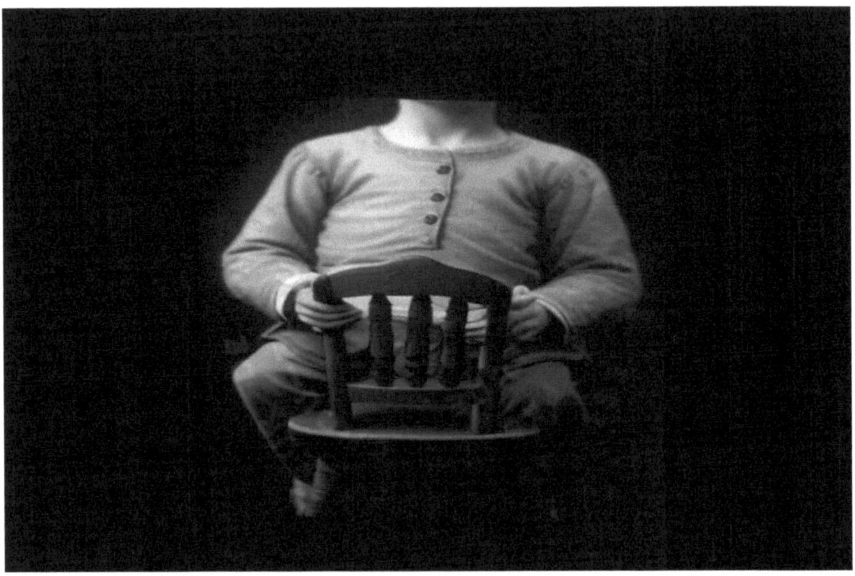

Fig. 10 *MetaAnémic Cinéma*, Alexandre [16]. *Source* [16] https://vimeo.com/798325068

before plus the incoherent sentences invented by Duchamp for his movie: "Bains de gros thé pour grains de beauté sans trop de bengué, kanji, japanese language, japanese people". This allowed us to see how the machine interpreted text in its form ("Kanji") and content ("Bains de gros thé pour grains de beauté sans trop de bengué"), and even showing us how these two realities combine. This was also an exercise about the theme of authorship where we explored what it means to be original or the true author of a work in a contemporary sense. *MetaAnémic Cinéma* [11] can be seen as a work that fits into a remix aesthetic and a meta-creativity trend, relating to Artificial Intelligence and machine learning [15].

To conclude this section, it is important to say that the process of experimenting with ideas, techniques, processes and materials will continue in the coming months. At a certain point this phase will end, and the knowledge acquired until then will be focused on the objects related to Fernando Gonçalves Lavrador—objects such as the poem *Meditação Sobre a Terra com Sede*. In this way, we expect to be able to create audiovisual artifacts that can merge the texts and words of Lavrador, with the learning experiences acquired before within the medium that encompasses visual, moving and interactive typography.

4 In Lieu of a Conclusion

The first half of this paper reflects a fragment of an investigation around typography in a modern and contemporary artistic context and a theoretical exercise about the use of this discipline in audiovisual and digital environments. The knowledge acquired so far is, in fact, a fraction of what is proposed to be a larger study that will explore authors, artistic trends, works and genres that were neglected in this text, e.g., ASCII art, PO.EX, clipoemas, typewriter art, the work of Rui Torres, which, we hope, will be discussed in future papers. The continued study will allow us to absorb more influences, understand new theories and get to know new methods and tools. All the theoretical inputs—as well as the knowledge acquired from works produced so far—will contribute to refine the design of future artifacts.

Alongside this theoretical approach, a study is already being carried out on the collection itself, whose documental mass is quite extensive. Because of this, it is important to define and limit the objects to be considered for future treatment: our focus will only be on documents of a more personal nature (e.g., poems, correspondence, postcards, memoirs, drafts, etc.). This decision happened because we considered these materials more adequate to the project's objectives, since all the cited elements are texts written by and for Fernando Lavrador (written both by hand and by machine).

The processes described above—bibliographical research and the study of the documentation—will result in the creation of new digital artifacts based on the collection. This phase will also be marked by some experimentation and constant adaptation, and it will involve a practice-based research methodology [3]. This type of methodology proposes a research process that combines artistic writing and production.

This is the direction that we intend to carry out from here on. The creation of future digital artifacts will seek to disseminate and enhance an important legacy not yet studied or shown to the general public. The textual richness of this unexplored collection is great. Therefore, there are many opportunities to act creatively on the visual composition of the texts, extracting new meanings and interpretations. Even so, it is important to emphasize that we are in an intermediate stage of a project that will expand and focus on other genres, categories and areas, in a quest to find new ways to (re)interpret the textual legacy of Fernando Gonçalves Lavrador.

References

1. Bachfischer G, Robertson T (2005) From Moveable Type to Moving Type – Evolution in Technological Mediated Typography. AUC Academic and Developers Conference 2005, Hobart, Tasmania, Australia. https://www.academia.edu/9374875/From_Movable_Type_to_Moving_Type_Evolution_in_technological_mediated_Typography
2. Betancourt M (2019) Asemic typography in kinetic design. Semiotica 231:245–257
3. Candy L (2006) Practice Based Research: A Guide. Creativity and Cognition Studios Report, Australia. https://www.researchgate.net/publication/257944497_Practice_Based_Research_A_Guide
4. Commit Global: Subtitling vs Dubbing: List of preferred method per country. Commit Global (2021). https://webcasts.td.org/uploads/assets/9888/document/Commit_Global_Subtitling_vs_Dubbing.pdf
5. Geiger N (2020) Discourse, Figure, d'après Lyotard. In: Valente AC (coord.) Avanca | Cinema 2020, pp 40–45. Edições Cine-Clube de Avanca. https://www.researchgate.net/publication/349717199_Discourse_Figure_d'apres_Lyotard
6. Gunaratne SA (2001) Paper, Printing And The Printing Press: A Horizontally Integrative Macrohistory Analysis. Gazette 63(6):459–479. https://doi.org/10.1177/0016549201063006001
7. Harris D (2003) The Calligrapher's Bible. Quarto Publishing, United Kingdom
8. Helfand J (2001) Screen: essays on graphic design, new media, and visual culture. Princeton Architectural Press, USA
9. Ivanova E (1998) Una battaglia a/di colori. La battaglia di Tripoli di F. T. Marinetti. Italian Culture 1(16), 143–156. https://doi.org/10.1179/itc.1998.16.1.143
10. Marinetti FT (2009) Les mots en libertés futuristes. Inter 103:18–25
11. Martins A (2023) MetaAnémic Cinéma [Video]. Vimeo (2023, February 13). https://vimeo.com/798325068
12. Martins A (2022a) [o][p][r][e][s][s][ã][o] [Video]. Vimeo. https://vimeo.com/771213662
13. Martins A (2022b) como se constrói uma casa [Video]. Vimeo. https://vimeo.com/764465418
14. Martins A (2022c) Muted Words [Video]. Vimeo. https://vimeo.com/764464638
15. McCarthy S (2020) Digital Typography at Stanford. The Journal of Design, Economics, and Innovation 4(6):546–560. https://www.researchgate.net/publication/347906526_Digital_Typography_at_Stanford
16. Navas E (2023) The Rise of Metacreativity: AI Aesthetics After Remix. Routledge, United Kingdom
17. Neitzel AA, Bridon J (2013) Poesia digital: reflexões em curso. Literatura y Lingüística (27), 111–133 (2013). https://bit.ly/3k6Pt9a
18. Palos KI, Campos GB, Silva AC (2019) Typography in Interactive Poetry: Gestures and Their Contributions to Reading Multiplicities. In Kurosu, M. (Ed.) Human-Computer Interaction: Design Practice in Contemporary Societies (pp.212–227). Springer (2019). https://www.researchgate.net/publication/334364277_Typography_in_Interactive_Poetry_Gestures_and_Their_Contributions_to_Reading_Multiplicities
19. Quelhas V (2012) Dynamic typography dynTypo > Understanding typography as a multimedia expression [Paper presentation]. V Congreso Internacional de Tipografía, Valencia (2012). https://www.researchgate.net/publication/303685968_Dynamic_typography_dynTypo_Understanding_typography_as_a_multimedia_expression

20. Rebelo S, Martins T, Bicker J, Machado P (2018) Typography as Image: Experiments on Typographic Portraits [Paper presentation]. 9th Typography Meeting, Tomar (2018, November 17–18). https://www.researchgate.net/publication/329351953_Typography_as_Image_Experiments_on_Typographic_Portraits
21. Reis J (2015) Jan Tschichold e o seu opus magnum: a Nova Tipografia enquanto plataforma fundadora de uma abordagem intervencionista e visual da escrita. Revista Gama 6(3):111–117. https://repositorio.ul.pt/handle/10451/18288
22. Sitney PA (1979) Image and Title in Avant-Garde Cinema. Essays in Honor of Jay Leyda 11:97–112. https://doi.org/10.2307/778237
23. Smith D (2001) Has History Been Too Generous to Gutenberg? New York Times (2001, january 27). https://www.nytimes.com/2001/01/27/arts/has-history-been-too-generous-to-gutenberg.html
24. Temple R (1986) The Genius of China: 3,000 Years of Science Discovery, and Invention. Simon and Schuster, United States of America
25. Toschi C (2020) Futurism and the Birth of Modern European Typography. In Berghaus, G. (ed.) International Yearbook of Futurism Studies, vol.10 (pp. 200–215.). De Gruyter (2020)
26. Uyan Dur BI (2021) Virtual Reality art and immersive experimental typography. e-Journal of New Media, 5(3), 219–233. https://doi.org/10.17932/IAU
27. Warde B (1955) The Crystal Goblet: sixteen essays on typography. The Sylvan Press, United Kingdom

Design Strategies and Challenges

Subscription Design

Gavin Perin

Abstract The professional opportunities and ethical implications of Artificial Intelligence (AI) on design are yet to reveal themselves fully. Nevertheless, Architecture's thirty-year project against postmodern iconography exemplifies how pre-existing concerns precondition AI's impact on design. Admittedly, there were pronounced conceptual and procedural differences in the methods developed during this period. However, a common belief in a pragmatic, instrumental approach to production has instigated a fundamental change in how one translates ideas into objects. This change, coinciding with the increasing use of the digital interface, now sees drawing as something one subscribes to rather than a set of techniques one uses to transcribe ideas. Contextualizing AI within a more extensive history of Architecture's experience with the digital environment, this paper will describe how pragmatism in Digital Architecture and Activism has encouraged designers to believe digital drawings enable an axiomatic translation of abstract ideas into real objects. At the same time, the most rudimentary of AI processes already contest this belief by presenting the disciplinary oddity of a digital technique based on images. AI not only reinvigorates architecture's often-problematic relationship to the semiotics of depictive imagery. It also problematizes issues of precedent and model by hiding how its algorithms select and combine sources. In raising questions around authorship and legitimacy, AI questions this thirty-year effort to recast disciplinary agency through instrumental drawing methods. Crucially, AI makes claiming authorship an issue again, while also asking whether axiomatic drawings can ever achieve prescribed socio-political goals.

Keywords Artificial intelligence · Design · Pragmatism · Representation

---

G. Perin (✉)
University of Technology Sydney, Broadway, NSW 2007, Australia
e-mail: Gavin.Perin@uts.edu.au

N. Martins et al. (eds.), *Perspectives on Design and Digital Communication IV*, Springer Series in Design and Innovation 33, https://doi.org/10.1007/978-3-031-41770-2_18

1 Introduction

Occasionally, one comes across a meme of a group of people with their heads buried in a newspaper. The byline says something along the lines of "All this technology is making us antisocial" [1]. The meme, in implying we are the source of this 'behavior', tacitly assumes the technology is benign. As the dissemination of misinformation through social media or the more mundane experience of dodging distracted phone users shows, this assumption often contradicts experience. The medium alters the terms of engagement, showing, at the very least, that technology finds alternative ways to manifest human behavior.

The newspaper meme echoes a more profound philosophical conversation around what—or who—controls technology. As a rudimentary internet search of news items related to Artificial Intelligence shows, it inspires the same arguments as the meme. These divergent opinions are less interesting than the sheer quantum of commentary. The extent of this commentary reinforces a growing sense that Artificial Intelligence (AI) is the digital era's defining technology because, unlike purpose-made professional software, it promises unheralded access to knowledge and expertise that had previously been germane to the professions. However, our experience with social media suggests that AI will not democratize production. The exact nature of this knowledge and expertise still relies on authored algorithms that mediate how ideas are formed and communicated. Ultimately, it is not the platform's content but the algorithms that manage our relationship with the world.

As a nascent technology, there are obvious risks in predicting any socioeconomic and political effects that stem from AI's impact on design. On the one hand, these risks operate at a personal level since predictions better reveal the author's unspoken biases rather than divine the future. On the other hand, universalizing AI's impact on—and through—different creative practices invites crude generalizations that diminish each practice's uniqueness. If avoiding the first issue involves placing AI within a more extensive digital design history, then confining this history to a disciplinary case study goes some way in addressing the second. The choice of Architecture for such a case study makes sense not only because of the unique complexities of production and the weighty responsibilities associated with designing inhabitable environments. It also reflects the implications of Robin Evans' [2] observation that architects never work "directly with the object of their thought". In making this observation, Evans effectively identified Architecture as one of the most mediated design practices. Left to operate on objects through drawings, architects are free to invent or combine techniques sourced from the existing representational stable or appropriated from other practices. Moreover, these representational choices must mediate—in no prescribed order—all manner of environmental, social, political, financial and material issues. Ultimately, as demonstrated in architecture's first two 'digital turns', the unique representational choices used to negotiate these complexities and responsibilities highlight many of the critical issues surrounding the digital translation of ideas into objects.[1]

It is a historical coincidence that Architecture's experience with the digital toolset began with a criticism of postmodernism's uncomfortably close relationship to capital.[2] This relationship not only questioned what an authentic and legitimate mode of production looked like. It also caused a great deal of introspection around architecture's social agency. For a discipline so dependent on drawing, the digital 'virtualization' of the toolset amplified these anxieties, especially since the most innovative digital techniques used software designed for other practices. Today, text-to-image AI software takes this virtualization process a step further. Professionally, AI questions whether the technical capacity to draw is enough to establish disciplinary expertise. Creatively, it questions the ontological and epistemological basis of architectural production. In raising these issues, AI poses fundamental ethical questions about how architectural production deals with algorithms that hide how data is selected and recomposed.

The recent popular interest in AI obscures that this technology represents a critical endpoint in our collective engagement with the digital interface. For Architecture, the furtiveness of AI's algorithms merely completes a progressive distancing of architects from the 'real' drivers of production. This estrangement, beginning with digital architecture's unquestioning faith in the toolset and consolidated by more recent attempts to neutralize digital technology, has fundamentally shifted how one translates ideas into objects. Admittedly, practices have oscillated over the last thirty years. However, whether drawings generate objects or illustrate disciplinary agency, a shared belief in pragmatism has instrumentalized the digital interface. If, previously, drawing manually emphasized the need to transcribe ideas through different mediums, the instrumentalization of the digital interface encourages one to 'buy into' what the algorithms deliver. This acquiescence to the tool instigates a form of exchange based on subscription. Crucially, by superseding methods that transcribe ideas through various mediums, subscription ensnares drawing within a consumptive mode of production. Since moral arguments alone are unlikely to stop the adoption of AI, the real challenge involves engaging with the technology without participating in the consumptive production model it tends to favor.

[1] 'Digital turns' references Mario Carpo's books *The Digital Turn in Architecture 1992–2012* (2012) and *The Second Digital Turn: Design Beyond Intelligence* (2017).

[2] For clarity, the term postmodern also concurs with those critics that placed deconstructivist architecture under the postmodern banner. While these practices have significant differences, one must concede that both privileged a semiotic understanding of architectural form.

2 From Transcription to Subscription

The manipulation of perspectival geometry in Piranesi's Carceri d'invenzione, Mies's collages, Cedric Price's schematic diagrams, and Bernard Tschumi's cinematic mappings illustrate how technical inventions have enriched the history of architectural representation. However, when it comes to the digital interface, digital architecture's failure to establish radial new modes of architectural representation partly reflects how digital techniques augmented long-established representational forms.

These established representational forms tend to translate ideas into objects through processes of transcription. Or, put another way, the design act operates through a technological transformation of data through the medium. Transcribing form through orthographic, paraline or perspective drawings focuses on providing a definitive account of the final object. While the viewer is expected to overlook the medium, the drawer's capacity to achieve this outcome requires a fundamental understanding of the medium and the conceptual scaffolding of the techniques used. Like subversive drawing practices listed above, the drawer is not interested in making the viewer conscious of how the medium has been adapted and manipulated. The viewer might register that the drawing is masterful or unique, but the primary ambition is to bring attention to the object.

The shift from transcription to subscription begins with the digital interface's impact on drawing as an embodied experience. The phenomenological undercurrents of the arguments against digital drawings often lament the loss of a tangible engagement between the body, the instrument and the medium. Whereas drawing manually involved coordinating body, eye and medium, moving between two-and three-dimensional views within the digital environment foregrounds the eye's importance. On the one hand, designing objects in Computer-Aided Design (CAD) and Computer-Aided Modelling (CAM) software has reduced the need to organize space, material and program in two dimensions. CAM software in fact alters past processes where two-dimensional drawings inform three-dimensional decisions. While one might now choose to begin in the plan view, there is a fundamental difference between refining an object in two dimensions and simply extruding preliminary line work into three dimensions. It is not that these past practices were inherently better. Denigrating a new technology by romanticizing older mediums and techniques overlooks how these past mediums and techniques transcribed architectural ideas through the Euclidean geometries embedded within the manual toolset. Nevertheless, the incapacity to impart order, scale, and materiality also leads to drawings that paradoxically decrease the importance of the embodied, experiential qualities of the object.

On the other hand, mediating lines through a keyboard and mouse both dispense with the corporeality demanded of the manual toolkit. Drawing a line using three clicks of a mouse and hitting the return button is very different from inscribing a line on paper.[3] Equally, the compass provides a reliable and repeatable embodied experience of the circle's Euclidean geometry. In effect, the manual drawing toolkit

[3] For clarity, the three clicks describe the process of selecting the 'tool' and then identifying the two endpoints of the line.

demands a more active, embodied relationship between drawer and medium. It is not just that splines only ever approximate the circle's geometry. Drawing a line and circle in the digital environment involves the same number of actions and physical movements. The only difference between a line and a circle is the selection of the tool in the menu.

This loss of corporeality—both in an understanding of the object's experience and in physical intimacy in the act of drawing—has of course been used to support claims that the digital interface ruptures the relationship between the author and the object. At the same time, Architecture has always relied on others to manufacture the toolset. The transition from the drawing board to the screen automatically alters several conditions that mediate the translation process. The most apparent difference being the digital environment's preference for splines recasts the geometric principles governing the depiction of form. This geometric shift becomes significant when exploiting the representational capacity to manipulate form as an integral facet of the design process. In these more explicit uses of spline geometry, the ability to manipulate form is more important than worrying about the algorithms operating behind the graphic user interface (GUI).

The other significant change accompanying the shift to the screen is that it dispenses with the fixed, standardized nature of the manual toolkit. In many respects, sustaining disciplinary sovereignty leveraged the standardized-yet-specialized nature of the disciplinary toolkit. The accessibility to an array of digital tools makes incursions across professional boundaries easier. By eroding the boundaries separating professions and the boundaries between professionals and their clients, the digital interface questions any residual belief in disciplinary autonomy. However, exploiting these circumstances involves seeing beyond the information used in orthographic projection, paraline drawings and perspectives. By extension, the digital toolset challenges disciplinary autonomy and expands the representational toolkit while eroding the embodied nature of drawing.

The use of the term subscription not only captures how 'exchange' in the digital economy increasingly involves signing up for things. It also recognizes that professionals subscribe to software to deliver professional services and platforms to promote this work. Digital technology is unique because it enables individuals and groups to use these sites as platforms to receive ideas and disseminate beliefs. The increasingly politicized nature of social media platforms also shows how 'signing up' now overlaps with subscribing to certain beliefs and using these platforms to share these beliefs. It is not just that individual and collective identities are formed and confirmed by what we subscribe to, but the platform is now the instrument and venue to articulate one's stance. For all these reasons, subscription registers a fundamental change in the exchange between programmers and practitioners and between practitioners and viewers.

Despite the computer's ubiquity, only around twenty-eight million people are professionally familiar with the algorithms driving the digital economy [3]. This figure is alarming enough, given that programming could be considered a general rather than a professional skill that should be taught at primary and secondary levels of education. Even more worrying is how few of the close to four million architects worldwide can appreciate how algorithms influence architectural decisions [4]. This degree of illiteracy is significant given that the pervasiveness of software represents a universal shift in the translation process. Ultimately, the toolset's ubiquity has a more significant bearing on production. If computers have shifted the basis of the translation process, one could argue that this ignorance of how the tools operate indicates that subscription is now the primary conceptual basis of translation. This circumstance reflects how this ignorance relies on an absolute trust in the toolset. Subscription, therefore, describes the tendency to accept passively various digital tools and platforms to receive and interpret information, and to disseminate opinions and ideas.

3 The Postmodern Problem

The shift in translation from transcription to subscription owes much to the disciplinary critique of postmodernism that began in the late 1980s. The ensuing focus on depicting objects diminished the role other representational techniques and mediums had in situating architectural objects within a larger cultural milieu. Crucially, many leading 'progressive' practices within the next generation of leading Anglo-American disciplinary figures rejected postmodernism semiotics with a more pragmatic approach to production. In this shift, imagery, words, and text were, at best, supplementary and, at worst, disruptive to architectural production. Despite pronounced formal, procedural, and representational differences between these practices, a common suspicion of imagery and theory prioritized decidedly instrumental drawings. Today, drawings depict effects.

Like other critiques of the time, architectural critic Alan Colquhoun placed the postmodern problem within a more comprehensive history. Positioning postmodernism as the logical endpoint of the two-hundred-year-long project, Colquhoun [5] argued that capitalism's ability to co-opt postmodern architecture continued modernist production's estrangement of makers and users. Postmodern iconography not only failed to reconnect architecture to a broader audience. In privileging a semiotic interplay of words, text and graphics, postmodernism had also inadvertently limited procedural innovation, restricting its formal 'language' to disciplinary precedents drawn from the Western Canon. Today, one rightly frowns on this form of cultural imperialism. However, this semiotic interplay was problematic even in less enlightened times because encultured beliefs were more important than the object's corporeal effects. In robbing objects of their material qualities, postmodern sampling and refashioning of past forms used imagery that displaced materiality as the basis for production. The fact that these images increasingly parodied their referents meant postmodern iconography pre-empted social media's tendency to make the technical capacity to disseminate images more important than an image's message.

It is difficult to dispute that postmodernism's corrupted symbolism used images to transform makers and users into producers and consumers. Much like Baudrillard's [6] idea of the hyperreal, modernism's dematerialization of its subjects had initiated a "liquidation of all referentials". Ultimately, postmodernism's genius was to fabricate a self-sustaining economy of signs that, increasingly devoid of an actual referent, preserved this separation. For the purists, postmodernism's general shift towards pictorial representation obscured the architectural drawing's traditional function to describe form. This problem, which never transformed into a concrete critique, tacitly recognized that the emblematic reduction of form to a symbol relied on pictorial representation to construct narratives. While architectural drawings still focused on the object, telling this larger story involved increasingly elaborate drawings full of all sorts of supporting figures. The abstract concepts represented in these images existed beyond the tectonic realities of the proposed object, drawing attention to the medium. The real problem with postmodern images was that their narratives supplanted the drawing's primary task of legitimizing architectural decisions by validating objects through their physical attributes alone.

Unsurprisingly, Colquhoun's 'solution' wistfully revisited an earlier modernism when production still had a genuine material basis. While this nostalgia posed its own issues, identifying them would be to 'miss the point'. In this framework, architectural drawings simply resolved real tectonic questions. Materiality's advantage was that architectural representation had nothing to do with pictorial representation. As an antidote to postmodern iconography, drawings had one job—to depict the physical impact objects have in and on the world.

4 Post-critical Solutions

In the period following postmodernism, pragmatism has followed that general "philosophical tradition that—very broadly—understands knowing the world as inseparable from agency within it" [7]. However, over these thirty-odd years, pragmatism has shifted from theoretical concerns to practical applications. This change represented a shift from an "articulate language [that] rest[ed] on a deep bed of shared human practices that can never be fully 'made explicit' " to an outcome-focused practice that aimed to "contribute directly to social progress" [7]. The basis for this transformation crystallized in the mid-2000s during the Post-Critical movement's ascendency. It is vital to note that the Post-Critical's import did not come from having shaped events but from sensing that something had shifted. It concretized the desire that first appeared in digital architecture for an authentic, politically aware mode of production. Only after the Global Financial Crisis (GFC) did pragmatism's basic credo of placing forming before form expose how digital architecture's theoretical pragmatism had neither the technological nous nor the sociopolitical relevance to legitimize production [8].

While sharing Colquhoun's concern of postmodern semiotics, this credo of pragmatism avoided a Miesian fetishizing of materiality and detail. In a somewhat unexpected move, avoiding tectonics made it theoretically possible to recast materiality and production as parts of one system. While approaches varied, each bore the influence of Deleuzian materialism. As an authority on Deleuze's oeuvre, Brian Massumi perhaps offers the best framework to place Deleuze's writings within a larger pragmatic philosophical tradition. Rather than being caught within the intricacies of Massumi's work, the most expeditious explanation of this version of materialism situates Deleuzian Affect within William James's pragmatic 'radical empiricism'. In this framework, James' notion of relations as the "givens of experience" makes them integral to the sensate conditions that form experience [9]. Therefore, any 'relation' between an act and the conscious identification of an experience sees a feeling as the result instead of the cause of action [9]. Accepting relationality explains the conditional nature of the context from which things form. The critical point is that Deleuzian Affect represents a "point of emergence" [9]. In this paradigm, forms emerge when abstract-but-real immanent relational conditions concretize.

Before the GFC, there was a common belief that architecture's susceptibility to capitalism resulted from postmodernism's attachment to "linguistic games" [10]. An alternative narrative of production focused on *forming* instead of *meaning* leveraged Deleuzian materialism to usurp capitalism's hold over disciplinary agency by negating the capacity for authorship. Today, this might be considered a capricious approach, but at the time, it was strategic. Ambitious to correct the errors of a previous generation, a new generation of thinkers was also wary that presenting radical new design methods risked being accused of replicating the ideological mistakes of the modernist avant-garde.[4] In particular, digital architecture's promotion of formal novelty demanded a type of pragmatism that removed ideology by diminishing authorship. Architects ceding authorship to 'the process', allowing Deleuzian materialism to deliver an ideologically free avant-gardism.

Procedurally, this attack against semiotics in the 1990s adopted animation software to give symbolic expression to Deleuzian Affect. In particular, Greg Lynn's invention of the animated diagram seemingly avoided semiotics using a drawing method that made the object's final form contingent on a complex field of contextual forces. In effect, procedural autonomy defied *meaning* since the drawing, not the architect, authored form. Undoubtedly, this new drawing demonstrated the opportunities of heeding Robin Evans' [2] call to explore architecture through the medium. It also showed the computer's potential to extend architecture's representational arsenal into methods that processed contextual information. At the same time, digital architecture's commitment to drawing Deleuzian affect demanded a smooth and transparent translation process. Nevertheless, it is essential to note that digital architecture's need for axiomatic drawings was both a theoretical principle and a defensive tactic aimed at diffusing any criticisms of these new representational methods. Consequently, axiomatic thinking had no choice but to place absolute trust in the algorithms to process the information. Whether the algorithm was on the screen or hidden behind a GUI, one never questioned what was 'under the hood'.

[4] This claim rests on two critical publications by Greg Lynn. Published in 1998 and 199, respectively, the books *Folds, Bodies & Blobs: Collected Essays* and *Animate Form* leveraged Deleuzian Materialism to attack [10, 11]. This generational contest was not only evident in Lynn's writings. Lynn's fight was against a preceding generation of architect-theorists. Robert Somol and Michael Speaks instigated a similar generation dispute over theory.

Alternatives to this drawn representation of affect adopted more conventional representational methods. In many instances, real-world affects resulted in an array of different techniques. At one extremity Robert Somol and Sarah Whiting's essay, 'Notes Around the Doppler Effect and Other Moods of Modernism', saw Affect as a style of authorship. This essay remains an oddity; comparing De Niro's 'method acting' to Robert Mitchum's innate 'coolness', Somol and Whiting [12] attempted to illustrate the perils of method over character. Sylvia Lavin [13, 14], who also spoke of 'moods', prioritized types of experiences. Recycling Walter Benjamin's 'shock effect', Lavin's notion of affect focused on objects that solicited unexpected experiences. Lavin effectively presented an alternative basis to legitimize formal novelty since form must be unusual enough to engender a particular type of experience. However, hindsight suggests that Michael Speaks' reading of affect matches a mode of pragmatism now seen in contemporary activist practices. Sympathetic to the technologies of production Speaks [15, 16] leveraged an argument against autonomy to call for architecture to appropriate new techniques and technologies drawn from outside the discipline.

The GFC drew a line under these forms of theoretical pragmatism that attempted to 'solve' architecture's relationship to capital by legitimizing alternative modes of production. Accordingly, Post-Critical thinking was an intermediatory step between theoretical pragmatism and its more practical counterpart. The systemic socio-political inequities exposed by the GFC and amplified by the climate emergency provided real-world issues on which to apply these technologies. In many ways, both digital architecture's abhorrence of semiotics and the Post-Critical attack on theory tried to materialize worldly effects to bestow authenticity and honesty to architectural objects. Nevertheless, this desire for authenticity also traced a shared wariness of the criticisms made of past avant-garde practices. This wariness instigated a reactive posture that was defined by anxiety around authorship. Consequently, dealing pragmatically with capital involved neutralizing politics at a personal level.

These different accounts of affect were captured in an internal disciplinary dispute over postmodernism and authorship. Post-Critical thinking marks a moment when the post-postmodern pragmatism transformed from an abstract account into its more normative, practical cousin. For the following generation, the digital toolset was, at best, a tool and, at worst, suspect. Reminiscent of Colquhoun's version of materiality, the practical nature of pragmatism today establishes legitimacy through service to others. Operating through a social lens, agency preferences modes of practice over technique. As laudable as this approach sounds, the contemporary return to a finely detailed and crafted orthogonal form recalls methods that value a masterly arrangement of space, program and material. If activism's antagonism to the digitizing of architectural production privileges politics as a social issue it is because real crises demand urgent action rather than new techniques. Drawings demonstrate pragmatic responses based on a problem–solution paradigm.

Somewhat counterintuitively, the practices bracketing Post-Critical thinking inverted the relationship between theory and pragmatic purity. The presumedly ideologically free digital drawing aimed to produce pragmatically pure practical objects, while activism's practical pragmatism uses conventional drawings to describe highly theorized and politized objects. In both instances, a theoretical and physical understanding of materialism incorporates alternative visions of pragmatism where drawings never lie. Irrespective of the genre, pragmatic materialism uses objects as the ultimate symbol of effecting agency in the world.

The shift from curvilinear form to its more restrained counterpart traces the impact Post-Critical pragmatism had on production. Whether contextual transformation involved drawn simulations or formal solutions to real-world socioeconomic or political issues, design prized contingency. Objects might materialize as novel forms or solve immediate, explicit problems, but politically speaking, pragmatism authenticates objects by avoiding authorship long enough to 'read the lie of the land'. If, for Lynn, animation software simply suspended authorship until the diagram had produced form, for others digital tools provided the systems and data required to make informed design decisions. Yet in all approaches, framing the object's politics demands pragmatic design methods that produce objects as an explicable, quasi-natural consequence of context. The differences in formal expression belie how, in the digital era, production is inherently instrumental.

5 Pragmatic Materialism and the Politics of the Axiomatic Drawing

The demand for alternative design methods rightly recognized that postmodernism had become trapped within an ever-decreasing spiral of self-referential parody. Rejecting postmodernism's often esoteric semiotic games promised to free production from an endless recycling of forms adapted from a fixed staple of precedents. However, Post-Criticality was a pivot point in thirty-year history of design methods trapped within reoccurring psychological preoccupations that have troubled architecture since postmodernism's end. As an overriding descriptor of practices over this period, pragmatic materialism represents various attempts to legitimize new modes of architectural production while avoiding any claims of ideological bias. Digital media complicated this project by altering how architecture used drawings to exercise agency and achieve a sense of authenticity. This challenge was made more difficult by insisting that drawings provide fidelity between the object and its graphic representation. This semantic approach to drawings creates new ontological conceits and epistemological inconsistencies. The discursive efforts to exercise agency without any trace of ideology only amplify these conceits and inconsistencies.

In the introduction to the 2011 publication, *Performalism: Form and Performance in Digital Architecture*, the editors Neuman and Grobman reassert digital architecture's relevance by arguing that its methods uncouple agency from ideology. In order to support this idea, the editors make two telling claims. The first is that performalism offers "a modality for performative architectural existence" capable of mediating outdated modernist dogmas and addressing more profound questions concerning the conditions of experience, the body, objecthood and politics [17]. The second claim, which comes as a response to Christopher Hight's contribution, argues that, unlike performativity, performalism has a capacity for political action [17]. This rearguard action aimed at repatriating digital architecture clearly failed. Despite the evolution from the animated diagram to parametrics to algorithms, digital architecture remains a marginal practice. This reality is not a problem around a wider disciplinary failure to attain relevant skill sets. The rolling crises following the GFC simply made the excessive nature of digital form increasingly anachronistic. Ironically, the solution to the problems of postmodern semiotics had produced an even more spectacular type of architecture. Worse still, the attempt to depoliticize the digital avant-garde through drawings that 'naturalized' the translation process adopted a posture that paralleled those neoliberal arguments that legitimized production by bastardizing a bastardized account of Darwinian evolutionary theory. Ultimately, digital architecture's failure to alter architecture's relationship to capital illustrated the limits of agency and that drawings could not depoliticize objects.

These issues haunted many of the dominant practices during the 1990s and 2000s. Lavin's current interest in the archive suggests the limited capacity of formal novelty to effect change. One must also admit that Somol and Whiting's refashioning of authorship was always more rhetorical than instructive. Illustrating 'cool' was easy; manufacturing it was far more problematic since only very select individuals could exercise this style of authorship. These collective failures of an abstract, theoretical type of pragmatism originated from framing politics through issues of ideology and authorship. In reducing more significant collective political problems to an issue of personal ideology, the advocacy of these methods ignored that capitalism not only continues to appropriate spectacular objects but also—as social media demonstrates—that only the quantum of content matters.

These failures contextualize why many practices treat CAD and CAM software as a digital drawing board. While often a default response, Activism's criticism of digital technologies transforms this neutering of the technology into a political imperative. Accordingly, the theories of affect and radical empiricism have shifted towards the left-leaning politics of many contemporary Activist practices. Moreover, a particular approach to materials means Activism aligns formally and politically with the Marxist undertones of Colquhoun's recalibration of production, labor, and materiality.

Crucially, activism's validation of objects refashions Neuman and Grobman's performalism. However, unlike digital architecture, performing an 'organic', 'bottom up' process values a particular form of practice rather than drawn representations of 'affective' conditions. Nevertheless, these more participatory design methods still employs drawings that must envision yet-to-be-built objects. Activism may not necessarily be concerned with representational innovation, but drawings are essential in demonstrating that people have been listened to. Moreover, activism's desire to erase the semiotic gap between the sign and the thing it signifies exhibits a similar axiomatic tendency. Like digital architecture, drawings legitimize action by vouching for the social or political integrity of both the author and object. Unlike digital architecture, political agency precedes drawing.

Despite wanting to effect change, activism challenges capitalism by withdrawing from capitalist production. This withdrawal is illustrated in Pier Vittorio Aureli's advocacy of Agonism, which celebrates rather than mediates conflict. Such practices have little time for Deleuzian synthesis. Instead, political agency operates through 'critical' projects as an "alternative to the one imposed by capitalist reality" [18]. Nevertheless, the increasingly ubiquitous promotion of 'collaborative' open office space shows that a project's political impact extends beyond the intentions embedded in the drawing. It is awkward enough that the media ignores the problematic politics of the open office. Or that these projects make space for the gig economy by reframing labor through the lens of entrepreneurism. The reality is that these spaces are the spatial expression of a deeper societal trend that transforms public service into a private practice. Ultimately, the media's promotion of such projects by offices like Dogma[5] and Assemble Studio never really contest, let alone sit apart from, the system they wish to dispute.

This brief critique of the collaborative space highlights the value of returning to a much older attack on the idea of the 'critical' project. In a succinct and scathing criticism of Krzysztof Wodiczko's *Homeless Projections*, Rosalyn Deutsche [19] draws attention to "The cynicism inherent in the use of a homeless person as a foil for the aesthetic merits" of an artwork. In its contemporary form, the ethical problem at the core of the 'critical' project comes from revering a supposedly simpler time that reduced politics to a dispute between capital and labor. Worse still, the pretense of operating at the margins of capitalist production ignores the resilience of specific capitalist tropes and methods.

Interestingly, Assemble Studio's work with ceramic manufacturing processes transforms the issues behind the collaborative workspace into the realm of materiality. Originating as a community workshop associated with the 2014 Yardhouse Project in Stratford, this practice has matured into a homewares business with production centered around the Granby project in Liverpool. Crucially, this trajectory echoes the developmental arc of William Morris's wallpapers, where an 'authentic' craft-based industry ends up servicing an exclusive niche market. While Assemble Studio makes no grandiose claims about the social benefits of this line of work, they never contest the belief that this commercialization of a cottage industry is an authentic social enterprise. At a literal material level, this activism also sees entrepreneurism as a valid measure to correct capitalism. In linking the mode of production to social good, Assemble Studio establishes authenticity by valuing material through its social impact rather than any intrinsic material properties. Like the numerous projects that celebrate sustainability by making a feature of Cross Laminated Timber (CLT), 'authenticity' comes from what material signals. Metamorphosed into a sign of real sociopolitical action, the performance of a material becomes a secondary concern. If the 'critical' project merely offered a representation of a problem as a solution, Assemble Studio's interest in ceramics fabricates a representational solution to a problem.

[5] Aureli is a partner of Dogma.

This semiotic reading of materials is not without precedents. In its architectural form, phenomenology imparted an almost sacred capacity for materials to evoke universally specific feelings [20]. Curtis [21] substantiates this argument when linking Scandinavian materiality to National Identity. Framption's [22] 'critical regionalism' simply follows this model by seeing Identity as a semiotic consequence of a material's geographic specificity to humanize international modernism. In this attempt to make modernism local Identity is constructed through a universal acquiescence to an image based on shared histories originating from unique geographical conditions. Activism merely alters this relationship by making meaning an extension of the author's abstract political preferences. This contemporary reading of materiality ignores shared histories, preferring instead to construct meaning around individuated, idiosyncratic readings. This process also involves an ambivalence toward physical properties, meaning materials embody principles that are as abstract as they are indisputable. Paradoxically, these individuated acts of assigning abstract political concepts to materials contradict claims that activism involves pragmatic design decisions. Materiality now extenuates the gap between sign and signified.

Epstein Jones's [23] essay "Little People Everywhere: The Populated Plan" indicates how far an activist mindset is willing to uncouple the signified from the sign. Arguing that drawings should "reflect the diversity we wish there to be in those worlds", Epstein Jones [23] abandons any expectation that the object fosters diversity. It is enough that the figures in the drawing depict diversity. It is not a coincidence that MOS's paraline drawings published in this paper treat architecture as a stage set. While this call for more realistic representations is welcomed by those of us tired of drawings populated by distorted silhouettes or runway models, the drawing is just as idyllic. The image validates the object by presenting an idealized depiction of a diverse, active, and harmonious community. In an update to Juhani Pallasmaa's poetic advocacy of drawing by hand, the 'little people' legitimize design decisions. Like Pallasmaa's advocacy for hand drawing—on the basis that the embodied experience of drawing creates an intimacy between the author, media and object–an equivalency is made between drawings that capture a 'real' embodied experience and those that 'evidence' the project's (and author's) credentials. In both instances, the image assumes a mystical agency that represents the author's idealized worldview.

Whatever the argument, such attitudes trap architectural representation into the same techniques and reassert existing disciplinary orthodoxies. The toolset has become increasingly mute because the drawing must represent real architectural solutions. It is this belief that means Activism underutilizes the range of tools and forms of data that could better inform real-world design decisions. In this sense, Activism's absolute rejection of the interface makes it possible to villainize digital tools by focusing only on their very worst effects. Instead, the expectation for the medium to not influence production means Activism amplifies a pre-existing notion of disciplinary exemplarity. If Activism neutralizes its own ideology, it is because its advocacy of supposedly collaborative, democratic design methods situates objects on the better side of a moral divide separating the left from the right.

Activism's focus on the object over technique and technology replicates the postmodern problem Colquhoun saw in modern production. Two critical steps create this circumstance. First, partisan intent legitimizes specific methods, allowing ideology to once again dictate and delimit how production should occur. The second critical move is that the distancing between objects and technologies uses materiality to represent, iconographically, partisan political beliefs. Morals may be acceptable, but digital architecture's advocates were rightly wary of designing according to personal beliefs. Clearly, there is no real need for another upmarket clay tile production facility. The tiles are important only because they are seen primarily as a political rather than a commercial activity. At the same time, this mode of production participates in a marketplace based on consumption.

Despite these conceits and inconsistencies, the procedural and intellectual differences between digital architecture and what followed never altered the expectation that drawings represented real-world objects and their real-world effects. In order to deliver on the author's promise, drawings had to provide a clean, axiomatic movement from idea to form. It made little difference whether the drawing simulated nascent contextual forces or embodied a deeply held political positions. If digital architecture aimed to depoliticize form by neutralizing the connection between authorship and ideology, then the drawing must faithfully translate the data into an object. In comparison, Activism's equivalency between the manual and digital interface highlights an active disinterest in representational innovation. Reinforcing the role played by existing modes of architectural representation, the drawing's primary role is to depict the effects of objects in the real world. For activism, the image is enough to validate its social good.

Since the 1990s, the computer has provided clear professional advantages by enhancing the capacity to draw things. In the intervening years, a pragmatic desire to effect a clean and authentic translation from idea to object has seen different genres deploy different methods to achieve indexical fidelity between drawings and objects. Unfortunately, instrumentalizing action in the real world also obliges one to see past the digital medium. This approach created two missed opportunities. First, presenting plausible procedural narratives to account for the images on the screen made it possible to coopt tools from outside architecture. However, this colonization of these tools has reinforced a false sense of disciplinary autonomy.[6] Second, the idea of the axiomatic drawing fails to acknowledge that the sign differs from the signified. Consequently, one never examines the digital interface to explore "the drawing's power as a medium… [by recognizing the drawing's] distinctness from and unlikeness to the thing that is represented, rather than its likeness to it" [2].

[6] The issue of disciplinary autonomy has reappeared over the last thirty years. While the digital drawing broke this myth of 'autonomy', it paradoxically never challenged disciplinary boundaries. Free to pick and choose knowledge from elsewhere, any disciplinary exchange of ideas was always to architecture's benefit. Despite Activism's collaborative reflex, architecture continues to reaffirm disciplinary silos and place itself atop a professional hierarchy. This new self-image, which sees architecture less as the mother of the arts and more godfather of the professions, has reasserted autonomy and disciplinary primacy.

6 Conclusion: Subscripting AI

AI comes at a time when architecture increasingly sees the computer as an instrumental yet dumb representational tool while scripting remains a marginal skill set. Within the academy and practice, the digital drawing either optimizes prosaic objectives or propagandizes design thinking. In architecture, the most rudimentary one-step AI process already presents a range of issues that pragmatic materialism believed it had resigned to the past. AI ironically presents the disciplinary oddity of a digital technique based on images. Somewhat annoyingly, the technology reminds architecture of its often-problematic relationship to the image. The reintroduction of the image into production is doubly confronting to digital design methods, given that Deleuzian materialism had supposedly dealt with these "linguistic games" [10]. Moreover, this reliance on digital imagery refashions postmodern appropriation and collage techniques. This obtuse return of precedent and model is problematic precisely because AI hides the algorithms used to select and combine sources. The resuscitation of these older techniques also coincides with the ethical and moral concerns around appropriating work without intellectual recognition or financial gain. After a thirty-year effort to depoliticize agency through the drawing, AI makes claiming authorship an issue again.

The exploratory work of Andrew Kudless, principal of Matsys, highlights how the shift from lines to text alters architecture's intellectual frameworks and extends its procedural reserves. While Matsys's work reinforces AI's dependence on recycling imagery Kudless's [24] digital processing pipeline differs from most digital processes because the content is generated and refined through multiple software programs. This pipeline is unique not only because it must sequence the roles of each software but also because it generates form using an image file format. Initially, one can afford to be agnostic about the file type. One might begin with a model file, but in the end, production moves quickly to an image file.

Crucially, the close relationship between image and text in this AI design process means that design depends on highly evocative descriptors. Herein lies a second oddity of AI. Namely that these descriptors rely on textual prompts. Text is central in envisaging material, formal and phenomenological effects. As Kudless [25] notes, reaching a final outcome involves negotiating how different AI software captures and processes imagery. However, these refinements are essentially curatorial and demand becoming more adept with word selection. Notably, AI's use of imagery reintroduces semiotics into digital forms of architectural production. Interestingly, this semiotics is akin to ekphrasis, given that production relies not so much on a verbal but textual "representation of a visual representation". This is an important procedural oddity since the most distinguishing aspect of Matsys's work is its atmospheric quality. 'Mood' returns through deliberate acts of authorship.

There is one other significant effect AI has on the discipline. If textual prompts are enough to generate all sorts of outcomes, AI's interface makes the drawing board skeuomorph obsolete. On the one hand, the importance of text ends any notion of drawing as an embodied act. On the other hand, AI changes the skill set because a command of language is now a fundamental attribute. If AI deals in image files, the need to convert these images into real objects means the technology poses no immediate threat to disciplinary sovereignty. However, using AI to generate code suggests that drawing does not guarantee this sovereignty.

These changes to production reinforce the opinion that AI represents the end of transcription. It does so partly because it replicates, professionally, the separation social media platforms establish between users and proprietors. Equally, artists' current concerns over AI's appropriation of their imagery illustrate the procedural and political concerns presented by the image's newfound currency. Nevertheless, as the meme cited at the beginning of this essay shows, one needs to be cautious of blaming technology for our current ills. It is also important to note that calls to halt the technology have come a little too late [26]. Worryingly for the discipline, the rhetorical fight against digital technologies in venues like *e-flux* wants to fight battles already lost. Worse still, these reactionary positions—which overestimate architecture's capacity to effect change—ignore that things have altered fundamentally. This stance replicates the errors of the last thirty years where solving neoliberalism ignores the technology driving its economy. Ultimately, such positions serve capital by inflating disciplinary agency while denying the discipline of the first-hand technical knowledge required to interrogate and challenge these politics.

The current tendency to villainize digital technology oddly ignores the partisan nature of each respective political position. There is also a failure to acknowledge how this politic participates in the polarized nature of present-day political discourse. One might approve of Activism's intentions, but this split between two battling moral codes cannot be framed as an old-fashioned skirmish between left and right. As Monbiot [27] and Malik [28] argue, contemporary politics have redrawn past affiliations. Frustratingly, these articles stop short of identifying the drivers behind this confluence of the fringes, but they observe how all players believe that they are on the 'right side' of history. In this world, subscription involves buying into belief systems. The act of picking sides merely involves an exchange where individuals consume rather than question ideology.

In this narrative of architectural production after postmodernism, the different intellectual, procedural and ideological forms used to legitimize work have, in some way, assumed a pragmatic understanding of materialism. If the GFC provided the socioeconomic and political causes to direct agency, the scientific realities of climate change offer the promise of some sort of technological solution. However, while the GFC and climate change are symptomatic of neoliberalism's inequities, these symptoms have been conflated. In this process, the understanding of power has been simplified, making it easy to package agency as an individual rather than collective responsibility. This tacit belief that change is an individuated activity is as equally problematic as believing that reforming the financial system or solving climate change solves everything all at once.

Pragmatic materialism made three critical errors in the attempt to uncouple architectural production from capital. First, an argument about capital fuelled a generational dispute. Second, leveraging politics to pathologize postmodern semiotics resulted in a more general propensity to overestimate disciplinary agency. If these two errors resulted from seeing architecture as an agent rather than a symptom of power, a third error arose over the nature of architectural production itself. Specifically, the trust invested in practical design methods demanded drawings that faithfully presented the as-yet-unbuilt object. This thinking stretches credulity, given the expectation that drawings of yet-to-be-built objects serve as concrete actualizations of abstract political ideas. Often constructed along partisan lines, these positions used drawings to present idealized relationships between individuals and between individuals and the world. These errors constructed a discursive imperative connecting agency to axiomatic, instrumental drawings. Despite an initial dissatisfaction with the semiotic logic driving postmodernism's interest in pictorial representation, there is little evidence that subsequent design methods have really altered this tendency. Digital architecture's abstract representational methods and activism's attempt to depict abstract political ideas have merely constructed their own idiosyncratic semiotic systems. The drawings may want to presume real action, but this symbolism uses the depictive logic of pictorial representation.

Ultimately, the inability to escape capitalist production reflects a widespread habit of seeing production as smooth and transparent. In this context, the belief that production could solve all sorts of ills repeats the same mistake Communism made about capital. As Bataille [29] noted, Communism's dispute with capital focused on who owned the tools rather than questioning the logic of accumulation that legitimized production. It is this complete faith in production that connects the practices operating over the last thirty years. It is no accident that the problem of postmodern production was primarily a North American dispute. Collectively, the validation of production speaks of another, more base form of North American pragmatism, where a liberalism—paranoid of institutions—always lurks below the surface.

The solution to this circumstance is not to trust institutions but to re-examine how one activates agency. This re-examination must acknowledge that power can co-opt the most authentic design acts at the same time as it withdraws into the background. Like modernism and postmodernism, pragmatic materialism overestimated architecture's capacity to alter entrenched socioeconomic and political relationships. The problem with early modern ideology was that reduced complex problems into universal solutions. Unwilling to question what a truly non-autonomous practice might look like, the ensuing notions of agency never doubted architecture's capacity—or even right- to effect change. Robbed of a longer historical perspective, the focus on the immediate problems avoided questioning whether pragmatism sustained power's hold over the effects of architectural objects.

It is worth noting that both Bataille [29] and Camus [30, 31] embraced actions that willingly usurped systems. It is worth repeating that 'subscription' best describes a moment where designing 'accepts the conditions' of the platform. If AI embodies a subscriptive translation process, then pragmatic materialism teaches us to approach design without an ideological faith in production. It is also worth remembering that Colquhoun's retrospective view of production failed to appreciate past instances where the representational toolset deliberately extenuated the connection between drawing and form. Believing materiality can solve capitalism's capacity to exploit images ignores how drawings are anything but neutral translational mediums. The initial understanding of AI in this essay has observed the increased procedural complexity in the design pipeline. This complexity opens more opportunities for translational slippages. If we fail to understand this condition, neutering or vilifying medium only divides action from the effect images have in mediating power in the real world.

References

1. VR Zone, Show them this photo if someone said technology is making us anti-social, Facebook, https://www.facebook.com/vrzone/photos/a.10150129751268010/10152417138358010, Last accessed 2023/03/06
2. Evans R (1997) Translations from Drawing to Building and Other Essays. MIT Press, Cambridge, Massachusetts
3. Vailshery LS (2023) "Statista: Number of software developers worldwide in 2018 to 2024". Statista, https://www.statista.com/statistics/627312/worldwide-developer-population/, last accessed 2023/03/06.
4. Ross, B.: "How Many Architects Are There in the World? Architizer, https://architizer.com/blog/inspiration/industry/how-many-architects-are-in-the-world/, last accessed 2023/04/05.
5. Colquhoun A (1988) Postmodernism and Structuralism: A Retrospective Glance. Assemblage 5:6–15
6. Baudrillard J (1994) Simulacra and Simulation. University of Michigan Press, Ann Arbor
7. Legg, C., Hookway, C.: "Pragmatism". The Stanford Encyclopedia of Philosophy, https://plato.stanford.edu/archives/sum2021/entries/pragmatism/, last accessed 2023/03/06.
8. Kwinter, S.: Who's Afraid of Formalism. In: Foreign Office Architects: Phylogenesis: Foa's Ark: Foreign Office Architects. Actar, Barcelona (2004).
9. Massumi B (2002) Parables for the Virtual: Movement, Affect, Sensation. Duke University Press, Durham & London
10. Lynn G (1999) Animate Form. Princeton Architectural Press, New York
11. Lynn, G.: Folds, Bodies & Blobs: Collected Essays. 2nd edn. La Lettre Volée Brussels (2004).
12. Somol R, Whiting S (2002) Notes around the Doppler Effect and Other Moods of Modernism. Perspecta 33:72–77
13. Lavin S (2000) The New Mood or Affective Disorder. Assemblage 41:40
14. Lavin S (2011) Design By Mood. Pidgin 10:24–33
15. Speaks, M.: Which Way Avant-Garde? Assemblage no.41, 78 (2000).
16. Speaks M (2006) Intelligence After Theory. Perspecta 38:101–106
17. Grobman, Y. J., Neuman, E., (Eds.): Performalism: Form and Performance in Digital Architecture. Routledge, London & New York (2011).
18. Aureli PV (2008) The Project of Autonomy: Politics and Architecture within and against Capitalism. Princeton Architectural Press, New York
19. Deutsche, R.: Krzysztof Wodiczko's Homeless Projection and the Site of Urban "Revitalization. October vol. 38, 63–98 (1986).
20. Norberg-Schulz C (1987) On The Way To Figurative Architecture. Places 4(1):18–23
21. Curtis WJR (1996) Modern Architecture Since 1900, 3rd edn. Phaidon Press, London
22. Frampton K (1980) Modern Architecture: A Critical History, 1st edn. Thames & Hudson, London
23. Epstein Jones D (2019) Little People Everywhere: The Populated Plan. Log 45:59–71
24. Kudless, A.: "(@matsysdesign)", Instagram, https://www.instagram.com/p/Clt1AWPJzMg/, last accessed 2023/05/14.
25. Kudless, A.: "(@matsysdesign)", Instagram, https://www.instagram.com/p/ClMREVvunbQ/, last accessed 2023/05/14

26. Olson, P.: “Analysis: Elon Musk Wants to Pause AI? It’s Too Late for That”, Washington Post, https://www.washingtonpost.com/business/2023/03/30/elon-musk-wants-to-pause-ai-progress-with-chatgpt-it-s-too-late-for-that/86a1487c-ceeb-11ed-8907-156f0390d081_story.html, last accessed 2023/03/30.
27. Monbiot, G.: “It’s shocking to see so many leftwingers lured to the far right by conspiracy theories”, The Guardian, https://www.theguardian.com/commentisfree/2021/sep/22/leftwingers-far-right-conspiracy-theories-anti-vaxxers-power, last accessed 2023/04/07.
28. Malik, K.: "It can be hard to distinguish the cultural claims of right and left. Just look at Qatar”, The Guardian.“ https://www.theguardian.com/commentisfree/2022/nov/27/it-can-be-hard-to-distinguish-the-cultural-claims-of-right-and-left-just-look-at-qatar, last accessed 2023/04/07.
29. Bataille, G.: The Accursed Share: An Essay on General Economy- Volume 1: Consumption. 1st edn. Zone Books, New York (1995).
30. Camus, A., Read (transl.): H., The Rebel. Penguin, Harmondsworth (1973).
31. Camus A., O’Brien, J. (transl.): The Myth of Sisyphus. Penguin, Harmondsworth (1976).

Data Representation with No-Code Augmented Reality Authoring Tools

Ana Beatriz Marques, Vasco Branco, Rui Costa, and Nina Costa

Abstract This study is framed within information design and focuses on data representation using augmented reality (AR) and its appropriation by designers. The need to investigate no-code AR authoring tools was motivated by the conclusions drawn during the development of "Floating Companies" AR prototype, which highlight the limited access of designers without programming skills to this technology. We intend to identify the limitations of no-code tools regarding development platforms, such as Unity, and represent data with no-code tools. The methodology used entails three phases—the collection and characterization of no-code AR tools; the review of its limitations regarding the development milestones in FLOC; and the proposal of data representation based on no-code tools. The AR tools landscape offers several free platforms which do not require programming skills. It was found that the analyzed tools do not support algorithmic data representation, which forces any representation to be designed manually in a customized way, presenting limitations regarding the amount of data, but also opportunities. The type of project that no-code platforms support falls within the concept of communicative visualization—a type of visualization that does not intend to deeply analyze the data, but rather to communicate and engage public.

Keywords Augmented reality · No-code · Data representation

1 Introduction

This study concerns data representation using AR and its appropriation by designers without programming background. The need to investigate AR authoring tools was prompted by conclusions drawn during the development of the AR prototype "Floating Companies" (FLOC), created for the exhibition of the Design OBS Project (FBAUL, 2021) [1]. FLOC is the prototype of an AR application for mobile devices

A. B. Marques (✉) · V. Branco · R. Costa · N. Costa
Department of Communication and Art, Research Institute for Design, Media and Culture [ID+], University of Aveiro, University Campus of Santiago, 3810-193 Aveiro, Portugal
e-mail: abmarques@ua.pt

N. Martins et al. (eds.), *Perspectives on Design and Digital Communication IV*, Springer Series in Design and Innovation 33, https://doi.org/10.1007/978-3-031-41770-2_19

that visualizes in hybrid space the design companies registered in Portugal in 2019, in the form of a set of spheres that float in space, justifying the name of the application. The project was framed by *Design OBS—Towards a Design Observatory: models, instruments, representations and strategies* [2], which collects and interprets data on the Portuguese design ecosystem to promote its knowledge and influence public policy.

FLOC was implemented using the Unity, one of the most widely used platforms for creating AR experiences, often involving the creation of scripts (code instructions). The path followed in the prototype development, presented below in a summarized way, represented a great effort, being the biggest barrier the need to learn and use programming concepts in C# without a specific background in that area [1]. Other authors have identified the problem of difficult access to the creation of AR experiences by non-technologists, namely [3–6]. Given this limitation, we identified the need to collect and investigate no-code AR authoring tools currently available, which can be used in a more constraint-free way, by designers who typically do not have a programming background and want to focus their attention on the application design.

Providing an access that presents fewer barriers for non-technologists to enter the creation of AR content is essential to develop the language in this medium. This study is relevant from two points of view—by presenting designers with tools aiming to simplify their introduction into AR content creation; and relevant in the sense that it contributes to the gradual exploration of AR by creators who will eventually find ways to adapt content to the medium, giving it its own identity. As MacIntyre et al. [7] refer, the history of new media has shown that any medium will only reach its full potential when it is made available to designers who, through their action, define the popular forms of the medium. Gutenberg invented the printed medium but not the novel, Edison invented the moving image but not film. Berners-Lee invented HTTP and HTML, but not the web. Sutherland invented AR technology, and the forms that AR content will take are still unknown. Design therefore involves imagining the new rather than finding a solution within a set, being concerned not with the way things are (deductive reasoning) but with the way they could be (abductive reasoning) [8]. While discussing the implications of recent technological advance for innovation and design theory, Verganti et al. [8] frame creativity as a process of problem-finding rather than problem-solving, which is synonymous of 'making sense of things'. The human role becomes that of understanding which problems/opportunities should be addressed and driving the continuous evolution of algorithms towards a meaningful direction. The core of this activity is therefore not problem solving, but problem discovery (pp.224–225). This notion of design for technological innovation applies in the context of this study, as the implications of making AR more accessible to designers do not focus on solving problems, but rather on formulating new problems that can drive data representation with AR in a meaningful direction.

The objectives of this study include mapping and characterizing the set of no-code tools currently available online; ascertaining the limitations and possible strengths in the design of data representations with the collected tools, compared to the work done during the development of the "Floating Companies" prototype in Unity (Sect. 2).

2 Floating Companies (FLOC) AR Prototype

FLOC involved the design, development, and user testing within the scope of the exhibition "Design Observatory in Portugal - Situation" (FBAUL, 2021). The project followed a practice-based methodology which aimed to investigate, from the design point of view, the strengths, and limitations that hybrid space presents for data visualization, particularly when there is no semantic relationship between that data and the space where it will be inserted.

FLOC represents information about the set of design companies in Portugal from a database curated by Design OBS [9]. In the application, each company in the sample corresponds to a sphere, whose color and diameter reflects its size (number of employees). Their positioning in space gives an approximate idea of the distribution of the sample across the Portuguese territory, but also of their relative performance (the vertical position of each company is an indicator of its profit per employee) (Fig. 1). On the ground plan there is a district map of Portugal under which the companies are located. By tapping on each sphere further information on that company can be accessed—name, district, and profit per employee. The visualization offers a second scene—the resume view—in which the number of companies per employee class in each district is represented based on a 3D sphere chart.

FLOC explored the inherent potential of AR to render abstract numerical values concrete by simulating them in the real environment. In the main view the aim was to give a concrete existence to the number, distribution, and attributes of design companies through the physical configuration of the virtual objects representing them. FLOC was developed using Unity game engine combined with Vuforia SDK, which allowed to use an area target that launches the experience through the recognition

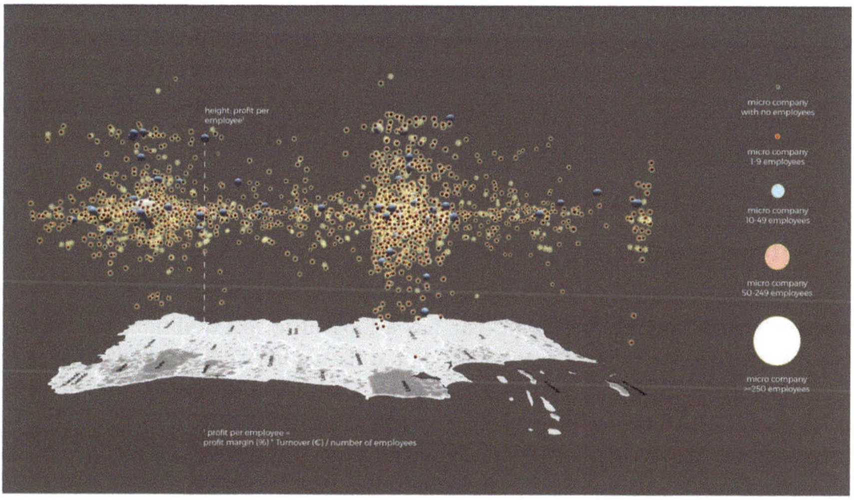

Fig. 1 Floating Companies App prototype. Arrangement of the virtual elements in space

of the exhibition room. Overall, the prototype development involved the following milestones:

1. Create and assign the prototype an area target to launch the AR experience with the recognition of the exhibition space.
2. Import the database on companies into Unity in CSV format and make it analyzable by Unity using a script.
3. Implement a script to analyze the database, generating a second database with the information on size, color, and location to be assigned to each sphere (prefab), based on the information on size, location, and profit of each company.
4. Implement a script to read the latest database and instantiate a sphere for each of the companies according to the location, size, and color information, within a bounding box.
5. Create a second scene - Resume View - where the total number of companies per district is represented. Here, instead of assigning a sphere to each company, the data are integrated, and each sphere represents the number of companies in a certain district.
6. Add User Interface elements (buttons to (1) the project description; (2) link to the original database; (3) the resume view and back).

   i. Attribute behavior to the spheres when they are tapped on - light effect and display of an information board (details-on-demand) with information extracted from the generated database.

Our team were composed mainly by designers without programming knowledge and the development process was initiated based on a tutorial on how to build a scatterplot graph from a database in Unity [10], being gradually adapted to the goals of the project over multiple development iterations. During this process we recurred to Unity forums, to its large community of users and often used excerpts from other existing programs. At a later stage, it was also necessary to call upon the expertise of a programming professional with knowledge in C#. This process was time consuming, laborious and at times restricted design options due to the limited programming knowledge.

3 Literature Review

Although visual representation of data has a long historical tradition, the vocabulary of the field is constantly evolving. For some authors data visualization is an area perfectly differentiated from infographics [11], for other authors infographics is a sub-area of data visualization [12, 13]. Iliinsky and Steele [11] distinguish infographics from data visualization in a synthetic way—infographics refers to a visual representation of data that is (1) manually-drawn (with a personalized treatment of the information); (2) specific to the data in question (and therefore difficult to recreate with different data); (3) aesthetically rich; (4) limited in terms of data (as each piece of information is manually coded). Data visualization, on the other hand, refers to a representation of data that is (1) algorithmically designed; (2) easy to regenerate with different data; (3) often aesthetically poor and (4) data-rich (as integration of large amounts of data is viable). On the other hand, Masud et al. [12] consider infographics as communicative visualization, ie. a type of visualization that is not used to convey a detailed data analysis, but rather to communicate a narrative. This type of visualization is used to raise awareness of a particular topic even if it is not intended to provide data analysis. Visual metaphors and illustration are visual strategies often used in communicative visualization.

Numerous examples of this type of visualization combined with AR come from journalism [14, 15], such as "Inflation Shrinking Ray" [16]; "EV Battery Breakdown" [17] or "Water footprint " [18]. In general, when creating AR applications, most content developers use the so-called game engines: software applications that allow teams composed of professionals from different contexts to collaborate in the creation of an application [19]. Many of the most well-known game engines, such as Unity or Unreal, require programming to create interaction, which hampers the exploration of these tools by non-programming designers [4]. Unlike game engines, high-fidelity prototyping tools, which will be covered in this article, are digital tools used by designers and software developers to address interface details without a full implementation [20].

In literature, authoring tools are classified according to a variety of criteria. Regarding the way content is previewed, they may be considered as 'digital tools', 'immersive tools' or both. Typically, digital authoring tools (e.g. Lens Studio) (Fig. 2) support a preview of the project using an emulator (a software that imitates another computer system) that needs to be deployed on the device for testing, while immersive authoring tools (e.g. Apple Reality Composer) (Fig. 3) allow editing while previewing the AR experience in the user environment [21]. Lee et al. [22] address the advantages of immersive authoring tools, which allow users to preview, experience and check first-hand the virtual content and its integration into the real environment.

Billinghurst et al. [23] collect and classify AR tools according to the required programming skills—from 'low level software libraries' requiring good development skills, to simple authoring tools for novice users with no programming knowledge. While libraries and low-level software provide a high level of flexibility while requiring programming skills, standalone authoring tools allow end users to easily create AR content with a minimum of programming knowledge, although the content created is quite simple. Due to the rapid evolution of this research field, the collection made in 2015 by Billinghurst et al. is no longer up to date with some of the software mentioned discontinued.

Dengel et al. [5] conduct a systematic literature review on AR Authoring Toolkits for educators without programming skills. Authors identify 69 different toolkits which they classify according to their accessibility, level of programming skills required, and supported interactivity. The methodology used is well documented, the time interval is wide—25 years—and the collection is recent and updated. However, due to the collection method based on bibliography only, the absence of some no-code

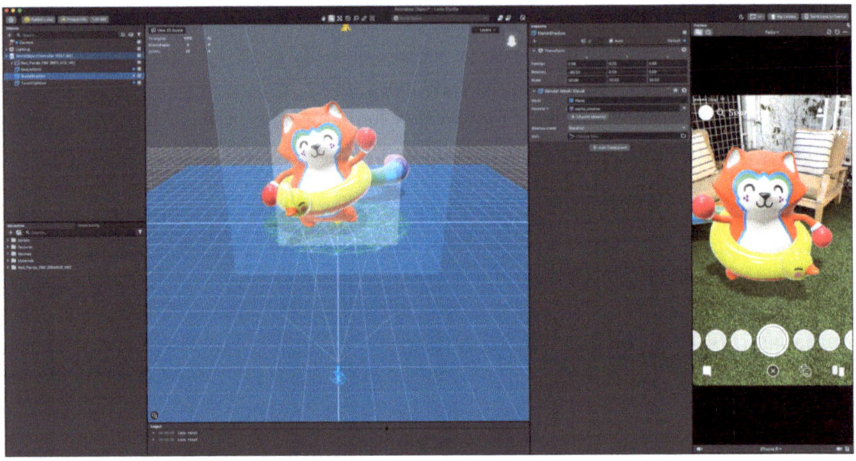

Fig. 2 Lens Studio AR emulator

Fig. 3 Reality Composer immersive environment

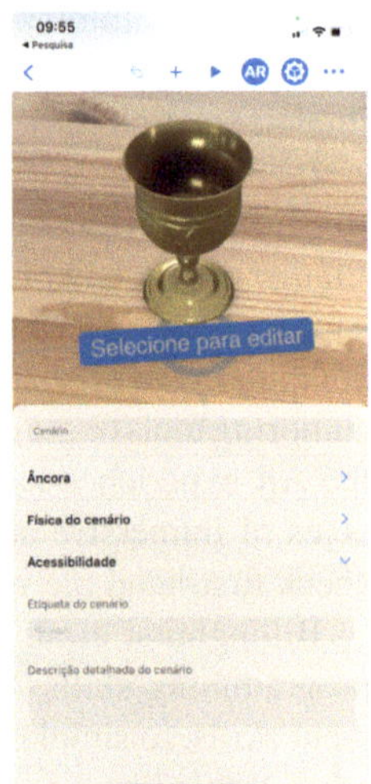

software available online, such as Reality Composer or Adobe Aero, was immediately noted. Furthermore, many of the collected tools are suitable for educators but are not sufficiently flexible for designers as they provide closed uneditable solutions. In systematic reviews on AR tools, other authors [24] considered the review supported only by academic literature potentially restrictive as it may exclude tools that despite not being academically validated are used by a community of practitioners.

Some authors develop and propose their own authoring tools [4, 6, 7, 25]. One of the best-known references is DART (Designer's Augmented Reality Toolkit) [7]—a tool launched in 2003 aimed at new media designers which has been widely used by a diverse population of creators during the following years. Gandy and MacIntyre collected the reflections of a group of DART users regarding the software's use and developed guidelines for building AR authoring tools for non-technologists [26]. The guidelines fall into the following categories: (1) access, (2) layered authorship, (3) collaboration with diverse teams, (4) making the process less painful and (5) the importance of community. Access concerns the technological barriers that non-technologists encounter when using AR authoring tools. Layered authoring is about providing a layered development environment that allows for an appropriate level of complexity, for instance a no-code tool that also provides access to scripting. Collaboration with diverse teams considers the need to integrate several specialized collaborators in the same project. Point 4 relates to the need to make the creation process easier. Finally, the importance of an active online community that conveys a sense of permanence is highlighted. Other authors present guidelines for the creation of an AR tool for non-technologists, such as Leiva et al. [4] who point out four design goals for a rapid prototyping tool: 1. create low fidelity assets through sketching; 2. position assets in space through direct manipulation; 3. capture user position changes through video; 4. create complex animations through direct manipulation. This is also the case of Schmalstieg & Höllerer [3] who argue that an AR authoring solution should address the following points: allow different possibilities of relating the application content to the real world (tracking modes); foster cross-platform compatibility; support a collaborative workflow and enable future reuse.

3.1 Description Categories

Categories for describing the authoring tools were outlined based on the literature review, combining points enunciated by different authors, but also based on the empirical knowledge accumulated throughout the development of previous AR experiences [1, 27]. **Accessibility**—The relevance of accessibility in 'new media' tools is referred to by Gandy and MacIntyre [26] in the technological sense, but also in the sense of presenting fewer barriers to entry in general. Technological accessibility relates to the need to provide a layered authoring environment that allows the user to access an appropriate level of complexity. Here, the criterion of accessibility has been extended to affordability, which can also be a barrier for designers looking to start exploring AR. **Tracking modes** and device compatibility—The support of different ways of relating the application content to the real world (tracking modes) and the need to foster cross-platform compatibility are two criteria pointed out by Schmalstieg and Höllerer [3] for a successful AR authoring solution. **Interactivity**—According to Billinghurst et al. [23] no-code authoring tools allow to easily create AR content with minimal programming knowledge but offer little flexibility. Therefore, it is important to ascertain whether the tools collected support some level of interaction and if this support depends on programming. **Type of authoring tool**—One of the important features in AR tools pointed out by Leiva et al. [4] is the possibility of positioning assets in space through direct manipulation, which is achieved through immersive or mixed authoring tools. **Publication**—The place of publication relates to accessibility, not accessibility for content creators, but user accessibility.

4 Research Question

Based on the literature review, we identified the absence of a comprehensive systematic review of AR no-code tools currently available online. This collection exists, but is limited to the academic literature, excluding many solutions that are currently available online for free. We also identified the inexistence of literature on the strengths and limitations of using these types of no-code tools for data representation.

The following research question is posed: What are the currently available no-code AR tools that are most suitable for designers without programming knowledge in data representation? As a secondary question, we intend to find out the advantages and limitations of using these tools in comparison with game engines, such as Unity. This study aims to fulfil three objectives: (1) To map and characterize the set of no-code tools currently available online. (2) To outline the limitations and strengths of working with the collected tools, using for comparison the work done previously during the development of "Floating Companies" in Unity. (3) To illustrate the type of visualization that can be designed with no-code tools, using data from Design OBS, similarly to the strategy previously followed [1, 27].

5 Methodology

This study can be divided in three parts—the first part (Methodology and Results) aims at collecting and characterizing the AR no-code platforms; the second part (Results) consists in verifying the limitations of those platforms regarding the tasks performed during FLOC development; the third part (Results) presents a data representation project built from a no-code tool, using Design OBS data regarding Portuguese design companies. The systematic collection of no-code tools and the identification of their limitations regarding development platforms are intended not only to map this universe, but also to qualify the use of these tools, identifying their advantages, but also the type of assignments for which they are not prepared.

The third part (Results) presents a data representation project named "Pencils" that showcases how data can be conformed through these tools. It also explores some challenges and complementarities between no-code AR apps, with full implementation tools, bringing AR closer to designers with no advanced coding competences.

The survey of no-code tools was carried out using the systematic review collection method. The collection process, as well as the characterization, are documented in the database "No-Code AR Platforms" [28]. To construct a comprehensive survey, this study uses a systematic literature review through web search as described by

Stansfield et al. [29], who suggest a three-stage process when constructing a systematic review through website search: (1) planning the search, (2) executing the search and (3) screening records for relevance and managing the results. 1. Planning the research involves justifying and informing decisions about: where to research, who is doing the research, and the timeframe and resources available for the review. 2. In the research, the main objective is to use each resource in a consistent and individually appropriate way. 3. finally, the results obtained must be filtered according to their relevance and according to previously established eligibility criteria.

5.1 Planning the Search

The search for no-code tools via web search, using Google's advanced search, was motivated by the need to gather a comprehensive collection that is often not documented in the academic literature. Given the topicality of the subject, no start was defined for the research time interval, which only has end date (all results until 26/10/2022 and all results until 27/10/2022). The collection entailed three advanced Google searches (Table 1). The aim of performing three searches from synonymous terms—'no-code augmented reality', 'augmented reality without coding' and 'augmented reality without programming'—was to encompass the maximum number of results and not to rely all the search on a single term, which could not be the most representative. One of the main challenges of research planning is to know which sites are most appropriate for the research [29]. In this context, it is intended to conduct a survey as complete as possible, so only ads, Facebook pages and LinkedIn profiles were excluded.

5.2 Executing the Search

Applied the exclusion criteria, advanced searches originated 30 web pages ('source' column of the 'Executing the search' sheet [28]), from which the AR platform names were manually extracted. In most web pages, platform names were extracted through a navigation along the URL. In the case of YouTube links, the platforms were extracted from the video content. In the case of publications on Reddit website or Twitter, only the search result post and its comments were considered. Of the total websites analyzed, two pages did not mention AR platforms, the remaining 28 pages pointed 38 names of supposed no-code AR platforms: Bundlar, Brand XR Studio, Geenee AR, Zapworks, Adobe Aero, AR Media, ARway, Aryel, Augmania, Minsar, PlugXR, Scapic, Spark Studio, XR + , Amazon Sumerian, Appypie, AR Code, Augm it! Blipparbuilder, Byldr, CanvasLogic, Envisage AR, Epigraph, Imersian, JigSpace, Lens Studio, MetaVRse, MyWebAR, Reality Composer, Scopear, ThreeKit, UniteAR, Vuforia Studio, Wintor AR, WorldCAST, XR Today, Hololink, Quartz Composer.

Table 1 Collection of no-code AR platforms based on three advanced searches using Google

Search nr	1	2	3
Search terms	'no-code augmented reality'	'augmented reality without coding'	'augmented reality without programming'
Time frame	All results until 26/10/2022	All results until 27/10/2022	All results until 27/10/2022
Google advanced search criteria	All these words: no-code augmented reality; this exact word or phrase: "no-code augmented reality"; language: english; region: any region; last update: anytime; terms appearing: in the title of the page; file type: any format	All these words: augmented reality without coding; this exact word or phrase: "augmented reality without coding"; language: english; region: any region; last update: anytime; terms appearing: in the title of the page; file type: any format	All these words: augmented reality without programming; this exact word or phrase: "augmented reality without programming"; language: english; region: any region; last update: anytime; terms appearing: in the title of the page; file type: any format
Results counting	23 results	6 results	4 results
Results counting (after excluding Facebook and LinkedIn pages)	20 results	6 results	4 results

5.3 *Screening Records*

For this study, only software that meets the following criteria were considered:

Type of tool: The platform must be an authoring tool. Therefore, it is necessary to verify that names extracted from each webpage correspond to actual AR tools - high fidelity and low fidelity tools are included. Excluded are platforms whose primary purpose is e-commerce and marketing. **Availability**: The platform must be available online, excluding platforms that have been discontinued or whose website cannot be accessed at the time of the search. **Accessibility**: The platform should be free or provide a free license (with no need for meeting with the team). Platforms with commercial licenses providing a free version, include the following typologies: Free limited version; Free use until x views; Free version for non-commercial use; Free version for personal and commercial use; Free version for testing and personal use. Excluded are platforms offering a trial version only for a limited period or offering a trial by appointment. **Required programming skills**: The platform should always provide forms of interaction without code. This includes tools which, in addition to the no-code feature, support coding for more complex behaviors. Tools relying entirely on programming skills are excluded.

The described criteria originated 12 platforms (Table A in annexes): Geenee AR; Adobe Aero; PlugXR Creator; Spark Studio; XR +; Blipparbuilder; Byldr; JigSpace; Lens Studio; Reality Composer; WorldCAST and Hololink.

6 Results

6.1 Comparison Between No-Code Tools

Affordability—Four platforms are completely free—Adobe Aero, Spark Studio, Lens Studio and Reality Composer. Remaining platforms are commercial but provide a free version that may: limit non-commercial use, limit features compared to the paid version, or require the use of a watermark. **Static/Interactive**—The following criteria analyze the capabilities of each tool without programming (for example, if a tool offers interactivity but only by scripting, it is considered static). Geenee AR, Jigspace and WorldCast platforms are considered static as they support a single predefined form of interaction. The remaining platforms offer varying forms of interaction—both in flexibility and in interaction implementation. Seven platforms support GUI-based interaction editing—PlugXR Creator; XR +; Blipparbuilder; Byldr; Lens Studio; Reality Composer and Hololink. Adobe Aero and Spark Studio base their interaction on a system known as 'drag and drop'. Geenee AR is the only platform requiring code to implement interaction. Interaction complexity among tools is highly variable—although Lens Studio and Blipparbuilder implement GUI-based interaction, Lens Studio presents a much wider set of interaction options, being highly flexible. Although Lens Studio and Spark Studio platforms allow creating AR experiences without writing code, they require some technical knowledge and familiarity with the development environment. These tools stand out as being those that present greater complexity but also greater flexibility. **Layered Authoring**—Six tools offer a layered authoring environment, simultaneously supporting coding: Geenee AR; Spark Studio; XR +; Blipparbuilder; Lens Studio and Reality Composer. All the analyzed cases use Javascript, except for Reality Composer, which is based on Swift. **Type of Authoring Tool**—Most tools are digital (9); Byldr is immersive; Adobe Aero and Reality Composer are considered digital and immersive simultaneously. **Tracking Mode**—Analyzed tools support a great variety of tracking modes (Table 2). The modes supported by most tools are surface tracking (10 cases); image tracking (9 cases); and face tracking (5 cases). The most versatile tool is Lens Studio followed by Geenee AR.

Device Compatibility—Most tools (Table 3) support mobile devices (11 out of 12), but there are also tools that support eyewear devices (3). Most tools support iOS operating system (11) and secondly Android (8). Most versatile tool in terms of device compatibility is XR + , covering Android and iOS mobile devices, but also eyewear devices (Oculus Quest, Oculus Go, Oculus Rift, and HTC Vive).

Publishing—Forms of publication diverge among platforms (Table 4). Most part (7 cases) allow for Web publishing—known as Web AR, which does not require the installation of an application and AR content is accessed through a link, QR Code or AR Code. Four tools allow the publication in the same platform used for the prototype creation; one tool allows for the publication in its own app; and two other tools allow for the integration of the prototype in an existing app. One tool

Table 2 Tracking modes enabled by each platform. * The software does not use tracking because it runs on AR/VR headsets

	Marker/ QR code	Image	Full body	Face	Shoulder	Upper body	Hands	Foot	Cats/ Dogs	Object/ Model	Content around user	Surface	World Mesh	Geo Anchor	Area Target	N/A *
Geenee AR	X	X	X	X			X					X		X	X	
Adobe Aero		X										X				
PlugXR Creator		X										X				
Spark Studio	X		X	X			X					X				
XR +		X		X			X	X		X		X				
Blipparbuilder		X									X	X				
Byldr																X
JigSpace												X				
Lens Studio		X	X	X	X	X	X		X			X	X			
Reality Composer		X		X						X		X				
WorldCAST	X	X										X				
Hololink		X														

Table 3 Device compatibility by platform

	Android	iOS	Windows	Oculus Quest	Oculus Go	Oculus Rift	HTC Vive	Magic Leap	Hololens 2
Geenee AR	X	X							
Adobe Aero		X							
PlugXR Creator	X	X							
Spark Studio	X	X							
XR +	X	X		X	X	X	X		
Blipparbuilder	X	X						X	
Byldr				X					X
JigSpace		X	X						
Lens Studio	X	X							
Reality Composer		X							
WorldCAST	X	X							
Hololink	X	X							

allows publishing in Web VR and another in AR Quick Look. There are even tools dedicated to social media content, such as Lens Studio and Spark Studio.

6.2 *Tools Limitations*

After collecting and characterizing the available tools, it was necessary to determine their limitations compared to the Unity game engine. To this end, a matrix was constructed summarizing the main milestones involved in the development of FLOC in Unity (Table 5) and each platform was then tested in the execution of each listed task. The following table presents the milestones of FLOC development (left) and the no-code tools supporting each task (right).

The milestones (M) below mentioned represent only the most important tasks included in the FLOC development:

(M1) Launch an AR experience in space using as area target—concerns the possibility of launching the virtual content through the recognition of an area, an interior space which in the case of FLOC was the exhibition gallery.

(M2) Regarding the creation of multiple scenes and the navigation between them was essential to integrate in FLOC a main view—composed by all the units of the database, and a secondary view (summary view) in which the same information was represented but in an integrated way. The existence of two views enabled the coexistence of two different representation paradigms and their comparison.

(M3) Refers to the software's ability to read and analyze a database, allowing objects to be instantiated directly from that database. This feature was essential, and

Table 4 Publishing/viewing mode by platform

	Web AR	Web VR	AR Quick Look	Platform used to create	Viewing Application	Integration into existing app	Instagram	Snapchat	Facebook	Facebook Lite	Messenger
Geenee AR	X										
Adobe Aero				X							
PlugXR Creator	X			X		X					
Spark Studio							X		X	X	X
XR +	X	X									
Blipparbuilder	X			X							
Byldr					X						
JigSpace	X										
Lens Studio								X			
Reality Composer			X	X		X					
WorldCAST	X										
Hololink	X										

Table 5 Matrix identifying the no-code tools that support important development milestones identified in FLOC

N	Development milestones with Unity (based on FLOC)	Geenee AR	Adobe Aero	PlugXR Creator	Spark Studio	XR +	Blipparbuilder	Byldr	JigSpace	Lens Studio	Reality Composer	WorldCast	Hololink
1	Launch an AR experience in space using an area target	X											
2	Create multiple scenes and the navigation among them	X		X	X	X	X	X		X	X		X
3	Instantiate objects directly through the reading and analysis of a database, with each database entry corresponding to an instantiated object												

(continued)

Table 5 (continued)

N	Development milestones with Unity (based on FLOC)	Geenee AR	Adobe Aero	PlugXR Creator	Spark Studio	XR +	Blipparbuilder	Byldr	JigSpace	Lens Studio	Reality Composer	WorldCast	Hololink
4	Attribute responsive behaviors to tap or click, including a change in the visual aspect of the object clicked and the display of a caption		X	X	X	X		X		X	X		
5	Import 2D images	X	X	X	X	X	X	X	X	X		X	X
6	Create a graphical user interface (screen canvas)				X					X			

of all the tasks it stands out as being the most relevant for the creation of FLOC, since it automated the representation of 2.714 companies, an action that would not have been feasible manually. Reading and analyzing the database also made it possible to assign a caption to each sphere (through tap) providing details about the selected company.

M4) Consists in assigning responsive behaviors to virtual objects. In FLOC this possibility materialized, for example, in the summary view, where, by clicking on the spheres, the visual aspect of that object changed, and a caption was presented—in this case a caption built manually and not by accessing the database. The importation of 2D images allowed, in the case of FLOC, to import the map of Portugal—an object that although simple, allowed to contextualize the experience geographically. The importation or creation of 3D models was also essential to the creation of FLOC, but is not included in the matrix since it is a task supported by all the tools analyzed.

M5) Finally, the possibility of creating a graphical user interface (screen canvas) supports the existence of buttons in the screen space, instead of virtual buttons. The advantage of the graphical user interface (GUI) in an AR experience is that it allows some buttons or menus to always be within the reach of the user who does not need to find them in the virtual space. In the case of FLOC, the GUI allowed, for example, to access the resume view and the database URL.

None of the tools under analysis can recreate all the tasks involved in FLOC. Milestones 4 and 5 stand out particularly for not being tackled using any of the tools analyzed. None of the tools included in this study allowed the analysis of a database to create or instantiate objects based on the stored data, a crucial step in the development of FLOC. In fact, Reality Composer allows creating two types of traditional AR charts from an imported csv database—bar charts and pie charts—but it does not allow, without coding, to program the instantiation of virtual objects directly from the database. There is a major limitation on the actions that can be performed from reading the database with Reality Composer, which is the only analyzed tool allowing to read a csv file. Without the ability to read, analyze and act based on the data stored in a database, it would not have been possible to obtain a working prototype due to the large amount of data involved. Except for the reading and analysis of a database, essential to algorithmic data representation, no-code tools allow to perform all remaining tasks involved in FLOC. Spark Studio and Lens Studio, both aimed at social media, stand out as the tools allowing to complete more tasks.

During the execution of the tasks described in Table 5, it was possible to observe less flexibility across all the no-code tools included in the study—a large part of the tasks can be executed, but in a predefined or closed way compared to Unity, offering much less creative freedom. On the other hand, they are far less complex tools and make the accomplishment of each task faster. Lower flexibility accompanied by lower complexity makes the whole AR creation process significantly easier and less frustrating or painful, as Gandy and MacIntyre point out, since it is quite straightforward to get a functional AR experience.

6.3 A No-Code Project—Pencils

Although Iliinsky and Steele [11] distinction between infographics and data visualization is too strict to fully comprehend this field, it offers a useful working basis for contextualizing AR data representation that is designed with no-code tools. Data representation in which the final visual representation is not done algorithmically is typically the type of visualization that no-code tools support, as they typically do not allow to read and analyze a database. However, this type of tool allows to develop the communication dimension mentioned by Masud et al. [12], which consists in exploring visual metaphors and storytelling to communicate results and to engage audience.

Pencils is a prototype intended to illustrate what an AR data representation using exclusively no-code tools can be. The project was developed to relate the information about the Portuguese design companies, specifically information about the revenue providing from sales and services (in K€) – the turnover—and information about the export percentage per NUT II (regions of Portugal: North, Center, MLA, Alentejo, Algarve, Madeira and Açores) in 2019. Table 6 synthetizes the data represented in this AR prototype, which was retrieved from the database about design companies registered with the code 7410 (design activity) in Portugal in the year 2019 [9], the same used in FLOC.

The software employed in the development of the project was Adobe Aero to create the AR experience and Cinema 4D to model and animate the 3D objects. Although Adobe Aero is not among the most versatile platforms in terms of tracking modes, device compatibility or publishing location, it is among the few platforms supporting both digital and immersive modes. Reality Composer also covers both modes, but its use on a computer is now integrated in Xcode—an iOS developer tool. By supporting digital and immersive modes simultaneously, Adobe Aero makes the AR design process very fluid, enabling the user to design in the digital environment and preview content in the immersive environment, switching between viewing modes very quickly. This feature resulted in a quick execution of the tasks listed in Table 5, which was decisive for its selection. The purpose of developing a project

Table 6 Data on Portuguese design companies represented in Pencils project

NUTS II	Year	Sales (K€)	Services (K€)	Turnover (K€)	Export (%)
North	2019	61,714	82,820	144,534	11,43
Center	2019	10,768	29,703	40,471	9,10
AML	2019	38,693	112,255	150,947	8,64
Alentejo	2019	997	7394	8391	10,85
Algarve	2019	5380	10,204	15,584	7,21
Açores	2019	384	939	1323	10,12
Madeira	2019	822	3589	4411	7,06

from data previously analyzed was to illustrate only the communicational phase, in which a visual metaphor is created.

In this case, the exercise was to find a way of compiling data into a fluid narrative using a significant visual metaphor, considering the subject matter. The iterative analysis and reflection were carried out together with the co-authors in periodic meetings, to design and progressively improve the experience, define possible paths from the communication point of view, such as the use of other visual metaphors as well as aspects from the design domain—size of the virtual objects, three-dimensional modelling, color, and aspects related to interaction. These meetings also served to reflect on the limitations of the AR experience in context and potential aspects to correct and improve, such as the distinction between services and products and the change of color, the use of other forms of target, the scale of the experience and other interaction possibilities.

6.3.1 Project Description

Pencils (Fig. 4) portrays the landscape of design companies in the seven regions (NUTS II) of Portugal—Norte, Centro, AML, Alentejo, Algarve, Açores and Madeira—concerning their turnover—similarly to FLOC application.

Each region is represented by a pencil whose physical configuration communicates information about the companies in that region (Fig. 5)—rubber represents

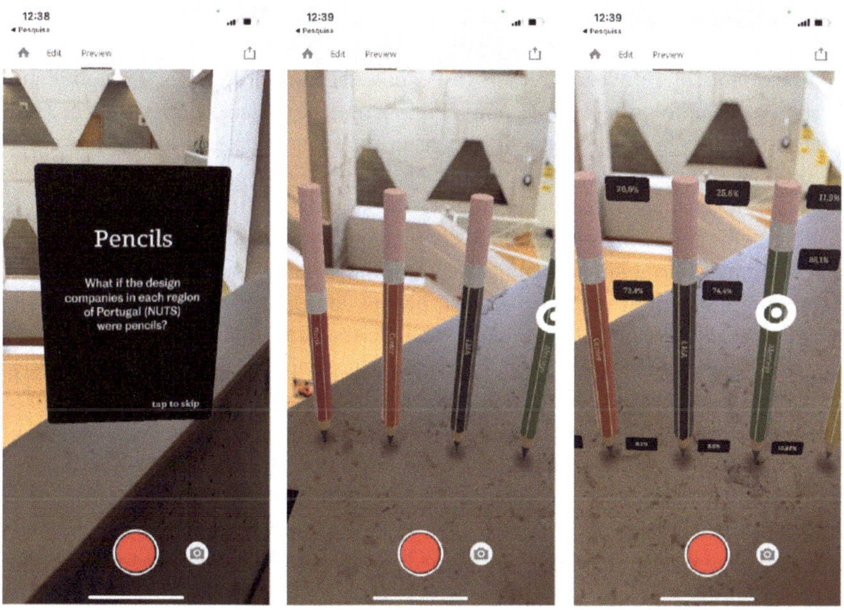

Fig. 4 Pencils screen captures

Fig. 5 Pencils caption

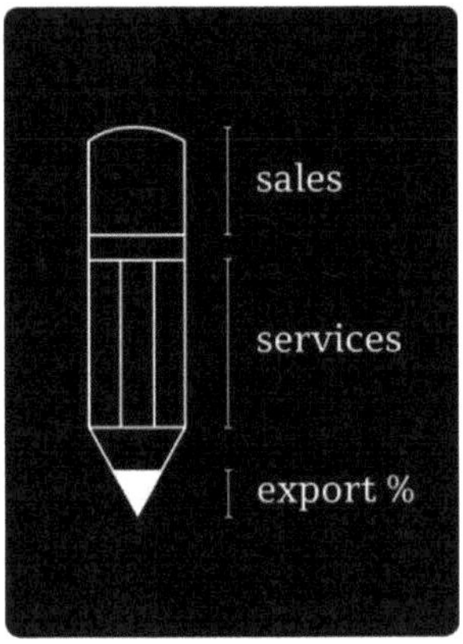

sales, wood represents services, and the charcoal bill represents the export percentage. The use of pencils as a visual metaphor to represent design companies was prompted by a second phase of user tests on the latest version of FLOC—which we named FLOC II, where one participant mentioned that visually the experience was a bit 'boring' but that she understood as it was an informational piece. Although it is really an informational piece, we tried to integrate the visual metaphor in Pencils as a way to emphasize the ludic and communicational character of the experience, since information and entertainment don't necessarily have to be incompatible, as the Immersive Journalism field has already demonstrated.

The AR experience can be accessed by QR code (Fig. 6) using an iOS device with Adobe Aero installed (free), from any location, as the target is a horizontal plane that can be a tabletop or even the floor. In Pencils, the user progresses through a short narrative that unfolds in several events (Fig. 7): first the project is introduced in a brief succession of panels; then pencils appear in the scene representing the percentage of product sales, services, and exports. In the third phase, percentage values are replaced by absolute values (€), showing the significant differences in turnover by region. In the following scene, pencils corresponding to the lowest turnover regions are combined into a single pencil, enabling the comparison with the two regions exhibiting the highest turnover. At last, pencils return to their initial configuration and are packed away in a box, closing the narrative, and enabling the user to access the URL where the databases supporting the experience are stored.

By aggregating multiple values into a smaller number of visual marks, Pencils fits into traditional data visualization. Differently, FLOC belongs to what [30] designate as immersive unit visualization—visualizations in which every data point is represented by a separate visual mark—while simultaneously offering a traditional, aggregated view of the data—the resume view. In Pencils we synthesize information and provide a data overview, while in FLOC there is no synthesizing exercise. Both projects aim to make a numerical abstraction concrete, but they do so in radically different ways and the tools per se seem to offer different forms of thinking and therefore visualizing. If we think about the distinction between infographics and

Fig. 6 QR code to access Pencils

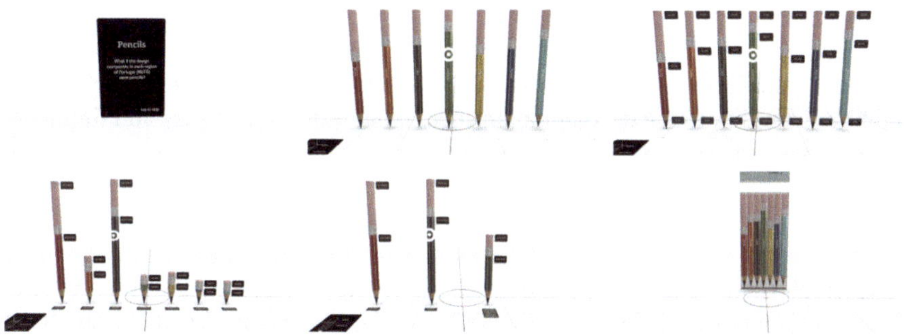

Fig. 7 Screen captures (Adobe Aero on desktop) showing Pencils narrative sequence

visualization criteria [11], FLOC fits better in data visualization and Pencils in infographics. FLOC supports a large amount of data, was designed algorithmically, and can be automatically reproduced with different data. Pencils was designed manually, with a personalized treatment of the information, and it would be laborious to recreate with different data. Due to the personalized treatment of information, the amount of data included is limited. On the other hand, in Pencils there was a greater investment in the design of the experience and narrative. While developing FLOC, the concerns with issues related to programming took away time and attention that could have been devoted to the design of the experience.

7 Discussion

No-code tools are no substitute for full implementation tools in data visualization as they are limited regarding the reading and analysis of a database, which are essential to algorithmically based visualization, without which the representation of large datasets is not viable. This limitation is more evident in the case of immersive unit visualization—where data is not aggregated, instead each unit is individually represented, making manual representation unfeasible or extremely time-consuming. Furthermore, no-code tools offer much less flexibility compared to more complex tools allowing to code and may also offer limitations in terms of scene size. Despite being less flexible than full implementation tools (for instance, no-code tools offer a limited number of interaction behaviors, which is not the case with development platforms relying on scripting), no-code tools allow to more quickly create visual metaphors capable of communicating a dataset. Their main advantage is the easy access to the creation of AR content, being a gateway for designers with no previous experience designing for AR. In fact, it is important to bring designers closer to all technologies, allowing them to broaden their field of action.

Adobe Aero low complexity, like other no-code tools, allows for quick design and brings fluidity to the development process, but it doesn't exactly add fluidity of thought. No-code tools automate development flows that are not automated in game engines which need to allow greater flexibility. For instance, when creating AR experiences with a no-code tool, the project evolves based on the tracking mode selection. In game engines, designers create their own path without a predefined development flow, which requires additional experience (they can even include several tracking modes simultaneously). No-code tools allow simple ideas to be quickly realized, but within a closed set of possibilities. Development tools and no-code tools are complementary and used at different moments. In any case, the following question arises: is it worth designing with these limited mechanisms or is it better to design using systems already mastered and ask translators (programmers) to then make these applications for a new support or medium? We consider that it is crucial that the design of an AR experience necessarily goes through contact with AR, even if it is with limited tools, because they allow to visualize the project in the hybrid space and this contact raises relevant questions related to the medium, which designing in other formats could not raise.

Still from the perspective of design, other question arises which is whether the designer should position himself aside of coding and entirely depend on someone who translates his thinking into a programming language. *Tradurre è tradire* (To translate is to betray), so it will always be desirable for a designer to think already from a digital grammar, but will no-code apps have a lexicon already broad enough for this design to be capable or creator of the new? Or is this loss greater than the translation of a more robust design proposal later translated by expert technicians? Although they are good gateways to the AR medium, no-code tools might excessively lead the project to a final standard solution. But then again, the more people engage with these tools, more solutions will arise.

Even today, most designers still draw manually, doodling (even if it's on an iPad), because there seems to be an immediate connection between thought and drawn images, and even because chance and serendipity plays an essential role in this process. Some students, try to do it directly on the computer, already with final applications, which somehow dulls their thinking—they think according to the limits of that machine, or let the machine manage their thinking. It is important that the no-code tools are assumed by designers as a gateway, and even as a form of rapid prototyping, but not as the exclusive place where the project emerges and is developed. Such modus operandi would conform any AR project to an always identical matrix, annulling the free exploration of new development paths—on which the construction of a proper language for this medium depends.

8 Conclusions

This paper aimed to identify, describe and compare AR no-code tools currently available online, outline their limitations regarding other development platforms like Unity and propose a project for data representation with AR using exclusively no-code tools.

By presenting alternative tools to game engines and development platforms that require programming knowledge, and by illustrating a possible use in the field of data visualization, we expect to make the appropriation of this technology by information designers more accessible, at least at an early stage of their introduction to the AR field.

This study also has limitations which in turn, point to future research directions. The absence in literature of criteria for an AR authoring tool specifically suited to the activity of designers would have been useful for the description and evaluation phases. As a future study, it would be interesting to compare the interpretation of Pencils with a traditional two-dimensional graph representing the same information, to ascertain on what specific criteria can AR surpass a traditional chart in terms of information communication.

Table A Authoring tools characterization

Platform	Static/ Interactive	Interaction Mode	Type of Authoring Tool	Tracking Modes	Compatibility	Publishing
Geenee AR	Static	N/A	Digital	Marker/ QR Code; Image; Full body Face; Hands; Surface; Geo Anchor (Beta); Area	Android; iOS	Web AR
Adobe Aero	Interactive	Drag and drop	Digital and immersive	Image; Surface	iPhone; iPad; iPad Touch	Adobe aero app
PlugXR Creator	Interactive	Scripts editing via GUI	Digital	Image; Surface	iOS; Android	PlugXR App; WebAR; integration into existing app
Spark Studio	Interactive	Drag and drop	Digital	Marker/ QR Code; Full body; Face; Hands; Surface	Android; iPhone; iPad	Effects for Instagram; Facebook; Facebook Lite and Messenger
XR +	Interactive	Scripts editing via GUI	Digital	Image; Face; Hands; Foot; Object/ Model; Surface	iPhones; iPads; Android; Oculus Quest; Oculus Go; and the combo Firefox/ Chrome + Oculus Rift/ HTC Vive on desktop computers	Web AR and Web VR (for smart glasses or VR headset)
Blipparbuilder	Interactive	Scripts editing via GUI	Digital	Image; Content appearing around the user; Surface	Android; iOS; Magic Leap	Web AR and Blippar App

(continued)

Table A (continued)

Platform	Static/ Interactive	Interaction Mode	Type of Authoring Tool	Tracking Modes	Compatibility	Publishing
Byldr	Interactive	Scripts editing via GUI	Immersive	N/A	Hololens 2; Oculus Quest	Byldr Marketplace
JigSpace	Static	N/A	Digital	Surface	iPhone; iPad, Mac and Windows computers	Web AR
Lens Studio	Interactive	Scripts editing via GUI	Digital	Image; Full body; Face; Shoulder; Upper body; Hands; Cats and dogs	iOS and Android devices	Snapchat filters
Reality Composer	Interactive	Scripts editing via GUI	Digital and immersive	Image; Face; Object/ Model; Surface	iPhone; iPad	Reality Composer; AR Quick Look; Integration into apps
WordCast	Static	N/A	Digital	Marker/ QR Code; Image; Surface	iOS; Android	Web AR
Hololink	Interactive	Scripts editing via GUI. Very limited interaction options	Digital	Image	iOS; Android	Web AR

Acknowledgements This work is financed by national funds from FCT – Fundação para a Ciência e a Tecnologia (Foundation for Science and Technology), I.P., under the PhD grant «SFRH/BD/144937/2019». This article is a result of the project Design Obs. Para um Observatório de Design em Portugal: Modelos, Instrumentos, Representação e Estratégias, (Towards a design Observatory in Portugal: models, instruments representation and strategies) supported by Lisbon Regional Operational Programme (LISBOA 2020) and the Competitiveness and Internationalisation Operational Programme (POCI-01-0145-FEDER-032445), under the PORTUGAL 2020 Partnership Agreement, through the European Regional Development Fund (ERDF) and FCT—Fundação para a Ciência e a Tecnologia (Foundation for Science and Technology).

References

1. Marques AB, Branco V, Costa R, Costa N (2022) Data Visualization in Hybrid Space—Constraints and Opportunities for Design. In: Martins N, Brandão D (eds) Advances in Design and Digital Communication III. DIGICOM 2022. Springer Series in Design and Innovation. Springer, Cham, pp 3–15
2. Branco V, Costa R, Costa N, et al (2018) designobs.pt. https://designobs.pt/. Accessed 21 Mar 2022
3. Schmalstieg D, Höllerer T (2016) Authoring. Augmented Reality: Principles and Practice. Addison Wesley, Crawfordsville, Indiana, pp 471–488
4. Leiva G, Nguyen C, Kazi RH, Asente P (2020) Pronto: Rapid Augmented Reality Video Prototyping Using Sketches and Enaction. In: Proceedings of the 2020 CHI Conference on Human Factors in Computing Systems. ACM, New York, NY, USA, pp 1–13
5. Dengel A, Iqbal MZ, Grafe S, Mangina E (2022) A Review on Augmented Reality Authoring Toolkits for Education. Front Virtual Real 3:https://doi.org/10.3389/frvir.2022.798032
6. Gasques D, Johnson JG, Sharkey T, Weibel N (2019). What You Sketch Is What You Get: Quick and Easy Augmented Reality Prototyping with PintAR. https://doi.org/10.1145/3290607.3312847
7. MacIntyre B, Gandy M, Dow S, Bolter JD (2005) DART: a toolkit for rapid design exploration of augmented reality experiences. ACM Trans Graph 24:932–932. https://doi.org/10.1145/1073204.1073288
8. Verganti R, Vendraminelli L, Iansiti M (2020) Innovation and Design in the Age of Artificial Intelligence. J Prod Innov Manag 37:212–227. https://doi.org/10.1111/jpim.12523
9. Costa N, Branco V, Costa R, et al (2021) Design Companies in Portugal (detailed view, 2018, SABI)
10. "mbs278" (2017) Beginning Data Visualization in Unity: Scatterplot Creation – Methodology blog. http://sites.psu.edu/bdssblog/2017/04/06/basic-data-visualization-in-unity-scatterplot-creation/. Accessed 27 Mar 2023
11. Iliinsky N, Steele J (2011) Designing Data Visualization. O'Reilly, Sebastopol, CA, USA
12. Masud L, Valsecchi F, Ciuccarelli P, et al (2010) From Data to Knowledge – Visualizations as Transformation Processes within the Data-Information-Knowledge Continuum. In: 2010 14th International Conference Information Visualisation. IEEE, pp 445–449
13. Segel E, Heer J (2010) Narrative visualization: Telling stories with data. IEEE Trans Vis Comput Graph 16:1139–1148. https://doi.org/10.1109/TVCG.2010.179
14. Marques AB, Branco V, Costa R (2022) Narrative Visualization with Augmented Reality. Multimodal Technologies and Interaction 6:105. https://doi.org/10.3390/mti6120105
15. Ikonen P, Uskali T (2021) Augmented reality as news. In: Uskali T, Gynnild A, Jones S, Sirkkunen E (eds) Immersive Journalism as Storytelling. Routledge, Oxon, pp 147–160
16. New York Times (2022) Inflation Shrinking Ray. https://www.nytimes.com/interactive/ 2022/04/20/business/inflation-shrink-ray.html?shared=1. Accessed 27 Sep 2022
17. New York Times (2022) EV Battery Breakdown. https://www.nytimes.com/interactive/ 2022/08/19/business/electric-vehicle-battery-made.html. Accessed 27 Sep 2022
18. Deutsche Welle (2020) Water footprint: Augmented Reality shows water cost of beef. https://www.dw.com/en/water-footprint-augmented-reality-shows-watercost- of-beef/a-52218248. Accessed 29 Sep 2022
19. Cappannari L, Vitillo A (2022) XR and Metaverse Software Platforms. In: Roadmapping Extended Reality. Wiley, pp 135–156
20. Li J, W. Tigwell G, Shinohara K (2021) Accessibility of High-Fidelity Prototyping Tools. In: Proceedings of the 2021 CHI Conference on Human Factors in Computing Systems. ACM, New York, NY, USA, pp 1–17
21. Billinghurst M, Nebeling M (2022) Rapid prototyping of XR experiences. ACM SIGGRAPH 2022 Courses. ACM, New York, NY, USA, pp 1–124
22. Lee GA, Kim GJ, Billinghurst M (2005) What you experience is what you get (WYXIWYG). Commun ACM 48:76–81. https://doi.org/10.1145/1070838.1070840

23. Billinghurst M, Clark A, Lee G (2015) A Survey of Augmented Reality. Foundations and Trends® in Human–Computer Interaction 8:73–272. https://doi.org/10.1561/1100000049
24. Freitas G, Pinho MS, Silveira MS, Maurer F (2020) A Systematic Review of Rapid Prototyping Tools for Augmented Reality. In: 2020 22nd Symposium on Virtual and Augmented Reality (SVR). IEEE, pp 199–209
25. Seichter H, Looser J, Billinghurst M (2008) ComposAR: An intuitive tool for authoring AR applications. In: 2008 7th IEEE/ACM International Symposium on Mixed and Augmented Reality. IEEE, pp 177–178
26. Gandy M, MacIntyre B (2014) Designer's Augmented Reality Toolkit, Ten Years Later: Implications for New Media Authoring Tools. In: Proceedings of the 27th annual ACM symposium on User interface software and technology. ACM, New York, NY, USA, pp 627–636
27. Marques AB, Branco V, Costa R (2021) Minard Revisited – Exploring Augmented Reality in Information Design. Springer, Cham, pp 79–89
28. Marques AB (2023). No-Code Augmented Reality Platforms. https://doi.org/10.5281/ZENODO.7640671
29. Stansfield C, Dickson K, Bangpan M (2016) Exploring issues in the conduct of website searching and other online sources for systematic reviews: how can we be systematic? Syst Rev 5:191. https://doi.org/10.1186/s13643-016-0371-9
30. Ivanov A, Danyluk K, Jacob C, Willett W (2019) A Walk Among the Data. IEEE Comput Graph Appl 39:19–28. https://doi.org/10.1109/MCG

Gaps, Goals, and Actions for the Development of the AfterLab: A Laboratory for Design Applied to the Dance Floor Experience

César Lugo-Elías and Pedro Cardoso

Abstract Clubs—like those devoted to disco, techno and house music—have worked as spaces for design, artistic and technological experimentation, creating a beneficial relationship between the discipline of design and the activity of clubbing. However, this relationship has caught little attention from design researchers, overlooking the possibilities that the club has to offer as a space for design experimentation and research. As an effort to foster these advantages, this chapter proposes the creation of the AfterLab as a laboratory for design—understood as the application of art and technology to our lives—applied to the dance floor experience. This proposal is divided in two parts: the first analyzes the economic, social, and aesthetic dimensions of clubbing; and the second exposes the research questions, goals, formats and initial actions proposed for the AfterLab, as well as some ethical guidelines. Once the AfterLab becomes a reality, it can benefit designers, artists, and professionals related to the area. Furthermore, the AfterLab can advance and deepen the relationship between design and clubbing, while preserving and improving local club cultures. The present text is shared with the scientific community in the hope of generating feedback and igniting synergies of collaboration.

Keywords Dance floor experience · Designing club culture · Fabrication lab

1 Introduction

The AfterLab is an emerging project consisting in a laboratory for design applied to the dance floor experience in the city of Porto, in northern Portugal. This paper presents this project by exposing two main arguments. The first argument reviews

C. Lugo-Elías (✉)
Unexpected Media Lab, (ID+) University of Porto, Research Institute for Design, Porto, Portugal
e-mail: c.lugoelias@gmail.com

P. Cardoso
DigiMedia, University of Aveiro, Aveiro, Portugal
e-mail: pedroccardoso@ua.pt

N. Martins et al. (eds.), *Perspectives on Design and Digital Communication IV*, Springer Series in Design and Innovation 33, https://doi.org/10.1007/978-3-031-41770-2_20

the established relationship between clubbing and design, where the contemporary dance floor has worked as a *petri dish* for artistic and technological experimentation, becoming one of the most exciting design artefacts developed in the last decades of the past century. The second argument showcases the need for the AfterLab, as a third space outside the academia and the club, where scientists, creators, activists, practitioners, and even clubbers and club owners can meet and align their interests for the benefit of their area of interest.

The AfterLab should work as a channel for transferring knowledge, ideas and skills among practitioners and people interested in learning technology and design topics related to the dance floor experience, such as DJing, coding, fashion, and the creation of immersive spaces and experiences. Hopefully, the AfterLab will improve the current lack of representation of women, immigrants, POC and members of the LGBTQ + in the local and regional clubbing scene, creating a diverse, healthier, and overall, more resilient scene—while improving literacy in matters of technology and Design.

This chapter has two main sections, in the upcoming second section, the economic relevance of club culture is presented, as well as the role that the dance floor has played in the creation of social utopias—far beyond preconceptions of drug consumption, plain hedonism and the seek of excess—and the relevance of the aesthetic output of the dance floor experience as a subject for design research.

A third section proposes the creation of the AfterLab as a way to foster the relationship between design and clubbing by using the dance floor as a laboratory. This section presents the research questions and goals of the laboratory as well as the initial proposed formats and actions. The section closes with some relevant ethical guidelines that have emerged during the writing process of this chapter. The fourth section offers a summary, underlining how the laboratory could help ethnographic-based research in clubbing by providing researchers with the experiential element of organizing a dance floor experience.

1.1 Working Definitions

Before continuing, it is necessary to give some basic definitions, starting with *club culture*, defined by the club commission of the city of Berlin[1] [3] in its 2019 study as: "the phenomenon of people meeting in clubs or in similar spaces (e.g. open-air concerts, warehouse raves or festivals) characterized by a program focused on music, restricted access of a certain nature to create a protected space with its own rules, and a community to listen to music, dance and socialize. Club culture is a subsector of the cultural economy in its own right." (p. 8). Here it is necessary to underline that club culture is attached to a specific space—the club— which "is the primary essential resource for club events" (p. 9), mainly because the club experience "is characterized by a high intensity of bodily physicality. An event is always experienced in a physical

[1] *Clubcommission e.V. Berlin.*

space together with other people who are present in a very analog way" (p. 9). This definition is open to a large variety of clubs and musical styles—including those devoted to music played live by bands, like jazz and rock music.

Yet, without challenging this definition, we need to point out that the current text works in a specific scene within the club culture: starting with the music, the AfterLab is more related to clubs characterized by a program consisting of electronically produced music. Likewise, the AfterLab is more aligned with clubs offering a dance floor experience consisting of dancing to music mixed by a DJ as part of a *DJ set*[2] or produced in situ by a producer as part of a *live set*—in which case, the use of analogue and electronic instruments is not excluded. On the other hand, the AfterLab is focused on a specific way of consuming music, which has dancing as the most common way of objectification and embodiment, but it remains open to other ways.

Furthermore, the AfterLab has a very specific area of action and study within the club: the dance floor. The dance floor is generally the most prominent architectural space inside a club, an area where the sonic energy contained in the music interplays with the kinetic energy contained in the attendees and is released in the act of collective dancing ([7], p. 41, [18], p. 163). The dance floor is then where music, DJ, audience, movement, sound-reproducing machines, and light and projection technologies became mediators of a communal aesthetic experience ([6], p. 19) which is customized to a specific social scene, space, and moment.

It is also necessary to underline that, for the *Clubcommission Berlin*, given the *high intensity of bodily physicality* of the club experience, virtual attendance of a club event is not a possible option ([9], p. 9, my emphasis), something that is understandable given the aim of this commission of influencing public policy regarding the protection of the analogue space of all clubs in the city of Berlin. However, for the AfterLab, the virtual space may be valid mainly because one of the aims of the AfterLab is the exploration of different technologies and their application to the dance floor experience, therefore, to exclude technologies like virtual reality would make little sense.

In the upcoming section, the dance floor will be analyzed under the economic, social, and aesthetic dimensions, following the structure of the above-mentioned study of the *Clubcommision.*

2 Economic, Social and Aesthetic Dimensions of the Dance Floor

The history of the dance floor reflects not only the cultural and social history of the societies that have danced on it, but it can also reflect economic aspects, as well as the history of technological innovation applied to the aesthetic experience of the dance floor. In this section, the economic, social, and aesthetic dimension of the club

[2] This tradition of *dancing to records* can be drawn front the 1970s discotheque to the contemporary house and techno clubs and festivals.

cultures will be explored, with the aim of exposing the relevance of clubbing for our society in general and for the discipline of design in particular—before advancing to the third section, where we propose the creation of the AfterLab as a way of fostering these values.

2.1 Economic Dimension

Clubs like those devoted to electronic music[3] are valuable from an economic standpoint because they are an essential part of our urban life and economies, mainly as independent cultural activities embedded in local subcultural scenes, produced thanks to private entrepreneurship and ingenuity. "It takes business thinking to organize an event. Anyone who cannot assemble the necessary resources and does not have a feel for what people want to experience will quickly fail" ([4], p. 18). This entrepreneurial aspect is not limited to booking artists and investing in opening a venue, but it can be seen as a process not unlike producing a play in theatre:

> The moments leading up to the event [...] the back-and forth email conversations about which bands should play, which DJs should spin, what people were planning on wearing, who was going to be on the guest list and what time everything should begin. In effect, this was all a process of staging nightlife not unlike the way a play is staged for the theatre. ([23], p. 62)

These events are then organised through internal procurements markets [4] or *scene markets*, in which a wide range of local people gets jobs beyond the artists—i.e., bartenders, administrators, managers, bouncers, drink providers, graphic designers, sound and light technicians. This industry also produces revenues for international companies who produce event-related equipment—such as the sound system, DJing and lighting devices—and even alcohol brands have found their way into the underground, where there is always a barrel of beer served through a beer tower exhibiting a beer logo. Admitting that it is extremely difficult to calculate exactly how much revenue the clubbing industry reports to those companies, it seems fair to suggest that in terms of marketing and the creation of customer loyalty, clubs can work as a showcase for brands. On the other hand, some authors have suggested that "nightlife is not only an international, multi-billion dollar industry—a billion-dollar industry (sic.) in New York City alone" ([23], p. 62), while other entities have evaluated their value considering their influence on other sectors like tourism, as in the case of Berlin:

> The sales market for club culture is the event market. The market volume is relatively small compared to other sectors. If clubs make profits at all, these profits are generally small. Only a few institutions in major cities make big profit margins – and the gap between them and

[3] With the term *electronic music*, we encircle genres such as disco, house, techno, electronica, experimental, and other genres produced mainly electronically in a studio and played back on the dance floor. We prefer this term—*electronic music*—over the other commonly used term *electronic dance music* or *EDM*, giving the connotations that this term has in reference to the commercial *dance music* of the end of the 1990s.

> other actors is considerable. However, the knock-on effects for other sectors are immense. The main benefactor is tourism, and other sectors such as retail and transport enjoy positive effects too. When examining the figures in Berlin, 168 million euros of direct revenue [here, year 2019] can be compared with a staggering 1.48 billion euros for tourism (see pp. 29–37). In addition, club culture is a location factor and an indicator for the attractiveness of a city, especially for the creative classes. ([3], p. 13)

In the Portuguese context, some data is offered by the Portuguese Association of Music Festivals—*Associação Portuguesa de Festivais de Música*—in their last pre-pandemic report of 2018, reporting an increase in the number of festivals in Portugal from 210 festivals in 2015, to 311 in 2018. An increase of ca. 100 new music festivals in three years, attending >2.7 million spectators in that year (APORFEST, p. 2), which is a considerable number in a country with a population of only ca. 11 million inhabitants. Now, this data represents all festivals, including those devoted to rock and pop music, which are among the main consumed music genres in festivals, and can only reflect the festival market, not the market of the urban clubs. However, we should not forget that outside the summer festival season, urban clubs became the main alternative to consume music over the weekends, and festivals can be seen as a manifestation of the fandom for music that is weekly fostered in clubs around the country. In other words, the increase in festivals reflects the willingness of Portuguese society to spend money on cultural activities related to music throughout the year.

On an empirical note, the corresponding author has been working and researching in the clubs scene of the city of Porto since 2016, where it is possible to report an increase in the number of tourists navigating through the different clubs of the city, with some of them mentioning the clubbing scene as a reason to come to Porto. This is especially true for tourists from the northern Spanish region of Galicia, who feel attracted by the international DJs presented in clubs such as the Gare and the techno parties organized by the Neopop festival[4] at larger venues such as the Hard Club, or the Super Bock Arena.

In sum, the economic dimension of clubbing is there, and even if we can't calculate it, it has become part of the tourist economy of the city. Another, maybe more obvious dimension of club culture is the social aspect, which will be described in the next section.

[4] The Neopop festival takes place once a year in the city of Viana do Castelo, northern from Porto in Portugal. It is one of the most prestigious festivals devoted exclusively to techno music, given its yearly program presenting high profile international and local DJs and music producers. The duration of the festival variates from three days—over a single weekend—and up to five days. The festival is organized by the events production company Made of You [https://www.madeofyou.pt/] who also organizes events like The Bpm Festival Portugal, Brunch Electronik Lisboa, and elrow Portugal. During the rest of the year, Neopop organizes several techno events in the city of Porto, hosted at large venues like the Hard Club and the Super Bock Arena—this last one with a capacity for ca. 8000 ppl.

2.2 Social Dimension

Despite the economic relevance of club culture, to the "untrained eye, nightlife is exclusively dedicated to decadence and the pursuit of pleasure" ([23], pp. 63–64). This is an effect of the calculated suspension of everyday rules, resulting from a combination of consumption of substances, the shared experience of dancing and even the spatial organisation of clubs [8], where corridors, corners, staircases, and darkness choreograph and organize anonymity, allowing the indulgence of sensual, sexual and drug-related activities [2]. This side of clubbing and night culture can be deemed as the *seek for excess*, meaning the deliberate renunciation of reason, turning away social norms as part of a process of *controlled loss of control* ([4], pp. 15–16, [10], p. 72). Yet, rather than negative, this aspect of clubbing has been demonstrated to be of great social potential, placing the club experience as a medium to explore the own personality—a process that does not necessarily must be aided by drugs. In this sense, there is little gain in terms of academic research, especially design research, to continue attributing exclusive agency [6] to the use of drugs as part of the rave and club culture.

On the contrary, underground ballrooms and clubs like those devoted to house and techno have been security zones for POC and members of the LGBTQ + community ([7], p. 12), especially transexual and transgender individuals [10, 14] who turned clubs from environments of controlled excess and indulgence, to secure spaces protected from the social norms where the exploration of utopian social and gender egalitarianism became possible [18].

In this sense, we agree with the suggestion that partying could be seen as a way of social and political participation [25], either intentionally or by default. Now, this social value is not always appreciated by society at large, but it has achieved great changes for all the above-mentioned minorities, and more importantly, the social component of clubs is the way they maintain themselves:

> The club visit's social component (meeting friends, experiencing something) is therefore the main reason for going to clubs. It is therefore central for these institutions to create suitable rooms and opportunities to remain attractive for visitors and thus to generate and maintain the economic basis for the club's success. ([17], p. 60)

Offering spaces to socialise is then one of the most important values of clubbing, a role that has turned more important and took more relevance in the post-COVID era, when people became more aware of the importance of social connections. Even if the tendency is the digitalization, clubs remain an important space for social life to thrive in an analogue form, allowing meeting mind-alike people to get to know each other.

2.3 Aesthetic Dimension

According to Drevenstedt [4], aesthetic considerations regarding club culture may first study the musical style attached to each scene (p. 17), and it can also consider the semi-public atmosphere of the club which offers a safe space for artists to try new music possibilities in front a selected audience (p. 16). This creates perfect conditions to have exchanges between artists and people, having the creative output as the basis for any dialogue [23], p. 71), while the dance floor works as the material context to inspire and get inspired.

Further aesthetic considerations regarding clubbing muss encircle fashion and make-up. For scholar Stanfill [26], the dynamics of inclusion and exclusion inherent to club cultures have rendered these spaces as key sites of fashionable display, where sartorial taste may not only be a matter of securing access and conformity to social codes—i.e., as a matter of subcultural capital as Thornton [27] may point out—but also a matter of spectacle and self-expression (p. 259). The urban nightclub can be seen as a *sensorium* [17, 28] for which the experience of getting dressed for the night constitutes an important element in the embodiment of the experience, in other words, we dress for the dance floor. Here:

> make-up—a largely overlooked area of design history—has been a powerful aspect of bodily adornment[5] in club culture. Make-up symbolises and expresses identity in these spaces [challenging] social conventions and, on many occasions, triggered real and lasting social change. ([22], p. 273)

Another aesthetic dimension of relevance is architecture, where clubs have served as spaces for the experimentation of environmental ideas. As previously documented [21], artists like Andy Warhol [11–13] and architects like Arata Isozaki [24] have used clubs as a "phenomenological-spatial apparatus" ([5], p.132) merging time, place and event through technology and media. Similarly, underground clubs and raves emerging "in locations where architects and town planners have not been involved. Deserted warehouse halls, damaged buildings, empty sites, and derelict areas in industrial estates offer ideal prerequisites for stimulating the visions and fantasies of pioneers in terms of locations for club culture" ([4], p. 17). In such spaces, clubs have been spaces where the aesthetic of the provisional, the emerging, the second-hand, and the *lost and found*, have emerged as part of an aesthetic narrative of the *possible* rather than the *finished* ([18], p. 58, [24], p. 118). Therefore, it is not surprising that clubs have become a working environment for diverse artists: "In particular, clubs serve as a workshop laboratory and presentation space for the visual arts" ([4], p. 17) and in general, clubbing has been seen as a site-specific performance [2] and as a loosely scripted work of art, a relational space of creation and curation ([23], p. 62).

There is also a technological component to this aesthetic dimension. The dance floor has been a territory where technology has always been a protagonist. As scholar Vitos [28] points out, the dance floor experience is located between audience and

[5] For Miller [22], under bodily adornment we can consider, dress, make-up, hair, piercings, tattoos, and jewellery (p. 273).

mediating technologies, becoming a product of the use, appropriation, and creative engagement with diverse technologies, discovering new applications and affordances (p. 65). An example of this can be found in the technics of DJing and creative music production,but also in the creation and experimentation with sound systems, a tradition that can be traced to the 1980s at the legendary club Paradise Garden in NYC, where the sound system—known as the Levan Horns, after the DJ resident Larry Levan—took legendary status for its sound fidelity and accurate and powerful bass [15]. Furthermore, clubs have been pioneers in the experimentation of lighting, projection and even olfactory technologies, with historic examples that can be dated back to the late 1960s when the club *Maddox* opened near Barcelona, for which fog, projection and cloud machines were designed [5], or the club *Cerebrum* in NYC in the 1980s, where the intention was to recreate a psychedelic multisensory environment [16] including the use of scent and olfactory technologies as part of the clubbing experience.

This relationship between design—understood as the application of art and technology to any aspect of our lives—and clubbing can also be explained by the exceptional characteristic of clubbing as a security space, where controlled excess is not only allowed but celebrated while bringing together new ideas, aesthetic proposals, and specific audiences under the same roof. The AfterLab pretends to tap into this well-established relationship to foster further design, art, and technological experimentation, as will be discussed in the upcoming section.

3 A Proposal for the AfterLab

The previous section presented a glimpse of the value and potential of the club culture, making emphasis on its economic, social, and aesthetic dimensions. Yet in the context of the city of Porto, these economic, social, and particularly aesthetic elements have caught very little academic attention; existing very few scientific works concerning aspects such as the aesthetic output, the structure of the audience, management aspects, and the impact of club culture in the local economy. With the intention of contributing to closing this gap, this section draws a proposal for the creation of a laboratory for studying and experimenting with the aesthetic dimension of clubbing in particular; and in general, to facilitate the study of the social and economic dimensions of clubbing.

At the core of this laboratory is the study of the intense experience on the dance floor, understood as a design-mediated experience. Here, Design is defined as (a) the output of the application of art and technology to any aspect of our lives; and (b) the discipline capable to mediate among art, technology, and the human body through the creation of artificial environments while triggering emotions and organizing experiences [19]. The dance floor is then defined as: the space—analogue or virtual—where the aesthetic output of the applied art and technology gets organized and embodied through design [20] and then objectified in every possible subjective realm—i.e., dancing, moving, socializing, flirting—with the aim of exploring the own or collective aesthetic experience and identity. The initial area of action of the laboratory is the city of Porto, Portugal. The laboratory does not aim to be a club, but to be a space *after* clubbing, where topics could be reflected on, and new experimentation could take place before the next clubbing experience. This is the first reason to name the laboratory the *After*Lab, the second reason is to link the AfterLab directly with the previous research carried out by the corresponding author on the local scene of the after-parties [18]. Therefore, by carrying the preposition *after* in the name of this project, we link the two projects semantically and conceptually.

The upcoming subsection exposes the research questions, goals, formats, and initial actions proposed for the laboratory, as well as some ethical guidelines we recognize as important. These subsections achieve the main goal of the whole text, which is to present and share these ideas with the scientific community to start a dialogue about the possibilities that such a laboratory could bring to the discipline of design, and to any other discipline involved in the study of the dance floor experience.

3.1 Research Questions

The AfterLab aims to answer the following key research questions:

1. How can we foster new dynamics between the discipline of design and the activity of clubbing for the benefit of both areas?
2. How the experimentation on the dance floor experience could advance our understanding of the subjectivities of design, art, and technology?

Additionally, the AfterLab can help to answer these secondary questions:

- How to enrich the cultural landscape of the city of Porto through design activities related to cubbing, while benefiting local populations, adding value to the local cultural and creative economy, as well as to other economic activities like tourism?
- How to tackle problems of representation in the club culture affecting marginalized populations—like members of the LGBTQ + communities, POC, immigrant populations, and individuals who may have little access to professional training in design, art, and the use of technologies?
- How to preserve and develop club culture, a culture that privileges the much-necessary social contact, the creation of security spaces for marginalized people to socialize, as well as activism and the exploration of the personal identity?

3.2 Goals

The AfterLab has two main goals, one is research-oriented, while the other is creation-oriented:

1. The research-oriented goal is to advance and promote design research projects related to the dance floor experience, especially regarding its aesthetic dimension.

3. The creation-oriented goal is to experiment and apply all possible technologies and creative proposals to be introduced as part of the design and production of the dance floor experience.

Additionally, the AfterLab follows these secondary goals:

- To establish connections among all stakeholders of the local club culture by creating a third space—analogue and/or digital—outside the academia and outside the club, where scientists, creators, practitioners, and even clubbers and club owners can meet and align their interests for the benefit of their own area of action—here please see the upcoming section *Formats*.
- To establish a space for artistic creation and technological experimentation in the production of the dance floor experience, with an emphasis on the transfer of knowledge—here please see the upcoming section *Actions*.

3.3 Formats

More than a single physical space, the AfterLab is thought to be a multi-format effort, a sum of actions that could and should take place in different spaces, both analogue and digital. However, as an established cultural entity the AfterLab aims to take some specific formats: a fabrication lab, an artistic workshop, a gallery, a forum of ideas, and a documentation site:

- As a fabrication lab, it should provide space and technology to test creative ideas within the laboratory. In the same way, it should provide access to events—even organizing them—clubs, and festivals for these ideas to be applied in real settings.
- The AfterLab can also take the format of a workshop—or a series of workshops—where artist and practitioners can share their knowledge.
- As a forum of ideas, the AfterLab organizes or co-organizes academic forums and informal meetings between researchers to interchange their ideas and present their research while documenting the research landscape regarding clubbing.
- As a documentation centre, the AfterLab can work as a website accessible for a wider audience.

3.4 Actions

To achieve our goals and in concordance with our formats, the following general actions are proposed for the AfterLab:

- To create the sensorium, where ideas and models could be formed through practice, a physical space where to apply technology and creative ideas related to the dance floor experience. In this space, audio and recording technologies could be tested; DJing techniques could be improved and taught; new technologies such as virtual realities could be explored. Yet, while on the long term one action could be to count with an own space, the initial sensorium(s) could be created inside the clubs of the city during the weekdays or in schedules when they are not being used.

- To create a digital repository, a website, facilitating access to literature and documentation of projects regarding clubbing such as books, music, journal papers, films, magazines, video footing and photography.
- To organize forums for practitioners—i.e., club owners, promoters, designers, DJs, music producers, coders, and visual and new media artists—to interact and discuss common issues among them and with academic researchers.
- From the interaction between these actors and stakeholders, to organize active collaborations to influence cultural policy at a local level, regarding the conservation of club culture.
- To promote the mentor principle [10], encouraging and giving a framework for the interchange and socialization among practitioners, beginners, and newbies, sharing important information about the scene, its history, and values,helping to rejuvenate the club scenes while creating social and professional networks that could benefit all parts (pp. 79–80). Here one target group can be constituted by underrepresented groups in the art-technology and club scene, especially women, members of the LGBTQ + community, POC, and financially disadvantaged people from all paths of life.

3.5 Some Ethical Guidelines

Along with the production of the present text, a series of ethical considerations to be applied to the AfterLab have emerged and are documented here to leave a first draft for a list of ethical guidelines to be observed along the activities of the Afterlab:

- Like a club, the AfterLab must be a safe space, free from any form of discrimination based on skin tone, gender, sex, nationality, musical or sartorial taste, ways of dancing or embodiment of the dance floor experience.
- The AfterLab is a space for free thinking and speech.
- The AfterLab should remain also a free space for experimentation, mistakes, errors, and learning, while promoting the mentor principle, encouraging contact among practitioners, artists, and academics with newcomers to the scene, with the aim of disseminating knowledge on design, art, and technology.
- The AfterLab should respect all the current legislation and stay away from any illicit act.
- The AfterLab should respect intellectual property rights and avoid any infringement of copyright, patent, or trademark laws.
- The AfterLab should be mindful of privacy and data protection laws and ensure that all personal data collected during its activities are consensually collected and processed ethically and in accordance with applicable regulations.

- The AfterLab should consider the environmental impact of its activities and take measures to reduce its carbon footprint, minimize waste, and promote sustainable practices.
- The AfterLab should strive to be inclusive and accessible to all, including people with disabilities, by providing appropriate facilities, equipment, and services.
- The AfterLab should ensure that all its activities are transparent and accountable and that it operates in accordance with its stated mission and goals.

4 Summary and Final Thoughts

This text—especially along the second section—has shown how the dance floor experience falls under several areas of study and responsibility, like economy, social development, politics, cultural preservation, and activism; but also, under artistic practice and technology experimentation. The past third section has shown the general guidelines describing the AfterLab.

Similarly, as a way of conclusion, this short section reflects on the possible advantages that the AfterLab could bring to studies based on ethnographic methods [10]. Here, from a methodological perspective, the AfterLab offers the possibility to add a component of practice and intervention to research projects in related areas regarding club culture—i.e., design, art, technology, sociology, and psychology. This can expand works based on ethnographic observations and interviews—a common way of researching club cultures—not by replacing the ethnographic method, but by expanding it: "Approaching a party as a project of curation and creation expands the ethnographic repertoire with a unique and valuable way of knowing, one that transforms history, theory, and observation into real experience" ([23], p. 64). The AfterLab can also help to *decentralize* the text as the single way to report research, since text presents important limitations when used as a means of representation, especially when dealing with the volatile and ephemeral experience at the dance floor ([18], p. 195).

The contemporary dance floor could be easily seen as one of the most relevant design artefacts created during the second part of the last century, however digitalization and social changes—and even the COVID pandemic—have threatened its existence. The AfterLab is still an idea, born out of our previous research, but also out of the necessity of preserving club culture, helping it to thrive and deepening the beneficial mutual relation between design and clubbing. This text constitutes the first materialization of the AfterLab and is here for the scientific community to read it, provide feedback and ignite collaborations.

References

1. APORFEST: Annual report: festivais de música em Portugal [Annual report: music festivals in Portugal]. Associação Portuguesa de Festivais de Música (ed.). Lisboa (2018)
2. Biehl-Missal B (2016) Filling the 'Empty Space': site-specific dance in a techno club. Cult Org 25(1):16 31. https://doi.org/10.1080/14759551.2016.1206547
3. Damm S, Drevenstedt L (2019) CLUBKULTUR BERLIN. Clubcommission e.V. Berlin. Berlin (2019). https://www.clubcommission.de/club-culture-study/, last accessed 2023/03/13
4. Drevenstedt L (2020) Dimensions of Club Culture: Learning from Berlin. In: Jori A, Lücke M (Eds) The New Age of Electronic Dance Music and Club Culture, pp. 9–20. Springer Cham, Switzerland. https://doi.org/10.1007/978-3-030-39002-0
5. Esteve P (2018) Total space. In: Kries M, Eisenbrand J, Rossi C (eds.) Night Fever: Designing Club Culture, 1960-Today, pp 130–147. Vitra Design Museum, Weil am Rhein
6. Ferreira, P.: When sound meets movement: performance in electronic dance music. Leonardo Music J. (18), pp. 17–20 (2008). http://www.jstor.org/stable/25578110, last accessed 2019/04/19.
7. Fikentscher K (2000) "You Better Work!": Underground dance music in New York City. University Press of New England, Hanover
8. Goulding C, Shankar A, Elliot R, Canniford R (2008) The Marketplace Management of Illicit Pleasure. Journal of Consumer Research 35(February):759-771. https://doi.org/10.1086/592946
9. Garcia, L.-M.: An alternate history of sexuality in club culture (2014), https://www.residentadvisor.net/features/1927, last accessed 2019/01/31.
10. Garcia, T.: Das intensive Leben – Eine moderne Obsession [The intensive life – a modern obsession]. Frankfurt a.M.: Suhrkamp (2017).
11. Heiser, J.: Club culture and contemporary art: a relationship. In: Kries, M., Eisenbrand, J. ,Rossi,, C. (eds.) Night Fever: Designing Club Culture, 1960-Today, pp. 172–189. Vitra Design Museum, Weil am Rhein (2018).
12. Joseph, B.W.: My mind split open": Andy Warhol's exploding plastic inevitable. Grey Room 8(Summer, 2002), pp. 81–107 (2002). http://www.jstor.org/stable/1262609, last accessed 2019/04/27.
13. Lavin, S.: Andy Architect™—or a funny thing happened on the way to the disco. Log 15(Winter 2009), pp. 99–110 (2009). http://www.jstor.org/stable/41765265, last accessed 2018/07/16.
14. Lawrence T (2011) Disco and the queering of the dance floor. Cult Stud 25(2):230–243. https://doi.org/10.1080/09502386.2011.535989,lastaccessed2018/07/28
15. Lawrence T (2018) Dance floor transformation: counterculture, post-industrialism, and environmental design in New York City during the 1970s and early 1980s. In: Kries, M., Eisenbrand, J., Rossi, C. (eds.) Night Fever: Designing Club Culture, 1960-Today, pp. 88–97. Vitra Design Museum, Weil am Rhein (2018).
16. Lovell, J,: Cerebrum: modelling the mind through psychedelic and playful environments. J Design Hist 34(2), pp. 149–165 (2021). https://doi.org/10.1093/jdh/epaa054, last accessed 2022/01/18.
17. Lücke M (2020) Club Culture and Electronic Dance Music in Berlin: An Economic, Social, and Aesthetical Perspective. In: Jori, A. and Lücke, M. (eds.) The New Age of Electronic Dance Music and Club Culture, pp. 55–62. Springer Cham, Switzerland. https://doi.org/10.1007/978-3-030-39002-0
18. Lugo-Elías, C.: Reassembling the Dance Floor: Exploring interactions at after-parties in spaces of consumption of electronic music in the city of Oporto in the years 2016–2018. Doctoral thesis, University of Porto, Portugal (2021).
19. Lugo-Elías C., Cardoso P.: Design is in the House: Basic Concepts to Understand the Role of Design in the Creation of the Club Experience. In: Martins, N., Brandão, D. (eds) Advances in Design and Digital Communication II. DIGICOM 2021. Springer Series in Design and In-novation, vol 19, pp. 517–527. Springer, Cham, (2022). https://doi.org/10.1007/978-3-030-89735-2_42 last accessed 2022/02/15.

20. Lugo-Elías, C., Cardoso, P.: Design Radicalized as Experience: Disentangling and Reassembling the Dance Floor Experience to Understand the Relationship Between Design and Clubbing. In: Martins, N., Brandão, D., Paiva, F. (eds) Perspectives on Design and Digital Communication III. Springer Series in Design and Innovation, vol 24, pp. 91–103. Springer, Cham (2023a). https://doi.org/10.1007/978-3-031-06809-6_6 last accessed 2023/05/11.
21. Lugo-Elías, C., Cardoso, P.: Embodiment, Memory, and Consent: Strategies to Conduct a Research in Design Using Participatory Observation in Clubs. In: Martins, N., Brandão, D. (eds) Advances in Design and Digital Communication III. DIGICOM 2022. Springer Series in Design and Innovation, vol 27, pp. 30–39. Springer, Cham (2023b). https://doi.org/10.1007/978-3-031-20364-0_3 last accessed 2023/05/11.
22. Miller, J.: Visibility and Invisibility: Make-Up forthee Nightclub. In: Kries, M., Eisenbrand, J.,Rossi,C. (eds.)Night Fever: Designing Club Culture, 1960-Today, pp. 272–280.VitraDesign Museum, Weil am Rhein (2018).
23. Moore M (2013) Looks: Studio 54 and the Production of Fabulous Nightlife. Journal of Electronic Dance Music Culture 5(1), pp. 61–74. https://doi.org/10.12801/1947-5403.2013.05.01.04, last accessed 2023/03/20
24. Munuera L (2018) Discotecture: the bodily regime of archi-social exploration. In: Kries M, Eisenbrand J, Rossi C (eds) Night Fever: Designing Club Culture, 1960-Today, pp 116–129. Vitra Design Museum, Weil am Rhein (2018)
25. Riley S, More Y, Griffin C (2010) The 'pleasure citizen': Analyzing partying as a form of social and political participation. Young 18(1):33–54. https://doi.org/10.1177/110330880901800104,lastaccessed2023/03/20
26. Stanfill S (2018) The Nightclub as Fashion Catwalk. In: Kries, M., Eisenbrand, J., Rossi, C. (eds.) Night Fever: Designing Club Culture, 1960-Today, pp. 258–270.Vitra Design Museum, Weil am Rhein
27. Thornton S (1995) Club Cultures: Music, Media, and Subcultural Capital. University Press of New England, Hanover
28. Vitos B (2020) Researching Informal Musical Expertise and Mediations of Aesthetic Experiences Among the Berlin Techno Audience: A Postdoctoral Research Proposal. in: Jori, A. and Lücke, M. (Eds.) The New Age of Electronic Dance Music and Club Culture, pp. 63- 74. Springer Cham, Switzerland (2020). https://doi.org/10.1007/978-3-030-39002-0, last accessed 2023/03/20.

MIX
Papier aus verantwortungsvollen Quellen
Paper from responsible sources
FSC® C105338
FSC
www.fsc.org

If you have any concerns about our products,
you can contact us on
ProductSafety@springernature.com

In case Publisher is established outside the EU,
the EU authorized representative is:
Springer Nature Customer Service Center GmbH
Europaplatz 3, 69115 Heidelberg, Germany

Printed by Libri Plureos GmbH
in Hamburg, Germany